ESR Spectroscopy in Membrane Biophysics

Volume 27

A Continuation Order Plan is available for this series. A continuation order will bring delivery of each new volume immediately upon publication. Volumes are billed only upon actual shipment. For further information please contact the publisher.

Marcus A. Hemminga and
Lawrence J. Berliner

ESR Spectroscopy in Membrane Biophysics

Volume 27

Springer

Marcus A. Hemminga
Laboratory of Biophysics
Wageningen University
Dreijenlaan 3
6703 HA Wageningen
The Netherlands
email: marcus.hemminga@wur.nl

Lawrence J. Berliner
Department of Chemistry &
Biochemistry
University of Denver
2190 E. Iliff Avenue
F. W. Olin Hall, Room 202
Denver, CO 80208
USA
email: berliner@du.edu

Library of Congress Control Number: 2006936900

ISBN-10: 0-387-25066-2 e-ISBN-10: 0-387-38880-X
ISBN-13: 978-0-387-25066-3 e-ISBN-13: 978-0-387-49367-1

Printed on acid-free paper.

9 8 7 6 5 4 3 2 1

springer.com

CONTRIBUTORS

R. Linn Belford (app. 2)
Department of Chemistry
University of Illinois
600 South Mathews Avenue
Urbana, IL 61801, USA
(217) 333-2553
rbelford@uiuc.edu

Enrica Bordignon (chap. 5)
Experimental Physics —
Macromolecular Structure
Universität Osnabrück
Barbarastrasse 7
Osnabrück D-49076, Germany
+49 541 969 2664
enrica.bordignon@uos.de

Louise Brown (chap. 4)
Department of Chemistry and
Biomolecular Sciences
Division of Environmental and
Life Sciences
F7B 335, Macquarie University
Sydney NSW 2109, Australia
+61-2-9850-8294
lbrown@alchemist.chem.mq.edu.au

Mikolai I. Fajer (app. 1)
National High Magnetic Fields Laboratory
Kasha Laboratory of Molecular Biophysics
Florida State University
1800 Paul Dirac Drive
Tallahassee, FL 32310, USA
(850) 645-1335
mfajer@gmail.com

Piotr G. Fajer (chap. 4, app. 1)
National High Magnetic Fields Laboratory
Kasha Laboratory of Molecular Biophysics
Department of Biology, Florida State
University
1800 East Paul Dirac Drive
Tallahassee, FL 32310, USA
(850) 644-2600
fajer@magnet.fsu.edu

Jack H. Freed (app. 3)
ACERT, Cornell University
B52 Baker Laboratory
Ithaca, NY 14853, USA
(607) 255-3647
jhf@ccmr.cornell.edu

Marcus A. Hemminga (chap. 1)
Laboratory of Biophysics
Wageningen University
Dreijenlaan 3
Wageningen 6703 HA, The Netherlands
+31 317 482044
marcus.hemminga@wur.nl

Gunnar Jeschke (chap. 2, app. 4)
EPR Group
Max Planck Institute for Polymer Research
Ackermannweg 10
Mainz D-55128, Germany
+49 6131 379 247
jeschke@mpip-mainz.mpg.de

Karen Mattson (app. 2)
Bayer Material Science LLC
100 Bayer Road
Pittsburgh, PA 15205-9741, USA

Mark J. Nilges (app. 2)
Illinois EPR Research Center
University of Illinois
506 South Matthews Avenue
Urbana, IL 61801, USA
(217) 333-3969
mjnilges@uiuc.edu

Kenneth L. Sale (app. 1)
Biosystems Research Department
Sandia National Laboratories
7011 East Avenue
Livermore, CA 94551, USA
(925) 294-2143
klsale@sandia.gov

Arthur Schweiger (app. 6)
Laboratory of Physical Chemistry
ETH Zürich, HCI F 225
Wolfgang-Pauli-Strasse 10
Zürich CH-8093, Switzerland
[deceased 4 January 2006]

Alex I. Smirnov (chap. 6, app. 5)
Department of Chemistry
North Carolina State University
614F Cox
Raleigh, NC 27695-8204, USA
(919) 513-4377
Alex_Smirnov@ncsu.edu

Tatyana I. Smirnova (chap. 6)
Department of Chemistry
North Carolina State University
612B Cox
Raleigh, NC 27695-8204, USA
(919) 513-4375
Tatyana_Smirnova@ncsu.edu

Likai Song
National High Magnetic Fields Laboratory
Department of Biology, Florida State University
1800 Paul Dirac Drive
Tallahassee, FL 32310, USA
(850) 645-1335
song@magnet.fsu.edu

Heinz-Jürgen Steinhoff (chap. 5)
Experimental Physics — Macromolecular Structure
Universität Osnabrück
Barbarastrasse 7
Osnabrück D-49076, Germany
+49 541 969 2675
heinz-juergen.steinhoff@uos.de

Stefan Stoll (app. 6)
Laboratory of Physical Chemistry
ETH Zürich, HCI F 225
Wolfgang-Pauli-Strasse 10
Zürich CH-8093, Switzerland
+41 44 6334259
stefan.stoll@ethz.ch

Janez Štrancar (chap. 3, app. 7)
Laboratory of Biophysics, EPR Center
Jožef Stefan Institute
Jamova 39
Ljubljana SI-1000, Slovenia
+386 1 4773226
janez.strancar@ijs.si

FOREWORD

This book flows from the "Graduate Course: Advanced ESR Spectroscopy in Membrane Biophysics" held in Wageningen (The Netherlands) during March–April 2004, which was coordinated by one of the editors, Marcus A. Hemminga (Wageningen), Janez Štrancar (Ljubljana), and Heinz-Jürgen Steinhoff (Osnabrück). The course was organized under the auspices of the Graduate School of "Experimental Plant Sciences" (EPS). It was a phenomenal success and fitted well the needs of students who desired a thorough understanding of the theory as well as the applications of spin-label ESR in biophysics.

In the spirit of the graduate course, this volume is strongly enhanced by the enlistment of an international group of experts. We are confident that it will receive the same enthusiastic reception that was given by the students and research scientists who participated in the first graduate course and should serve as an excellent pedagogical tool.

Lawrence J. Berliner
Denver, Colorado

PREFACE

Membrane proteins comprise almost a third of all the proteins in an organism or cell. However, progress in determining their structure has been slow. Compared to soluble proteins, only a handful of membrane protein structures have been solved. Therefore, membrane proteins offer the greatest challenge in structural biology, and there is an urgent need to develop and apply new biophysical methodologies that are able to generate detailed structural information. Among modern biophysical techniques, site-directed spin-labeling electron spin resonance (SDSL-ESR) appears to show the highest potential to further develop the field. The objective of this book is to provide in-depth information about new advances of SDSL-ESR in membrane biophysics, emphasizing recent developments in the application to membrane proteins. Of course, the methodology is applicable to other protein systems as well, such as water-soluble proteins, proteins assemblies, and protein–nuclei acid complexes.

The contributions in this volume illustrate the current state of the art of SDSL-ESR and highlight new advances in high-field ESR and pulsed ESR, especially aimed at providing highly resolved structural information about proteins. In addition, background information is presented concerning the analysis of ESR spectra in complex situations.

Together with this book, a CD-ROM is bundled that contains background material not suitable for inclusion within the book and software to simulate and analyze ESR spectra. In this way, the CD-ROM covers the most recent innovations and advances in the field of ESR spectroscopy, and will also be helpful in training new students in the field.

We are very much indebted to the contributors to this book, who are experts in the field, and who spent a great deal of time and effort in producing the materials for the book as well as reviewing other chapter contributions.

Marcus A. Hemminga
Wageningen, The Netherlands

Lawrence J. Berliner
Denver, Colorado, USA

CONTENTS

Chapter 3

Advanced ESR Spectroscopy in Membrane Biophysics

Janez Štrancar

Chapter 4

Practical Pulsed Dipolar ESR (DEER)

Piotr G. Fajer, Louise Brown, and Likai Song

Chapter 5

Membrane Protein Structure and Dynamics Studied by Site-Directed Spin-Labeling ESR

Enrica Bordignon and Heinz-Jürgen Steinhoff

Chapter 6

High-Field ESR Spectroscopy in Membrane and Protein Biophysics

Tatyana I. Smirnova and Alex I. Smirnov

Appendices: Software Descriptions

Appendix 1

Molecular Modeling of Spin Labels

Mikolai I. Fajer, Kenneth L. Sale, and Piotr G. Fajer

Appendix 2

SIMPOW6: A Software Package for the Simulation of ESR Powder-Type Spectra

Mark J. Nilges, Karen Mattson, and R. Linn Belford

Appendix 3

ACERT Software: Simulation and Analysis of ESR Spectra

Jack H. Freed

2. MEMBRANE PROTEINS: PRODUCTION AND RECONSTITUTION CHALLENGES

High-resolution structural studies of membrane proteins by x-ray crystallography or by nuclear magnetic resonance (NMR) spectroscopy have traditionally been limited by technical and practical difficulties (Torres et al. 2003). High-yield expression, purification, and refolding of membrane proteins are still much more demanding than the corresponding work with soluble proteins, and it is generally difficult to crystallize membrane proteins from detergent solutions (Fernández and Wüthrich 2003). There are also more fundamental reasons why the success rate for the structural characterization of membrane proteins to date has been so poor. As a class, these proteins are very different. This is largely due to the very heterogeneous environment in which they are found. This protein environment includes the bulk aqueous environment outside the membrane, the hydrocarbon core of the membrane, and the thick interfacial regions between the hydrocarbon and water domains. The lipids that make up these membranes also induce certain global properties, such as a tendency to be non-planar. Consequently, in reconstituted protein–lipid samples it is often necessary to mimic this complex environment with a simpler model environment. This latter environment may or may not be adequate to support the native conformational and functional state of the protein.

3. SDSL-ESR

Since the conventional high-resolution techniques such as x-ray crystallography and NMR spectroscopy are often insufficient to study complex protein–lipid assemblies, new biophysical characterization techniques, such as Fourier transform infrared (FTIR) (Sonar et al. 1994; Vigano et al. 2000), site-directed fluorescence labeling (SDFL), fluorescence (or Förster) resonance energy transfer (FRET) (Heuck and Johnson 2002; Chattopadhyay and Raghuraman 2004), and electron spin resonance (ESR) (or electron paramagnetic resonance), are needed to advance the field. However, the possible structural resolution of these techniques is generally less than for x-ray crystallography and NMR spectroscopy. Among those techniques, site-directed spin labeling (SDSL) ESR provides both detailed structural and dynamic information, and therefore becomes very useful for the analysis of membrane proteins. SDSL was introduced and developed by Wayne Hubbell and coworkers (Hubbell and Altenbach 1994). It offers unique local probing of the environment of a specific amino acid residue in a protein or protein complex of interest. It is based on the spin-labeling technique, pioneered by Harden McConnell in the early 1960s. The general idea was based on the "reporter group" approach in which spectroscopic probes are placed in a biological system, such as an enzyme or a protein, to monitor spectroscopic variables to which the biosystem alone is transparent. In almost all SDSL studies nitroxide spin labels are used for this purpose.

SDSL combines site-directed mutagenesis (nowadays a standard tool in molecular biology and biosynthetic protein production) with ESR spectroscopy. In general, ESR is used to investigate paramagnetic centers in a molecular system. Only electrons whose spin is not paired with the oppositely directed spin of another electron give an ESR signal. With this technique, in general, information can be obtained about certain transition metal ions, free radicals, and free electron centers. By changing the genetic code, a specific attachment point in the protein is created for a nitroxide spin label that contains a stable free electron. When a nitroxide spin label is introduced into the protein, it is influenced by interactions with surrounding structures and the motion of the protein, which is reflected in the ESR spectra of the spin label. Alternatively, two nitroxide spin labels are attached at different sites in the protein to obtain distance measurements through analyzing the magnetic dipolar interactions between the radical centers.

SDSL-ESR is tremendously useful for studying membrane proteins, because they can be examined in their native membrane, in reconstituted lipid bilayers, or in other membrane-mimicking systems. Additional advantages for using SDSL include a relatively high sensitivity (nanomolar quantities) and effectively no molecular weight limitation. ESR-derived parameters of membrane proteins may be obtained at room temperature, under physiological conditions, preserving native lipid-protein interactions.

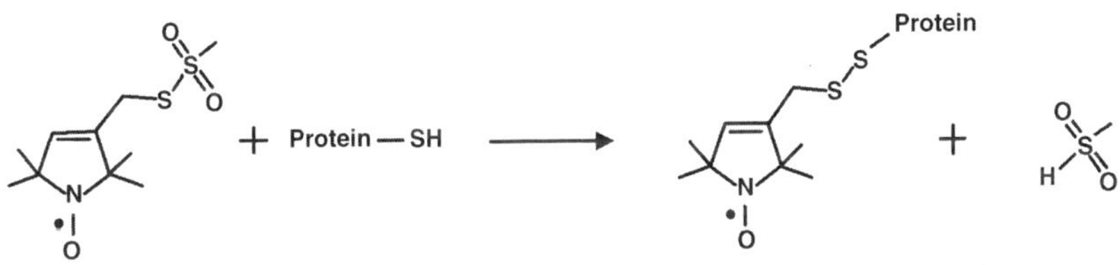

Figure 1. Reaction of the methanethiosulfonate spin label with cysteine to generate the disulfide-linked nitroxide side chain to a protein. This spin label is frequently used in SDSL of proteins. The dot indicates the free electron in the NO bond of the spin label that is stabilized by the adjacent methyl groups. The free electron has a strong anisotropic dipolar interaction with the nitrogen nucleus. Because nitrogen has a nuclear spin state of 1, the ESR line shape will basically consist of three independent line shapes arising from each of the three quantum states of the nitrogen nucleus. The electron is also sensing the anisotropic electronic environment of the chemical bond. The anisotropies of the interactions make the spin label sensitive to its motion and local structural constraints. See Mchaourab et al. (1996) for a detailed interpretation of the spectral line shapes.

4. CYSTEINE MODIFICATION

The standard labeling methodology for proteins by site-directed mutagenesis is to introduce a unique cysteine (Cys) residue at a desired location. After protein expression, the cysteine is chemically modified by reaction with a thiol-specific nitroxide spin label (a nitroxide-containing side chain, see Figure 1). This reaction is very specific. In site-directed labeling studies the question always is: what is the disturbance of the spin label on the protein under investigation? While some type of structural perturbation may occur, the evidence continually shows that the majority, including those at internal positions in a protein, does not result in disruptive structure–function changes. Thus in general, the effects are limited, but in every SDSL application this question should be addressed. Therefore, it is necessary to check the integrity and activity of the protein at each step of the procedure.

5. STRUCTURE AND DYNAMICS INFORMATION FROM SDSL-ESR

It has long been recognized that proteins are dynamic structures, and that the dynamics associated with backbone and side-chain mobilities is an essential aspect of protein function (Karplus and McCammon 1983; Parak and Frauenfelder 1993; Gumbart et al. 2005; Karplus and Kuriyan 2005). Concerning the structure–function relationships of membrane proteins, it should be noted that there are two physical aspects of a protein that are relevant for the determination of function, and SDSL-ESR enables one to correlate structure and dynamics directly to function:

1. Structure — Most information that connects protein structure to function occurs at the local molecular level.
2. Dynamics — The function of a specific protein is directly related to movement of the domains throughout its structure.

Table 1. Description of the SDSL-ESR Approach and Resulting Data

1. Use site-directed mutagenesis to place one or two cysteine residues at specific sites in the membrane protein
   - A zero-cysteine mutant may be needed
   - Check integrity and activity of the protein
2. Turn the spectroscopic data into structural and dynamic data
   - Structure
     - Distances between labeled sites (from 0.8 to 80 nm)
     - Local polarity (dielectric constant)
     - Proticity (propensity of hydrogen bonding)
     - Local order parameters
     - Topology and local protein conformation (secondary structure)
     - Location, tilt and orientation of trans-membrane helix segments
   - Dynamics
     - Motion of the nitroxide on the nanosecond time scale
     - Rotary diffusion of the protein
     - Side chain internal motions
     - Local backbone fluctuations
     - Conformational switching
     - Transient conformational states

Analysis of the ESR spectrum of the spin label provides a wealth of information about the local environment in the protein (see Table 1). With a sufficiently large set of labeled mutant proteins, global information on structure is obtained and, most importantly, changes in the structure during function can be followed in real time. Doubly labeled protein will enable one to map inter-residue distances and, ultimately, to generate reliable three-dimensional structures. Specifically, three key types of data are extracted from SDSL-ESR experiments (Hubbell et al. 1996; Hubbell et al. 1998; Hubbell et al. 2000). This multidimensional data set can be used to derive both global folding patterns and conformational changes in proteins:

1. Dynamics — The dynamics of the spin label and the membrane protein to which it is attached are revealed from the ESR line shape. Line shape analysis provides information about backbone dynamics, conformational changes, and local secondary structure (Hubbell et al. 2000; see also Bordignon and Steinhoff 2006). A semi-empirical approach is available for analysis of protein backbone dynamics that enables one to identify functional protein domains of membrane proteins (Columbus and Hubbell 2002). An example for analysis of helix dynamics of membrane proteins is shown in Figure 2.

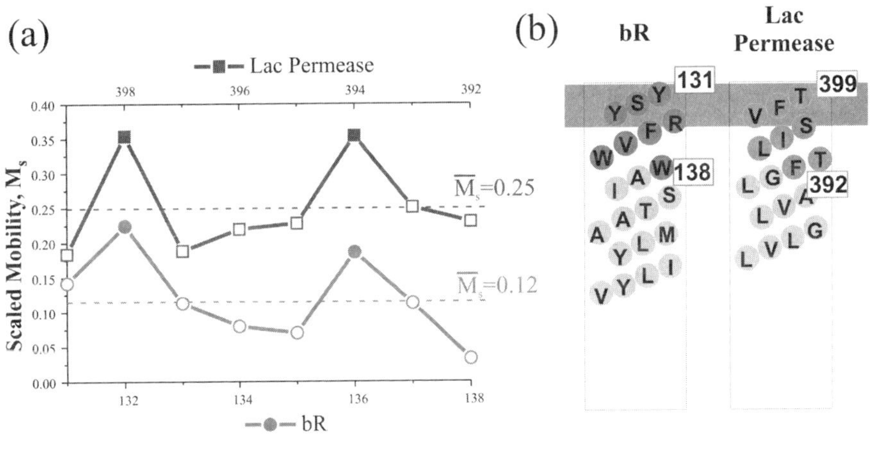

Figure. 2. Helix dynamics in the membrane proteins bacteriorhodopsin (bR) and lac permease (Columbus and Hubbell 2002). (a) The scaled mobility along 131–138 helix segment in bacteriorhodopsin (●, ○) and along helix segment 392–399 in lac permease (■, □). The mean scaled mobility for each helix, $\overline{M}_s$, which is based on differences in the line widths (Hubbell et al. 2000), is given by the dotted lines. The filled data points indicate lipid-exposed helical sites and the open data points are tertiary contacts and buried sites. (b) The helical segments of bacteriorhodopsin (bR) and lac permease, showing their orientation with respect to the membrane–aqueous interface (gray border). The SDSL-ESR data are consistent with the protein crystal structures showing that residues 132 and 136 face the bilayer in bacteriorhodopsin (1QHJ.pdb). For lac permease, residue 396 is buried in the protein core (1PV6.pdb). In fact, the SDSL work was published a year before the x-ray structure was obtained (Abramson et al. 2003).

2. Accessibility profiles — Collision parameters between the spin label and a paramagnetic species in the environment are used to generate solvent accessibility profiles (Perozo et al. 1998, 2002). These profiles are used to determine secondary structure elements and orientation within membranes and at membrane surfaces. Paramagnetic species are natural oxygen from the air that is preferentially located in the lipid bilayers (alternatively, oxygen may be added to the sample) and paramagnetic ions that are added to the sample and preferentially reside in the water phase (see Figure 3). Therefore, SDSL-ESR has been especially powerful in discerning membrane spanning vs. exposed segments, secondary structure (α vs. β vs. random structure), and helix–helix interfaces within the protein structure (Malmberg and Falke 2005; see also Bordignon and Steinhoff 2006).

3. Distance information — Measuring spin–spin dipolar couplings between two spin labels provides distance information. Various continuous wave and pulsed ESR techniques are available for distance measurements in proteins (Persson et al. 2001; Jeschke 2002). The distances between the two spin labels become constraints for three-dimensional structure determination. The dipolar interaction of the electron on the spin label is much stronger than that of a proton. Consequently, ESR can obtain measurements over much greater distances than NMR (see Bordignon and Steinhoff 2006).

6. PULSED ESR SPECTROSCOPY

Time-domain ESR represents a recent highlight of ESR development. In nuclear magnetic resonance (NMR), the use of pulsed techniques has been standard for decades, but most ESR spectrometers systems are continuous wave (cw) systems (i.e., using a fixed microwave frequency and sweeping the applied magnetic field). Pulse techniques, which involve irradiating the sample with a sequence of short pulses and then measuring the time-resolved response, offer the ability to resolve closely spaced lines and are good for measuring relaxation effects. To excite the whole spectrum at once, however, requires very short, very high-power pulses and requires system dead time (the time after the last pulse before the measurement can begin) to be minimized. System dead time often means that the signal from any broad lines has decayed before the measurement starts. Initially the focus of time-domain ESR, in particular pulsed ESR, was on solids, typically at low temperatures, but also the technological challenges were resolved for studies of biological systems under physiological conditions (Freed 2000).

In doubly spin-labeled proteins the distance between the radical centers can be investigated using frozen cw-ESR spectroscopy for intermediate distances (from 0.8 to 1.5 nm), where the exchange interaction is small and the dipolar term large (Hustedt and Beth 1999). For longer-range dipolar couplings (from 1.5 to 8 nm) pulsed ELDOR (electron–electron double resonance) (also called: PELDOR — pulsed-electron double resonance), or the pulsed DEER (double electron–electron resonance) variant (Jeschke 2002; Jeschke et al. 2004) can be applied. The high quality of the observed dipolar effects also results in distance distributions, which can be used to set upper and lower bounds in assessing macromolecular structure (see Fajer et al. 2006).

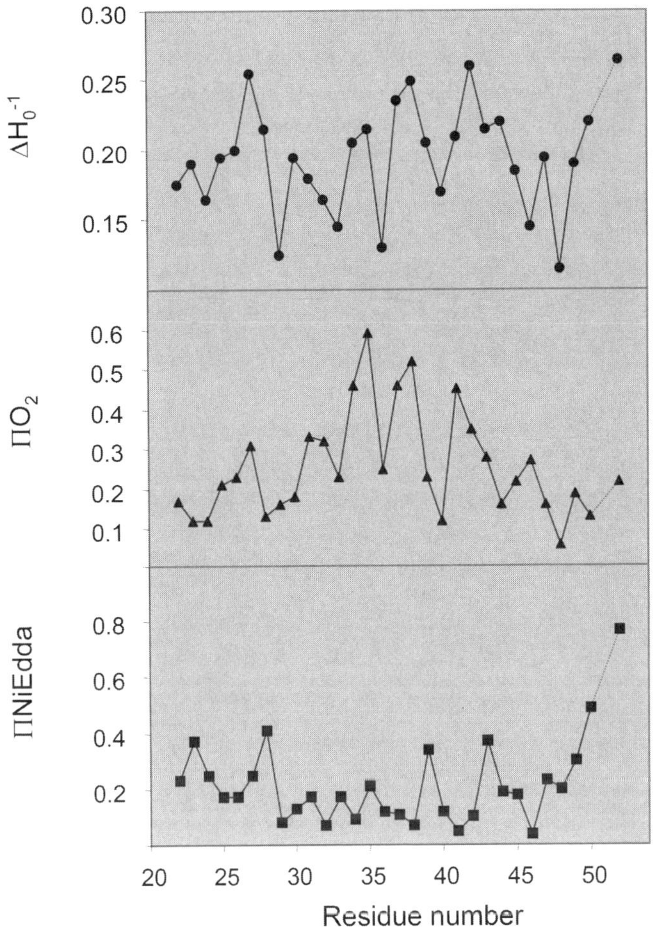

Figure. 3. Dynamics and accessibility profiles applied to structural analysis of the first transmembrane segment of the *Streptomyces* K^+ channel, a 160-residue membrane protein that includes two putative transmembrane segments and a pore region. Residue-specific environmental parameter profiles are shown for the mobility parameter ΔH_0^{-1} (top panel, filled circles), oxygen accessibility parameter ΠO_2 (center panel, filled triangles), and nickel (II) ethylenediaminediacetate (NiEdda) accessibility parameter ΠNiEdda (bottom panel, open circles). The ESR line shape reflects the local steric restrictions on the mobility of the spin label. This motional information is quantified using the inverse of the width of the central line $\Delta \underline{H}_0^{-1}$ (Mchaourab et al. 1996). A high O_2 accessibility (ΠO_2) is indicative of a membrane-exposed residue, whereas a high NiEdda accessibility (ΠNiEdda) is diagnostic of a water-exposed residue. All plots show a periodic behavior characteristic for an α-helical structure. After Perozo et al. (1998).

Table 2. Microwave Bands for ESR Spectrometers and Corresponding Magnetic Field Values for Nitroxide Spin Labels

Band name	Frequency (GHz)	Wavelength (mm)	Magnetic field (T)
X-band	9.5	31.6	0.34
K-band	24	12.5	0.86
Q-band	35	8.6	1.25
W-band	95	3.2	3.39
D-band	130	2.3	4.65

7. HIGH-FIELD ESR SPECTROSCOPY

Recent developments in superconducting magnet and millimeter-wave technology have led to significant technical advances in high-field/high-frequency ESR, offering now routine experiments at the W-band (Table 2); see Jeschke 2006 and Smirnova and Smirnov 2006. Up to now, ESR studies on biological samples were concentrated mainly on standard X-band frequencies (9.5 GHz). However, standard ESR (particularly in frozen solution samples) suffers from strong inhomogeneous line broadening, resulting in low spectral resolution (Figure 4). Over the past years a number of laboratories has met the technological challenge to construct mm and sub-mm high-field ESR spectrometers, thereby opening a promising new research area (Freed 2000; Steinhoff et al. 2000; Smirnov 2002; Möbius et al. 2005a). Nevertheless, high-field electron paramagnetic resonance (EPR) instrumentation development will remain a challenging task for experimental physicists. In summary, high-field ESR instrumentation in combination with novel theoretical methodologies leads to:

1. Increased signal-to-noise ratio (S/N), giving rise to the use of smaller samples and/or lower concentrations. Optimum concentration sensitivity appears to be attainable in the W-band region. However, at much higher microwave frequencies a compromise has to be made between the gain in resolution and loss of sensitivity that, for technical reasons, turns out to be unavoidable (Möbius et al. 2005b).

2. Improved spectral resolution by an enhanced *g*-factor resolution compared to ESR at conventional frequencies. This results in a better description of the chemical environment (local polarity) of the spin label. For example, changes in the protein microenvironment, including conformational changes, can be identified as potential molecular switches for membrane transport processes in proteins at work (Möbius et al. 2005; see also Bordignon and Steinhoff 2006).

3. A faster "snapshot" of the motional dynamics. Due to the fact that ESR spectroscopy at high frequencies is sensitive to faster molecular motions, new information about molecular motions in biomolecules can be obtained. These mo-

tions are associated with characteristic dynamic processes, such as protein or lipid motion and protein folding.

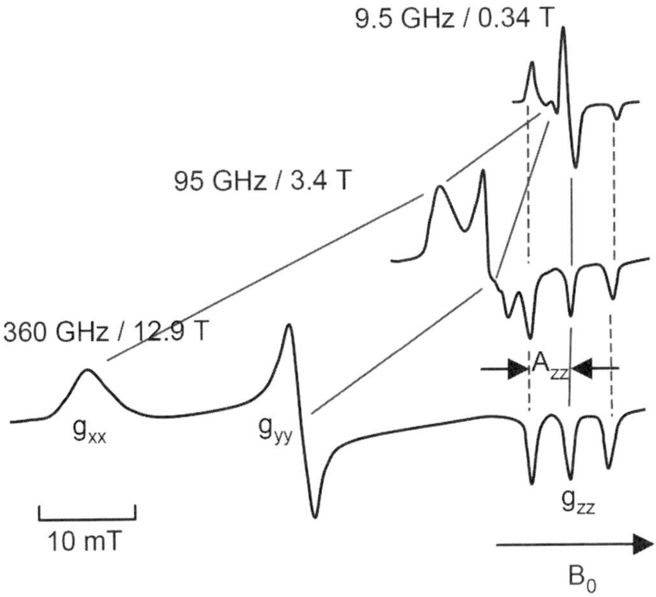

Figure 4. Dramatic gain in resolution of nitroxide spin probe powder ESR spectra by separation of the *g*-factor tensor components at high field. At the X-band (9.5 GHz) the ESR spectrum is dominated by the anisotropic components of the hyperfine interaction tensor, of which the component A_{zz} is most significant. At high field (360 GHz), the spectral features are dominated by the components of the *g*-factor tensor g_{xx}, g_{yy}, and g_{zz}. After Möbius et al. (2005).

8. MOLECULAR DYNAMICS SIMULATIONS

A further new development is the introduction of molecular dynamics (MD) simulations to study the conformational dynamics of membrane proteins. Molecular dynamics simulation numerically solves Newton's equations of motion on an atomic level of the protein–lipid system to obtain information about its time-dependent properties. There is a large body of evidence to show that the restriction of nanosecond dynamic disorder exhibited by a spin label attached to a protein is a direct reflection of the local structure and dynamics of the protein, due to nearby residues within the protein or to contacts made with other molecules (Steinhoff and Hubbell 1996; LaConte et al. 2002; Steinhoff 2002; Fu et al. 2003; Kim et al.

2004). When attached to a specific protein residue, probe restrictions are well modeled by an MD simulation reflecting the restrictions imposed by neighboring protein atoms (Stoica 2004). In the past decade, the calculation speed of computers has increased tremendously, and novel advances in software developments have been implemented, bringing molecular dynamics simulations now in the nanosecond time domain of ESR spectroscopy. Analysis of the dynamics and accessibility of the spin label side chains as well as determination of inter-spin distances and of the polarity in the vicinity of the spin label binding sites provide information for restraint modeling of protein structures and conformational changes (see Figure 5). For the purpose of studying proximity relationships within doubly spin-labeled variants, double electron–electron resonance (DEER) techniques have been developed and successfully applied (see Fajer et al. 2006).

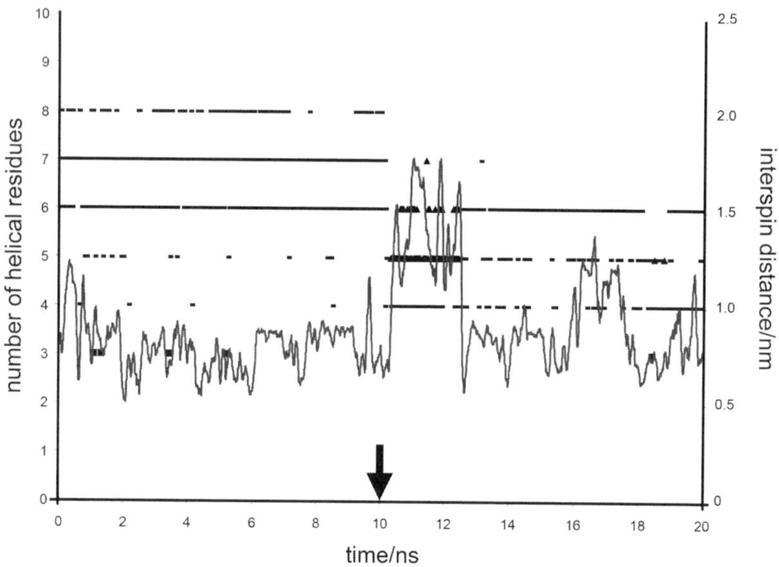

Figure 5. Molecular dynamics simulation of a doubly labeled MTSSL-labeled artificial decapeptide starting from an α-helix conformation (Vos et al. 2006). On the left axis the number of helical residues in terms of the 3_{10}-helix (■), α-helix (–), and π-helix (▲) is shown. Snapshots were taken every 10 ps. The right axis shows the interspin distance. Here a moving average with period 10 is used. During the first half of the simulation (0–10 ns), a restraining potential on the C_α atoms was applied to keep the peptide in an α-helix conformation. The arrow indicates the time (10 ns) at which the restraining potential was removed. Because of crowding of the data points on the horizontal axis, they are not shown (most of the time the number of residues in the 3_{10}- or π-helix is zero).

9. SPECTRAL SIMULATION AND ANALYSIS

Since the first applications of spin-label ESR spectroscopy in biosystems, computer simulation of the spectra was an important step to quantitate the motional and structural characteristics of the spin probe and its environment. The ESR spectra arising from many biological studies are complex and can be easily misinterpreted by superficial analysis. Many contemporary applications of ESR would not be possible without the pioneering work in theoretical and simulation methods developed by Jack Freed and his research group (Freed 1972). There is also a strong educational and training aspect of computer simulations in that one can systematically evaluate the dependencies of spectral features on the magnetic, dynamical, and structural parameters. So, one can learn a lot by simply simulating ESR spectra.

Recent new developments in analysis software are the use of automated spectral fitting procedures, based on genetic algorithms for ESR spectral parameter optimization (Filipic and Štrancar 2001; Stopar et al. 2005), as shown in Figure 6, and analysis of DEER (double electron–electron resonance) signals to obtain accurate inter-spin distances (Jeschke 2002).

10. COMPARISON WITH SITE-DIRECTED FLUORESCENCE LABELING

In many aspects, site-directed fluorescence labeling (SDFL) is comparable to SDSL. Also, SDFL involves introducing unique cysteine residues at defined sites in a protein, then attaching fluorescent labels to these sites to act as local reporter groups. The information these probes report on their local environment (for example, buried vs. solvent exposed) is then used to assess the structure and monitor dynamic changes in the protein. By systematically scanning through a protein region, SDFL can be used to obtain localized secondary structure information, map proximity in proteins, study protein folding, assess conformational changes in a protein's structure, determine membrane protein topology and insertion, and monitor protein–protein interactions (Spruijt et al. 1996; Spruijt et al. 2000; Koehorst et al. 2004; Mansoor and Farrens 2004; Sosa-Peinado and Gonzalez-Andrade 2005). In fact, by using the same cysteine mutants, both SDSL and SDFL can be combined to provide complementary information about the protein. In addition, SDFL can be used in combination with distance measurement techniques such as fluorescence (or Förster) resonance energy transfer (FRET) to determine structure (Vos et al. 2005). Measurements of longer distances are most effective in providing constraints on the structures of large, complex proteins. DEER and related techniques have several key advantages when compared to FRET:

1. Nitroxide spin labels are usually smaller than fluorophore molecules.
2. There is no orientational problem, since the electron–spin dipole moments are aligned by the magnetic field.
3. The same spin label is used at both sites rather than a distinct donor and acceptor.
4. The distribution in distances can be directly measured.

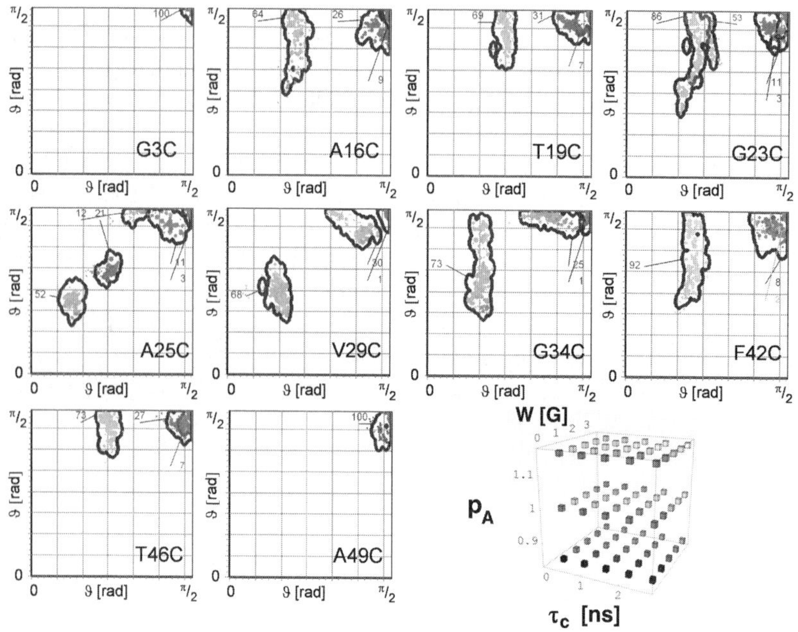

Figure 6. Analysis of complex ESR spectra of different spin-labeled M13 coat protein mutants reconstituted in DOPC vesicles in terms of polarity, spin-label mobility, and exposure to water/lipid phases. The parameter ϑ is the cone angle within the spin label attached to the protein, and parameter φ describes the asymmetry of the cone. The relative fractions of a group of solutions for the two spectral parameters are indicated for each spin-labeled mutant with a larger font, in addition to small gray letters that indicate the fraction of the subgroups, if they exist. The cube at the lower right bottom codes for the relative values of the correlation time τ_c, additional spectral broadening W, and correction for polarity p_A in their definition intervals (0–3 ns), (0–4 G), and (0.8–1.2), respectively. The light-gray component represents the asymmetry of the local environment consistent with a transmembrane topology of the protein. The other fractions arise from nonspecific spin labeling of the protein (Štrancar 2006). Reprinted with permission from Stopar et al. 2005. Copyright © 2005, American Chemical Society.

On the other hand, the fluorescence technique is much more sensitive than the ESR technique, so that greatly lower protein concentrations can be used. This is often an advantage, if the production of protein material is a limitation. Another benefit of fluorescence is that distance measurements using FRET can be carried out at room temperature, whereas distance analysis by ESR needs to be done at low temperatures (around 200 K).

11. FUTURE

As compared to isotope labeling studies as carried out by FTIR and NMR spectroscopy that are effectively non-disturbing, SDSL has the disadvantage of introducing a protein mutation in combination with covalently adding a molecular probe. The risk of this methodology is that an unwanted effect on the structure and function of the protein is introduced. The problems related to the disturbance of the labeling may be strongly reduced eventually by new insights arising from MD simulations and molecular modeling that can explicitly take into account the cysteine mutation and molecular properties of the spin label. In addition, new spin labeling strategies are under development (Becker et al. 2005). A promising general method is based on expanding the genetic code to include new "unnatural" spin-labeled amino acids. These amino acid analogues can then directly be introduced into the protein sequence at any desired place (Dougherty 2000; Hendrickson et al. 2004; Hahn and Muir 2005). Nevertheless, a major limitation of this method remains the charging of the tRNAs with unnatural amino acids (Lu 2005).

Spin-labeled amino acids with reduced residual side chain motion will provide better-defined orientations with respect to the protein backbone. This will simplify discrimination between protein backbone dynamics and its modulation upon conformational changes and residual side chain dynamics. Furthermore, the reduced residual motion and the defined orientation of the nitroxide side chain with respect to the backbone will provide reliable data of inter-spin distances and relative orientations of the nitroxide side chains. It may be expected that by using such sets of doubly spin-labeled engineered proteins, the determination of structural details and conformational changes with atomic resolution will be achievable (Becker et al. 2005).

As has been shown above, there is an impressive range of novel methodologies available for SDSL-ESR, all of which have emerged from recent instrumental and methodological developments. SDSL has dramatically advanced the type of questions biochemists and biophysicists can address using ESR spectroscopy. Given the highly flexible way of introducing spin labels at virtually any desired site in a protein, the relatively low quantities of material needed for ESR, and the possibility to work under physiological conditions, challenging questions related to the three-dimensional structure and structure function relationships can be tackled for membrane proteins. This field of protein research is emerging as one of the most important areas in structural proteomics nowadays, as the use of other biophysical techniques for the purpose of structure determination in these systems is generally limited. In this situation, SDSL-ESR comes out as a key methodology in advancing the field of structure determinations and linking structural changes to membrane protein function. Indeed, the future of SDSL-ESR for membrane protein research is bright.

12. ACKNOWLEDGMENTS

I gratefully acknowledge Wayne L. Hubbell for reading the manuscript and for providing Figure 2. Also, Laura Wessels helped a lot in providing useful comments on the manuscript.

13. REFERENCES

Abramson J, Smirnova I, Kasho V, Verner G, Kaback HR, Iwata S. 2003. Structure and mechanism of the lactose permease of *Escherichia coli*. *Science* **301**:610–615.

Arora A, Tamm LK. 2001. Biophysical approaches to membrane protein structure determination. *Curr Opin Struct Biol* **11**:540–547.

Becker CFW, Lausecker K, Balog M, Kálai T, Hideg K, Steinhoff H-J, Engelhard M. 2005. Incorporation of spin-labelled amino acids into proteins. *Magn Reson Chem* **43**:S34–S39.

Bordignon E, Steinhoff H-J. 2006. Distance measurements in membrane proteins. Chapter 5 in this volume.

Bowie JU. 2001. Stabilizing membrane proteins. *Curr Opin Struct Biol* **11**:397–402.

Chattopadhyay A, Raghuraman H. 2004. Application of fluorescence spectroscopy to membrane protein structure and dynamics. *Curr Sci* **87**:175–180.

Columbus L, Hubbell WL. 2002. A new spin on protein dynamics. *Trends Biochem Sci* **27**:288–295.

Dougherty DA. 2000, Unnatural amino acids as probes of protein structure and function. *Curr Opin Chem Biol* **4**:645–652.

Fajer PG, Brown L, Song L. 2006. Practical pulsed dipolar ESR (DEER). Chapter 4 in this volume.

Fernández C, Wüthrich K. 2003. NMR solution structure determination of membrane proteins reconstituted in detergent micelles. *FEBS Lett* **555**:144–150.

Filipic B, Štrancar J. 2001. Tuning EPR spectral parameters with a genetic algorithm. *Appl Soft Comp* **1**:83–90.

Freed JH. 1972. Electron spin resonance. *Annu Rev Phys Chem* **23**:265–310.

Freed JH. 2000. New technologies in electron spin resonance. *Annu Rev Phys Chem* **51**:655–689.

Fu Z, Aronoff-Spencer E, Backer JM, Gerfen GJ. 2003. The structure of the inter-SH2 domain of class IA phosphoinositide 3-kinase determined by site-directed spin labeling EPR and homology modeling. *Proc Natl Acad Sci USA* **100**:3275–3280.

Gumbart J, Wang Y, Aksimentiev A, Tajkhorshid E, Schulten K. 2005. Molecular dynamics simulations of proteins in lipid bilayers. *Curr Opin Chem Biol* **15**:423–431.

Hahn ME, Muir TW. 2005. Manipulating proteins with chemistry: a cross-section of chemical biology. *Trends Biochem Sci* **30**:26–34.

Hendrickson TL, de Crécy-Lagard V, Schimmel P. 2004. Incorporation of nonnatural amino acids into proteins. *Annu Rev Biochem* **73**:147–176.

Heuck AP, Johnson AE. 2002. Pore-forming protein structure analysis in membranes using multiple independent fluorescence techniques. *Cell Biochem Biophys* **36**:89–101.

Hubbell WL, Altenbach C. 1994. Investigation of structure and dynamics in membrane proteins using site-directed spin labeling. *Curr Opin Struct Biol* **4**:566–573.

Hubbell WL, Mchaourab HS, Altenbach C, Lietzow MA. 1996. Watching proteins move using site-directed spin labeling. *Structure* **4**:779–783.

Hubbell WL, Gross A, Langen R, Lietzow MA. 1998. Recent advances in site-directed spin labeling of proteins. *Curr Opin Struct Biol* **8**:649–656.

Hubbell WL, Cafiso DS, Altenbach C. 2000. Identifying conformational changes with site-directed spin labeling. *Nat Struct Biol* **7**:735–739.

Hustedt EJ, Beth AH. 1999, Nitroxide spin–spin interactions: applications to protein structure and dynamics. *Annu Rev Biophys Biomol Struct* **28**:129–153.

Jeschke G. 2002. Determination of the nanostructure of polymer materials by electron paramagnetic resonance spectroscopy. *Macromol Rapid Commun* **23**:227–246.

Jeschke G. 2006. Instrumentation and experimental setup. Chapter 2 in this volume.

Jeschke G, Wegener C, Nietschke M, Jung H, Steinhoff H-J. 2004. Interresidual distance determination by four-pulse double electron–electron resonance in an integral membrane protein: the Na^+/proline transporter PutP of *Escherichia coli*. *Biophys J* **86**:2551–2557.

Karplus M, Kuriyan J. 2005. Molecular dynamics and protein function. *Proc Natl Acad Sci USA* **102**:6679–6685.

Karplus M, McCammon JA. 1983. Dynamics of proteins: elements and function. *Annu Rev Biochem* **52**:263–300.

Kim N-K, Murali A, DeRose VJ. 2004. A distance ruler for RNA using EPR and site-directed spin labeling. *Chem Biol* **11**:939–948.

Koehorst RB, Spruijt RB, Vergeldt FJ, Hemminga MA. 2004, Lipid bilayer topology of the transmembrane alpha-helix of M13 Major coat protein and bilayer polarity profile by site-directed fluorescence spectroscopy. *Biophys J* **87**:1445–1455.

LaConte LEW, Voelz V, Nelson W, Enz M, Thomas DD. 2002. Molecular dynamics simulation of site-directed spin labeling: experimental validation in muscle fibers. *Biophys J* **83**:1854–1866.

Lu Y. 2005. Design and engineering of metalloproteins containing unnatural amino acids or non-native metal-containing cofactors. *Curr Opin Chem Biol* **9**:118–126.

Malmberg NJ, Falke JJ. 2005. Use of EPR power saturation to analyze the membrane-docking geometries of peripheral proteins: applications to C2 domains. *Annu Rev Biophys Biomol Struct* **34**:71–90.

Mansoor SE, Farrens DL. 2004. High-throughput protein structural analysis using site-directed fluorescence labeling and the Bimane derivative (2-pyridyl)dithiobimane. *Biochemistry* **43**:9426–9438.

Mchaourab HS, Lietzow MA, Hideg K, Hubbell WL. 1996. Motion of spin-labeled side chains in T4 lysozyme: correlation with protein structure and dynamics. *Biochemistry* **35**:7692–7704.

Möbius K, Savitsky A, Schnegg A, Plato M, Fuchs M. 2005a. High-field EPR spectroscopy applied to biological systems: characterization of molecular switches for electron and ion transfer. *Phys Chem Chem Phys* **7**:19–42.

Möbius K, Schnegg A, Plato M, Fuchs MR, Savitsky A. 2005b. High-field EPR spectroscopy on transfer proteins in biological action. *Acta Phys. Pol. A* **108**:215–234.

Parak F, Frauenfelder H. 1993. Protein dynamics. *Physica A* **201**:332–345.

Perozo E, Cortes DM, Cuello LG. 1998, Three-dimensional architecture and gating mechanism of a K^+ channel studied by EPR spectroscopy. *Nat Struct Biol* **5**:459–469.

Perozo E, Cortes DM, Sompornpisut P, Kloda A, Martinac B. 2002. Open channel structure of MscL and the gating mechanism of mechanosensitive channels. *Nature* **418**:942–948.

Persson M, Harbridge JR, Hammarstrom P, Mitri R, Martensson L-G, Carlsson U, Eaton GR, Eaton SS. 2001. Comparison of electron paramagnetic resonance methods to determine distances between spin labels on human carbonic anhydrase II. *Biophys J* **80**:2886–2897.

Smirnov AI. 2002, Spin-labeling in high-field EPR. *Electron Paramagn Reson* **18**:109–136.

Smirnova TI, Smirnov AI. 2006. High-field ESR spectroscopy in membrane and protein biophysics. Chapter 6 in this volume.

Smith WL, Garavito RM, Ferguson-Miller S. 2001. Membrane protein structural biology minireview series. *J Biol Chem* **276**:32393–32394.

Sonar S, Lee C-P, Coleman M, Patel N, Liu X, Marti T, Khorana HG, RajBhandary UL, Rothschild KJ. 1994, Site-directed isotope labelling and FTIR spectroscopy of bacteriorhodopsin. *Nat Struct Biol* **1**:512–517.

Sosa-Peinado A, Gonzalez-Andrade M. 2005. Site-directed fluorescence labeling reveals differences on the R-conformer of glucosamine 6-phosphate deaminase of *Escherichia coli* induced by active or allosteric site ligands at steady state. *Biochemistry* **44**:15083–15092.

Spruijt RB, Wolfs CJAM, Verver JWG, Hemminga MA. 1996. Accessibility and environmental probing using cysteine residues introduced along the putative transmembrane domain of the major coat protein of bacteriophage M13. *Biochemistry* **35**:10383–10391.

Spruijt RB, Meijer AB, Wolfs CJAM, Hemminga MA. 2000. Localization and rearrangement modulation of the N-terminal arm of the membrane-bound major coat protein of bacteriophage M13. *Biochim Biophys Acta* **1509**:311–323.

Steinhoff H-J. 2002. Methods for study of protein dynamics and protein–protein interaction in protein–ubiquitination by electron paramagnetic resonance spectroscopy. *Front Biosci* **7**:C97–110.

Steinhoff H, Hubbell W. 1996. Calculation of electron paramagnetic resonance spectra from Brownian dynamics trajectories: application to nitroxide side chains in proteins. *Biophys J* **71**:2201–2212.

Steinhoff H-J, Savitsky A, Wegener C, Pfeiffer M, Plato M, Mobius K. 2000. High-field EPR studies of the structure and conformational changes of site-directed spin labeled bacteriorhodopsin. *Biochim Biophys Acta* **1457**:253–262.

Stoica I. 2004. Using molecular dynamics to simulate electronic spin resonance spectra of T4 lysozyme. *J Phys Chem B* **108**:1771–1782.

Stopar D, Štrancar J, Spruijt RB, Hemminga MA. 2005. Exploring the local conformational space of a membrane protein by site-directed spin labeling. *J Chem Inf Model* **45**:1621–1627.

Torres J, Stevens TJ, Samsó M. 2003, Membrane proteins: the "Wild West" of structural biology. *Trends Biochem Sci* **28**:137–144.

Štrancar J. 2006. Advanced ESR spectroscopy in membrane biophysics. Chapter 3 in this volume.

Vigano C, Manciu L, Buyse F, Goormaghtigh E, Ruysschaert J-M. 2000. Attenuated total reflection IR spectroscopy as a tool to investigate the structure, orientation and tertiary structure changes in peptides and membrane proteins. *Biopolymers* **55**:373–380.

Vos WL, Koehorst RBM, Spruijt RB, Hemminga MA. 2005. Membrane-bound conformation of M13 major coat protein: a structure validation through FRET-derived constraints. *J Biol Chem* **280**:38522–38527.

Vos WL, Vermeer LS, Hemminga MA. 2006. Conformation of a peptide fragment encompassing the proton channel of vacuolar H^+–ATPase. In press.

White S. 2006. Membrane proteins of known 3D structure. http://blanco.biomol.uci.edu/Membrane_Proteins_xtal.html.

CHAPTER 2

INSTRUMENTATION AND EXPERIMENTAL SETUP

Gunnar Jeschke
Max Planck Institute for Polymer Research,
Mainz, Germany

Instrumentation aspects of ESR spectroscopy are treated thoroughly in a standard text (Poole 1997). However, a majority of ESR users are nowadays concerned with application work rather than development of instrumentation and methods. For such work, a comprehensive knowledge of instrumentation aspects is not required, yet a basic knowledge is needed to measure high-quality spectra and notice problems with hardware. The present chapter intends to present just this basic knowledge and to relate it to proper procedures for experimental setup. It complements Poole's book, and where more detailed descriptions are needed, they can probably be found there. For pulsed ESR and ENDOR, selected original papers on instrumentation are cited to take modern developments into account. High-field ESR beyond frequencies of 95 GHz, where no fully fledged commercial spectrometers exist, is not included here. This reflects my lack of experience rather than a judgment on the usefulness of higher frequencies for work on biological membranes and membrane proteins. Note also that I discuss instrumentation and setup procedures with particular emphasis on measuring spectra of nitroxide spin probes and spin labels, although it should not be too difficult to adapt the procedures to work on transition metal centers.

The chapter starts with a short section on continuous-wave (CW) ESR, as I feel that the physical intricacies of this experiment are often underestimated. This section also introduces basic knowledge on resonators and coupling and should be studied even if one is mainly concerned with pulsed ESR. The following section on basic pulsed ESR emphasizes the different optimization criteria for resonators compared to CW ESR and their different handling. It also discusses the question of proper echo integration, as sensitivity depends critically on the choice of the integration window. The section on pulsed ENDOR discusses the reasons for baseline problems and the fact that Davies ENDOR is better suited for large couplings and Mims ENDOR for small couplings. It also looks at the question of how to measure very small intramolecular hyperfine couplings that can provide distance information on a length scale between 0.3 and 1 nm. The final section on pulsed ELDOR stresses the fact that instrumentation for implementing ELDOR for pulsed work is actually simpler than for CW work. The discussion of experimental setup concentrates on distance measurements. Obtaining the full information from such measurements — reliably — depends more strongly on a precise setup than is the case for most other ESR measurements.

Readers should realize that it is impossible to give foolproof setup procedures that will work with any given spectrometer. When reading this chapter, they should aim to understand *why* these particular procedures are recommended, as such understanding is the basis for adapting them to different hardware. Finally, note that references are selected merely to illustrate thoughts that cannot be evolved in detail here. They are not comprehensive, as in a review, and are not intended to provide an overview of the history of ESR instrumentation.

1. CONTINUOUS-WAVE ESR

CW ESR (Weil et al. 1994) spectra are often analyzed for their line shape or fitted by least-square procedures. This is particularly true for spectra of nitroxide spin probes or labels. Such data analysis depends crucially on how reliable the experimental line shapes are. Investing ten minutes more in a careful experimental setup may save hours of work in data analysis.

1.1. How the Signal Arises

In the CW ESR experiment the first derivative of a field-swept ESR absorption spectrum is measured. The microwave frequency $\omega_{\mathrm{mw}} = 2\pi\, \nu_{\mathrm{mw}}$ during the experiment is constant. Thus sensitivity can be increased by optimizing the high-frequency components for a narrow frequency band and by placing the sample in a microwave (m.w.) resonator. The use of a suitable m.w. resonator and its proper coupling are crucial for obtaining correct CW ESR line shapes. An important characteristic of the resonator (or cavity) is its loaded quality factor:

$$Q_L = \frac{\omega_{mw} L}{R_0 n^2 + r}, \tag{1}$$

where L is the inductivity of the resonator, R_0 is the impedance of the line that transmits the m.w. power from the source to the resonator (usually 50 Ω), n is a parameter that characterizes the coupling between transmission line and resonator, and r is the impedance of the resonator. The impedance r varies between samples due to dielectric m.w. losses in the samples. It does not include the resonant m.w. absorption of the electron spins. The quality factor is related to the width $\Delta\nu$ of the resonator mode by

$$\Delta\nu = \frac{\nu_{mw}}{Q_L}. \tag{2}$$

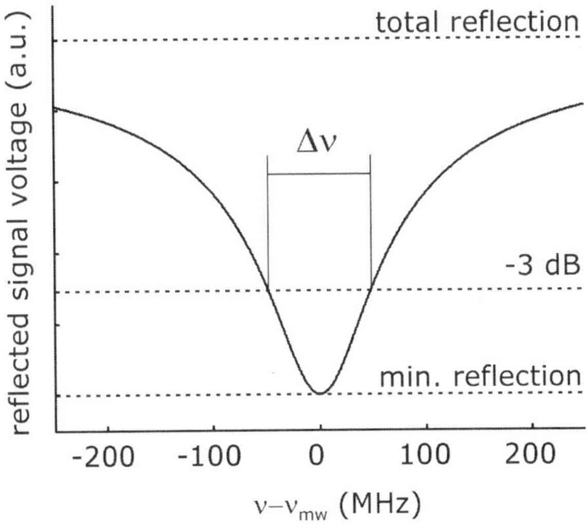

Figure 1. Idealized m.w. mode (dip) with $Q_L = 100$ at a frequency of 9.6 GHz. At the 3-dB points, the power that enters the resonator is half as large as on resonance ($\nu - \nu_{mw} = 0$). The 3-dB resonance band width $\Delta\nu$ in this example is 96 MHz.

The shape of the resonator mode is shown in Figure 1. Most CW ESR spectrometers use a reflection resonator (Figure 2). The m.w. power is transmitted through ports 1 and 2 of a circulator to the resonator and reflected power is transmitted from the resonator through ports 2 and 3 of the circulator to the detector (m.w. diode). With the electron spins off resonance, the reflected power should be zero. This is achieved by critical coupling of the resonator, i.e., the coupling is adjusted so that the condition

$$R_0 n^2 = r \tag{3}$$

is fulfilled. In this situation, all m.w. power that enters the resonator is converted to heat in the impedance r.

Now consider what happens during the field sweep when the electron spins become resonant with the irradiated m.w. frequency. The sample will then absorb additional m.w. power, which drives the ESR transition and is eventually also converted to heat via spin–lattice relaxation. In other words, the impedance of the resonator increases to $r + \Delta r$. This means that the condition of critical coupling, Eq. (3), is no longer fulfilled. Hence power is reflected from the resonator and transmitted to the detector.

Based on these considerations the dependence of the detected signal on Δr and hence on the amplitude $\chi''(\omega)$ in an ESR absorption spectrum can be derived (Feher 1957). The change in input voltage of the detector, ΔV_{refl}, is given by

$$\Delta V_{\text{refl}} = C\chi''(\omega)\eta Q_L, \tag{4}$$

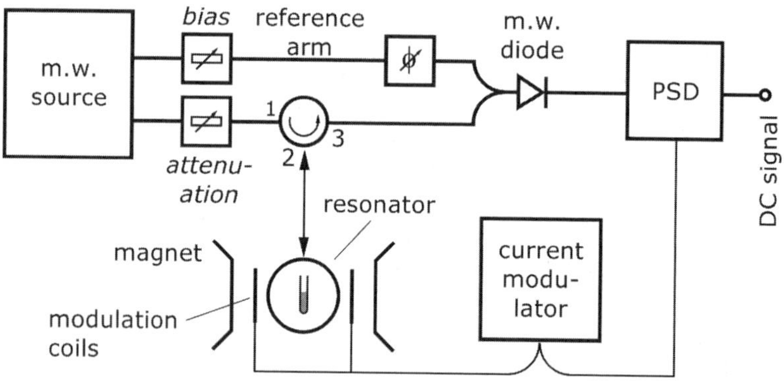

Figure 2. Simplified scheme of a CW ESR spectrometer.

where C is an instrumental constant and η is the filling factor of the resonator. The filling factor is the integral of the m.w. field amplitude B_1 over the sample volume divided by the integral of B_1 over the whole resonator.

It follows that sensitivity decreases if Q_L decreases due to dielectric losses. This is usually an important consideration in work on biological membranes or membrane proteins, since these samples contain water which has significant dielectric m.w. losses at all relevant ESR frequencies. As the dielectric losses are associated with the m.w. electric field, a good resonator should exclude the electric field from the sample volume. Furthermore, the part of the cavity where the m.w. magnetic field B_1 is concentrated should be filled as much as possible with sample. In a rectangular cavity, both these conditions can be fulfilled by placing the sample in a flat cell or in a bundle of capillaries such as a Bruker's AquaX cell. A disadvantage of flat cells and AquaX for work with precious membrane protein samples is the relatively large required sample volume. Loop-gap resonators, as available from Jagmar (Krakow, Poland) or in the U.S. via Molecular Specialties Inc. (Milwaukee, WI), have a lower intrinsic Q_L than cavities but very high η and allow for measurements with less than 1 μL of sample.

The reflected signal is usually detected by an m.w. diode (crystal detector). To measure correct line shapes the output signal of the diode needs to be linear to input voltage, i.e., to the square root of microwave power P_{mw}. At very low input power this is not the case. In order to detect the very small reflected power in the linear regime of the diode, it has to be biased by some reference power (Figure 2). Best results are achieved with a bias that puts the off-resonance detector current to 200 μA. Note that m.w. diodes are slowly destroyed by too much input power. One should avoid situations where the diode current is at the upper limit or exceeds it. Therefore, one should not touch the sample when in operate mode (this may shift the m.w. dip and lead to total reflection of the power) and decrease m.w. attenuation only gradually, as a large fraction of power may be reflected unless the resonator is critically coupled. Modern spectrometers may have protection circuitry that limits the input power of the diode. However, it is good practice not to rely on such protection.

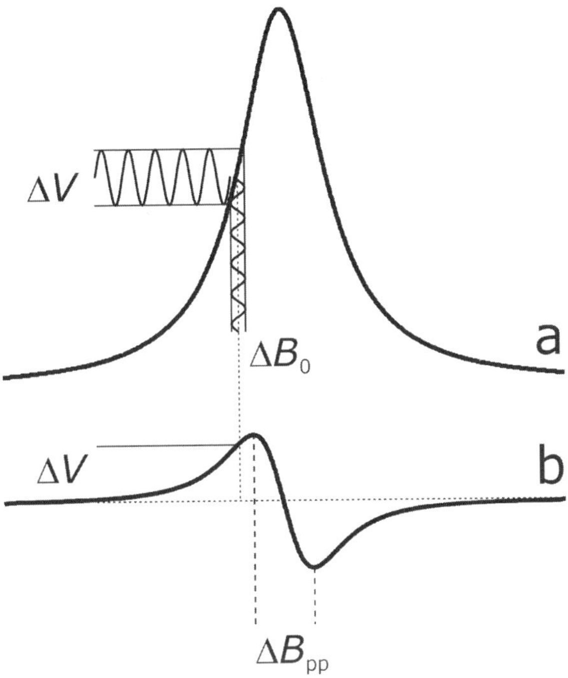

Figure 3. Detection of the first derivative of the absorption line by field modulation with amplitude ΔB_0. The signal after the m.w. diode (see Figure 2) oscillates with the frequency of the field modulation and with amplitude ΔV. The signal after the phase-sensitive detector (PSD) is a DC signal with amplitude ΔV. The field difference between the inflexion points of the absorption line is the peak-to-peak line width, ΔB_{pp}.

If one would just apply continuous m.w. power to the sample and sweep the field as described so far, the output current of the diode would be proportional to the absorption spectrum, not its derivative. The derivative is obtained by modulating the external magnetic field B_0, which leads to an oscillating absorption signal, and by detecting the amplitude of this oscillation with a phase-sensitive detector (PSD). The principle is depicted in Figure 3. The advantages of this additional step are a rejection of noise that does not oscillate with the modulation frequency and of non-resonant m.w. absorption that does not depend on the magnetic field. Furthermore, the derivative spectrum is better resolved than the absorption spectrum. As a disadvantage, signals that are much broader than the maximum modulation amplitude are hard to detect, because field modulation then causes only very small signal oscillations.

Generally, sensitivity increases with increasing modulation amplitude ΔB_0 as long as it does not exceed the line width. To obtain a very high-quality line shape for a strong sample, the amplitude of the field modulation should be five times smaller than the peak-to-peak line width, ΔB_{pp}. However, unless very small effects are to be detected in line shape analysis, the modulation amplitude can be selected as large as $\Delta B_{pp}/3$. If just the resonance field, but not the line width, is to be measured, the best sensitivity is obtained when ΔB_0 matches ΔB_{pp}. In Figure 3, the modulation amplitude was chosen 4.12 times smaller than the line width. For nitroxides, modulation amplitudes of 2 G in the solid state and 1 G in solution are usually a good choice. Two times smaller modulation amplitudes may be required for deuterated nitroxides or nitroxides that don't have protons, such as Fremy's salt. At high field, a modulation amplitude of 2 G is often tolerable even in solution.

The standard modulation frequency of 100 kHz may lead to line shape distortions by modulation sidebands for lines that are narrower than approximately 100 mG (Weil et al. 1994). Such narrow lines are hardly ever encountered with nitroxides. At high fields, it may nevertheless be advantageous to decrease the modulation frequency. This is because mechanical vibration caused by the field modulation (microphonics) can be a significant source of noise for the small and delicate high-frequency resonators. The optimum modulation frequency depends on the particular probe head and has to be found out experimentally.

1.2. Matching and Phasing with Automatic Frequency Control

Standard CW ESR spectrometers at frequencies up to Q band (35 GHz) feature automatic frequency control (AFC). By AFC small frequency shifts of the resonator mode due to temperature changes and due to the real part of the magnetic susceptibility of the sample $\chi'(\omega)$ are compensated for by automatic adjustment of the m.w. frequency. To measure with an almost critically coupled resonator, AFC is mandatory, as otherwise the magnetic susceptibility of the sample may carry the resonator from undercoupling to overcoupling (or vice versa), which leads to a sign change of the signal (Feher 1957). This effect can indeed be observed with very strong samples such as DPPH whose susceptibility upsets AFC. Adjustment of the m.w. phase is also simplified by AFC, as it keeps the frequency exactly at the center of the m.w. mode. If the mode is symmetric, as is usually the case for critically coupled resonators in this frequency range, the following procedure avoids any line shape distortions by phase errors or by nonlinear detection.

1. Go to tune mode. Switch reference arm off or set bias to zero. Find the resonator dip (m.w. mode) and shift it to the center of the display/oscilloscope picture. Adjust coupling so that the dip is as deep as possible. This is critical coupling to zeroth approximation.

2. Go to operate mode; set attenuation to 20 dB. If the diode current is at the upper limit, increase attenuation until it is in the range of the display. Adjust frequency so that AFC error is close to zero (lock offset display). Adjust coupling so that the diode current is as small as possible. *Slowly* decrease attenuation and readjust coupling, keeping diode current as small as possible. If you use the correct

sample cell and it is correctly placed in the resonator, it should be possible to decrease attenuation to 0 dB and still have the diode current at the lower limit. If you cannot get diode current to the lower limit at 20 dB, you should consider using a different sample cell or at least a tube with a smaller diameter. As a stopgap solution, you may go back to tune mode and pull the sample tube out until the dip is sufficiently narrow and deep, the restart at step 1. In fact, with resonators that are coupled from the bottom, such as in a Bruker's W-band spectrometer, pulling out the sample a bit may be a fairly efficient strategy. Once you have attained a small diode current at low m.w. attenuation, you have achieved critical coupling to first approximation. It is only a first approximation since the diode is rather insensitive at low input power.

3. Go back to 20-dB attenuation and tune mode. Switch on the reference arm and increase bias until you see significant changes in the displayed dip. Adjust phase so that the dip is as deep as possible and symmetric. Go back to operate mode and adjust AFC offset (lock offset) to zero. Adjust the bias so that the diode current is 200 μA (center of the display). Now finely adjust the phase so that the diode current is maximum. If you did everything right until here, the maximum should be only slightly above 200 μA. Readjust bias (attenuation of the reference arm) so that the diode current is again 200 μA. Note that this is a fine adjustment of phase that requires that you are already close to the correct phase. There are two phase settings where the diode current is maximum, corresponding to constructive and destructive interference of reflected and bias power. You need constructive interference, i.e., a deep dip in the tuning mode.

4. Now slowly increase attenuation to 40 dB, adjusting AFC offset if necessary. If phase and bias are correct, AFC should stay in the lock at 40 dB. Readjust bias to 200 μA if necessary. Slowly decrease attenuation while readjusting the coupling so that the diode current stays at 200 μA. Don't change the bias in this step. The highest power (lowest attenuation) at which the diode current is still 200 μA is the highest power at which you can measure reliable line shapes. Now go back to the attenuation at which you want to measure. If you fail with the 4th step, you may still measure a good line shape; the setting is just not absolutely perfect.

This procedure may seem tedious, but it is done easily and fast once learned. It guarantees that one can detect any problems with the sample, resonator, or spectrometer and obtain reliable line shapes under almost all circumstances. Some modern spectrometers have an autotune mode that is reliable as long as the mode is not broadened too strongly by m.w. losses of the sample. It is still useful to learn manual tuning, as autotune may not always work, particularly if specialized resonators are used.

At high frequencies or with samples that have strong dielectric m.w. losses in an imperfect resonator, it may sometimes be difficult to achieve critical coupling. If the diode current exceeds 200 μA with the reference arm switched off even at 20 dB attenuation and optimum coupling, one may want to measure with the reference arm off and the minimum attenuation that brings the diode current down to 200 μA. The diode is then biased by the m.w. power reflected from the cavity off resonance.

This way of operating an ESR spectrometer should only be used if proper tuning cannot be achieved.

1.3. Matching and Phasing without Automatic Frequency Control

High-frequency spectrometers often do not have AFC, or this AFC is rather unstable or ineffective. Without AFC, good line shapes can then only be obtained if everything else is very stable. The spectrometer electronics should be warmed up and the whole probe head should be at thermal equilibrium. It is therefore advisable to use temperature control with a stream of ambient temperature nitrogen even for room-temperature measurements. In any case, switch on the spectrometer and put in the sample at least half an hour before the intended measurement. Furthermore, it is usually difficult to adjust critical coupling with the same precision as at more conventional frequencies. In this situation, matching and phasing are better done with the following procedure.

1. Go to tune mode. Switch off reference arm or set bias to zero. Find the resonator dip (microwave mode) and shift it to the center of the display/oscilloscope picture. If you can adjust the frequency of your resonator, try to find a frequency where the mode is as symmetric as possible. Adjust coupling so that the dip is as deep as possible. This is critical coupling to zeroth approximation.

2. Go to operate mode. Adjust the frequency so that the diode current is as low as possible. This guarantees minimum reflection; the m.w. frequency is at the center of the dip. You may try to readjust coupling now, but note that this may shift the frequency of the mode. You thus have to readjust coupling and frequency together to minimize diode current.

3. Go to tune mode, switch on reference arm, and increase bias until you see changes in the dip. Adjust phase so that the dip is as deep as possible and as symmetric as possible. Go back to operate mode and adjust bias so that diode current is at 200 μA.

4. If your spectrometer has a setup scan feature, go to setup scan. Finely adjust the phase so that noise in the setup scan is minimized. If you don't have a setup scan feature, you can mimic this by doing short measurements of a baseline region of the spectrum and readjusting phase until noise is minimum.

The success of this procedure depends on symmetry and the frequency stability of the mode. Check for correct phase by integrating the spectrum after a baseline correction with a constant function. This baseline correction is required to cancel receiver offsets. If the phase is correct, the absorption spectrum obtained by integration also has a flat baseline at the zero level at its high-field end. Generally, spectra with correct phase are easy to measure at low field and hard to measure at high field. Spectra with wrong phase are difficult to fit with a least-squares procedure, so it pays to take care when performing the experiment.

1.4. Setting Receiver Gain and m.w. Power

Reliability of line shapes may also be affected by a too high receiver gain or a too large m.w. power. Too high receiver gain is easily recognized if the maxima and minima of the spectrum are cut off, but the line shape may already be compromised if the signal amplitude is close to the input limit of the analog-to-digital converter. Therefore, observe the receiver level display during the first scan of spectrum acquisition. If it comes close to the maximum or the minimum of the display, the receiver gain should be reduced. Lowering the receiver gain does not decrease sensitivity. If in doubt, measure with lower rather than higher receiver gain.

Reducing m.w. power usually does reduce sensitivity. Thus spectra should ideally be measured at the highest m.w. power at which saturation broadening is still absent. If signals are weak and spectra accumulated for a long time, it pays to adjust the power according to this criterion. Saturation broadening is absent as long as the signal amplitude still increases linearly with the m.w. field B_1, i.e., with the square root of m.w. power P_{mw}. Power is measured on the logarithmic dB scale. On this scale, 10 dB corresponds to a factor of 10, and 3 dB corresponds to roughly a factor of 2 (log 2 = 0.30103). Hence, a 6-dB decrease in attenuation corresponds to an increase in m.w. power by a factor of 4. In the linear regime, signal amplitude should thus increase by a factor of 2. To find the maximum non-saturating power for nitroxides one may start at room temperature with an attenuation of 20 dB, at 80 K with 30 dB, and at 10 K with 50 dB, and then increase or decrease attenuation in steps of 6 dB. Note that above a certain m.w. power level, the noise level may no longer be dominated by detector noise but by phase noise from the source. If this is the case, increasing m.w. power no longer improves signal-to-noise ratio.

2. BASICS OF PULSED ESR

2.1. Pulsed ESR Hardware

Pulsed ESR (Schweiger and Jeschke 2001) requires higher m.w. power than CW ESR, a pulse-forming unit (PFU), and a detection technique that does not depend on field modulation (Figure 4). Usually pulses are formed at powers of less than 1 W and are amplified later. A good pulse-forming unit (PFU) allows for separate power and phase adjustments of at least two channels. For some experiments, four independent channels are required. The power level of the whole pulse sequence can be adjusted with a main attenuator. At least the main attenuator should be calibrated, i.e., it should be possible to select its attenuation on a dB scale.

Pulsed ESR experiments depend on excitation bandwidths of at least a few megahertz — often tens of megahertz. The bandwidth of a resonator is defined as the frequency difference $\Delta\nu$ between its 3-dB points, i.e., between the upper and lower frequencies where the power in the resonator is attenuated by a factor of two

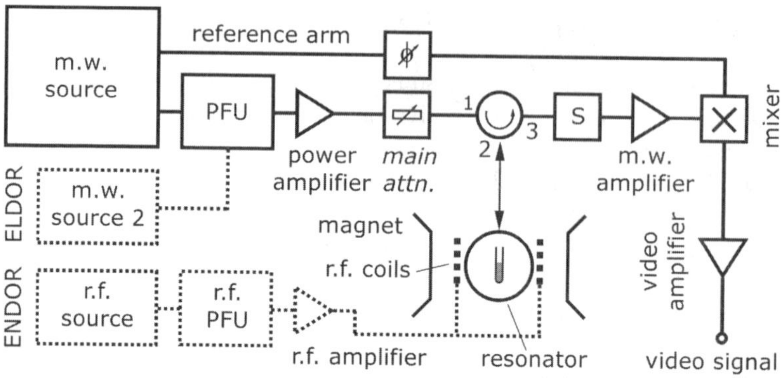

Figure 4. Simplified scheme of a pulsed ESR spectrometer. Optional components required for pulsed ELDOR or ENDOR are displayed as dashed parts.

compared to the center of the dip (Figure 1). A critically coupled high-Q resonator at X band ($Q_L \approx 5000$ or more), such as an m.w. cavity, a loop-gap resonator (Froncisz and Hyde 1982), or a dielectric resonator, has a bandwidth of less than 2 MHz according to Eq. (2). Such resonators have to be overcoupled ($R_0 n^2 > r$ in Eq. (1)). In Figure 1, overcoupling to $Q_L = 100$ is assumed, so that $\Delta\nu = 96$ MHz at $\nu_{mw} = 9.6$ GHz. At such low Q_L values, cavity resonators are not competitive with other resonator types that have larger filling factors. Loop-gap, split-ring, and dielectric resonators are well suited and similar to each other in performance. For resonators with intrinsically low Q_L, such as the bridged loop-gap resonator (Pfenninger et al. 1988), Q_L can be adjusted only within a small range. As finding the correct dip requires at least moderate Q_L and the measurements require small Q_L, such resonators may be more difficult to work with. Nevertheless, they are valuable for special applications where their transparency to radio frequency (r.f.) fields is important (Schweiger and Jeschke 2001).

The bandwidth of the resonator also influences the shape of the m.w. pulses, and thus the dead time t_d that passes after a pulse before the signal can be detected. In an ideal resonance circuit, signals rise as $1 - \exp(-t/\tau)$ and fall as $\exp(-t/\tau)$ (see inset in Figure 5b). The rise time τ is related to Q_L by

$$\tau = \frac{Q_L}{\pi \nu_{mw}}, \tag{5}$$

corresponding to $\tau = 3.3$ ns for $Q_L = 100$ at $\nu_{mw} = 9.6$ GHz. The actual dead time depends on pulse power P_{mw} and signal power P_{signal}, as at the time of detection the pulse power (ringing) should have decayed to a level significantly lower than P_{signal}. For the power of a 1-kW m.w. pulse to decay to less than 1 μW, the dead time in the above example would be $t_d = 68$ ns. The actual dead time can never be shorter than the one imposed by the bandwidth of the resonator, but it can be longer if m.w. power is reflected forth and back in transmission lines.

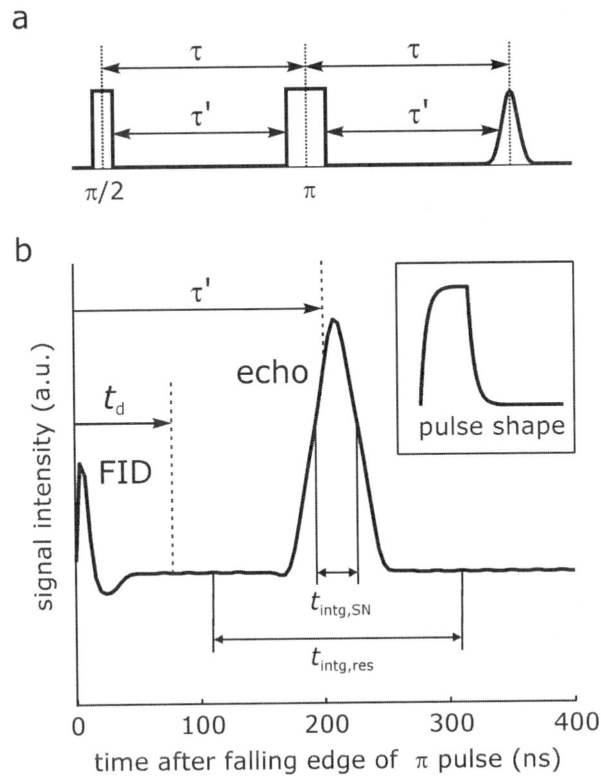

Figure 5. Precise definition of interpulse delays τ in echo sequences (a) and optimum integration windows (b) for reasonable field resolution in echo-detected field-swept EPR ($t_{intg,res}$ = 200 ns) and for maximum sensitivity in non-selective experiments such as ESEEM, ENDOR, or DEER ($t_{intg,SN}$ = 32 ns). The signal trace is a simulation for a two-pulse echo sequence with $t_{\pi/2}$ = 16 ns, t_π = 32 ns, τ' = 200 ns (τ = 224 ns) applied to a Gaussian ESR line with a full width at half height of 84 MHz (30 G). The inset in (b) shows the shape of a pulse with a length of 32 ns assuming rise and fall times of 4 ns.

To obtain maximum detection sensitivity and maximum m.w. field strength B_1 at given m.w. power P_{mw}, Q_L should not be lowered unnecessarily. The field strength is given by

$$B_1 = \sqrt{\frac{2\mu_0 Q_L P_{mw}}{V_C \nu_{mw}}}, \tag{6}$$

where V_c is the effective volume of the resonator. The field strength is related to duration t_p and flip angle β of the pulse by

$$\beta = F \frac{g\mu_B B_1}{\hbar} t_p, \tag{7}$$

where $F = 1$ for species with electron spin $S = 1/2$ and

$$F = \sqrt{S(S+1) - m_S(m_S+1)}\,, \tag{8}$$

if the pulse drives a single transition of spin $S > 1/2$ between the states with magnetic quantum numbers m_S and $m_S + 1$. If the pulse excites all transitions of a spin $S > 1/2$ simultaneously, $F = 1$.

Short pulses are necessary to achieve large excitation bandwidths,

$$\Delta\nu_{exc} \approx \frac{1.2}{t_p}\,, \tag{9}$$

and to avoid significant relaxation of the electron spins during the pulse. Here the excitation bandwidth $\Delta\nu_{exc}$ is defined by the full width at half height of the theoretical excitation profile of a rectangular pulse. A pulse with a width of 12 ns thus has an excitation bandwidth of ±50 MHz. Since 2.8 MHz corresponds to 1 G at $g = g_e$, the excitation bandwidth on a field scale is ±18 G. According to Eq. (7), such an excitation bandwidth requires $B_1 = 93$ G in our X-band example for a pulse with $\beta = \pi$.

From Eq. (6) it becomes clear that at given m.w. power a larger B_1 can be achieved with a smaller active volume V_c of the resonator. Using such a smaller resonator does not lead to a decrease of concentration sensitivity as long as the filling factor η can be maintained. In practice, miniaturization of resonators tends to decrease η somewhat as the ratio of wall thickness of the sample tube to resonator dimension increases. However, in work with membrane proteins absolute sensitivity is often more important as the amount of sample is limited. Absolute sensitivity increases with decreasing resonator size. If the filling factor remains the same, absolute sensitivity is approximately proportional to the inverse third power of the resonator dimension. In practice, the filling factor decreases somewhat with resonator size as sample container walls cannot be made proportionally thinner.

Working with overcoupled resonators has the disadvantage that much of the power is reflected. Even if pulses are short and duty cycles low, a reflected power of some hundred watts would damage any detector that is sensitive enough to measure signals that have only microwatts of power. Hence, the detection circuitry has to be protected either by a switch (S in Figure 4) or by a power limiter. A switch has to be closed well before high-power pulses are applied and has to remain closed until the pulse power has decayed down to the milliwatt level. This is usually done by control software and an auxiliary pulse from the pulse pattern generator. It is good practice to keep the main attenuator at 60 dB until the presence of the protection window in the signal trace has been verified by an oscilloscope or in a Bruker's SpectJet window.

In pulsed ESR spectrometers signals are detected by downconversion to video frequencies (between zero and 100 MHz) with a mixer. A mixer combines m.w. directly from the source (transmitted through the reference arm) with the m.w. coming from the sample. Video output is at the difference frequency. Such downconversion by a mixer causes a significant loss in signal-to-noise ratio if one of the input signals would have an amplitude below the noise level of the mixer. Therefore the m.w. coming from the sample needs to be pre-amplified (see Figure 4). The m.w. pre-amplifier usually has a fixed gain. Signal amplitude can be varied by changing the gain of the video amplifier that follows after the mixer. Video amplifier gain should be adjusted so that the maximum signal during the experiment is about 1/2 to 2/3 of the maximum input of the analog-to-digital converter. In rare cases the signal already saturates the m.w. pre-amplifier. Even with the lowest gain of the video amplifier one then observes an echo that appears flattened at the top or line shape distortions in echo-detected ESR spectra (a "compressed" top or even a cutoff near the maximum). For such strong signals the m.w. pre-amplifier has to be switched off.

2.2. Adjusting Pulse Amplitude and Phase

Before setting up a pulse ESR experiment one decides on the desired excitation bandwidths and hence lengths of all the pulses. For nitroxide spin labels the maximum feasible excitation bandwidth is smaller than the total width of the spectrum. The excitation bandwidth of the longest pulse in the sequence then decides what fraction of the spins contributes to the signal. Hence, sensitivity increases with decreasing duration of the longest pulse. On most spectrometers the rise time of the m.w. switches and the minimum Q_L of the resonator do not allow for well-defined pulses shorter than 8 ns. Depending on the resonator and the available m.w. power, the minimum length of a pulse with flip angle $\pi/2$ or π according to Eqs. (6) and (7) may be even longer. Experience tells what is feasible with a given spectrometer and probe head. If one decides that π pulses have twice the length of $\pi/2$ pulses, one can use the same pulse channel for both of them. In HYSCORE (Höfer et al. 1986) and PELDOR or DEER (Milov et al. 1984; Pannier et al. 2000) experiments, it is advisable that the π pulses have the same length as the $\pi/2$ pulses, so that different channels with separately adjustable power are required.

The number of required channels also depends on phase cycling (Schweiger and Jeschke 2001). In most ESR pulse sequences at most two phases ($+x$ and $-x$) will suffice for any given pulse. Single-frequency experiments for distance measurements such as double-quantum techniques (Borbat and Freed 2000) or the SIFTER experiment (Jeschke et al. 2000) require four phases ($+x$, $+y$, $-x$, $-y$) for certain pulses. In recent Bruker bridges (stripline bridges) there are preset channels for these four phases, which simplifies setup. Note, however, that such preset channels are optimized at a certain m.w. frequency. Depending on probe head and sample properties (dielectric permittivity) one may be forced to perform measurements at a frequency where the preset phases are slightly off. The following procedures, and pulse experiments on nitroxides in general, work well in the temperature range between 50 and 80 K. Lower temperatures lead to an increase in T_1 that forces slower repetition of the experiment without leading to an increase in T_2. At higher temperatures, T_2 may drop strongly in some matrices (e.g., glassy ethanol), so that a significant part of the signal is lost by relaxation. Generally, pulse ESR on nitroxides is feasible up to temperatures 10 K below the glass transition temperature of the matrix.

The first step in experiment planning is thus to decide which pulse length/flip angle/phase combinations are required to perform the experiment and to assign these combinations to the available pulse channels. After that the proper power level and phase for all those channels have to be set. Spectrometer electronics and the probe head should be in thermal equilibrium at that time, i.e., the probe head should have been cooled down to the target temperature and the spectrometer switched on 30 minutes before setup. For relatively short pulses up to 200 ns length setup is most easily done with an echo sequence: pulse(t_p)–delay(τ)–pulse($2t_p$)–delay(τ)–echo for $\pi/2$ pulses with length t_p and with a sequence: pulse($t_{p/2}$)–delay(τ)–pulse(t_p)–delay(τ)–echo for π pulses with length t_p. For channels with preset phases and no individual attenuators, power has to be adjusted with the main attenuator. If the channels have individual attenuators, it is advisable to set the main attenuator to 0 dB and to adjust pulse power with the channel attenuators. Before this is done, the m.w. frequency ν_{mw} should be determined and the magnetic field set to the expected maximum of the nitroxide spectrum. The maximum corresponds to $g_y \approx 2.006$, and the field can be computed by $B_0/\mathrm{G} = 356\ \nu_{mw}/\mathrm{GHz}$.

The following setup procedure assumes quadrature detection, i.e., both the real and imaginary part of the signal are observed. For a $+x$ channel, start with maximum attenuation (minimum power) and slowly decrease attenuation. An echo signal with at first arbitrary phase (components in both the real and imaginary part) should appear. Next, slowly decrease the attenuation until the echo amplitude stops increasing and eventually decreases. Adjust the video gain so that the echo is clearly visible. Now adjust the signal phase, preferably by using the main phase shifter that acts on the reference arm. The signal should be in the real part and positive, as shown in Figure 5b. The signal in the imaginary part should be symmetric about zero, i.e., it should look like a dispersion line. Once the phase is correct, fine adjust the power by searching for the maximum echo amplitude. This may unbalance the phase a little bit, so that the signal phase has to be readjusted.

This procedure has to be repeated for all required channels. However, in all channels except for $+x$ the channel phase shifter is used rather than the main phase shifter for phase adjustment. A $+y$ channel should have a maximum positive echo in the *imaginary* part, a $-x$ channel maximum *negative* echo in the real part, and a $-y$ channel maximum negative echo in the imaginary part. Setup of inversion pulses for Davies ENDOR (Davies 1974) or for pulse ELDOR experiments follows a different procedure (see below).

For soft pulses longer than 200 ns, the free induction decay (FID) signal that directly follows a single pulse is longer than the dead time, even if the ESR line is infinitely broad (Schweiger and Jeschke 2001). In this case, adjustment of power and phase is better done on this FID signal. Note that the FID signal reaches its maximum amplitude at flip angle π, not $\pi/2$, if the excitation bandwidth of the pulse is much smaller than the ESR line width. Further increase of the power does not lead to a smaller amplitude of the FID but to an easily recognizable deformation.

2.3. Echo Integration

Most pulse ESR experiments on solids use echo detection, i.e., echo intensity is recorded as a function of one or two variables, such as magnetic field, delay times in the pulse sequence, r.f in ENDOR experiments, or a second m.w. frequency in some ELDOR experiments. The echo amplitude is integrated over some window of length t_{intg}, which is at least as long as the time resolution of the detection circuit, but can be much longer. The center of the window should coincide with the echo maximum. Note that the actual delay τ between pulses is the time that passes between the *centers* of the pulses, not their edges (Figure 5a). The optimum width, t_{intg}, of the integration window depends on the type of experiment.

In recording echo intensity as a function of the magnetic field (echo-detected field-swept ESR), a too short integration window can lead to line shape distortions. The distortions result from an excitation bandwidth of the pulses that is larger than the width of the most narrow features in the absorption spectrum. As the echo amplitude (maximum intensity) in the time domain corresponds to the integral of the intensity over the whole excitation window in the frequency (or field) domain, the lines are broadened. To obtain the intensity in the frequency domain at only the resonance field, one has to integrate over the whole echo. In practice, $t_{intg} = 200$ ns suffices for nitroxides in the solid state. The window should be symmetric about the echo maximum (Figure 5b). A long integration window also helps alleviate line shape distortions that may arise for short windows due to nuclear modulation effects. For spectra with some very narrow features, a much longer integration window may be required. In this situation, symmetric placement would necessitate a very long interpulse delay τ in the echo sequence and would thus lead to relaxational losses. It is then better to work with short τ and start the integration window at the echo maximum. Note also that the echo maximum does not occur at time τ after the falling edge of the π pulse, but somewhat later (Figure 5).

In experiments other than field-swept ESR, contributions from spins in the whole excitation window are usually wanted. The integration window should then be as long as the longest m.w. pulse, as this maximizes the signal-to-noise ratio (Figures 5 and 6). If the integration window is selected four times as long as the longest pulse, one already needs six times as many accumulations to obtain the same signal-to-noise ratio.

3. PULSED ENDOR

3.1. Pulsed ENDOR Spectrometer

A basic introduction to pulsed ENDOR (Gemperle and Schweiger 1991) and more detailed descriptions of sophisticated experiments (Schweiger and Jeschke

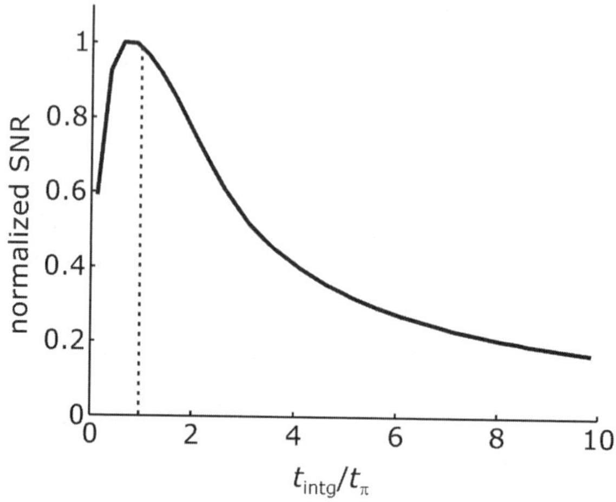

Figure 6. Normalized signal-to-noise ratio (SNR) as a function of the width, t_{intg}, of an integration window centered at the echo maximum (simulation). The simulation is the same as in Figure 5 with white noise added. The time axis is normalized to the π pulse length.

2001) can be found in the literature. To extend a pulsed ESR spectrometer to a pulsed ENDOR spectrometer, an r.f. source, a pulse former for the r.f. pulses, and a broadband high-power r.f. amplifier need to be added (see Figure 3). Furthermore, r.f. coils have to be added to the probe head. The r.f. source and amplifier should cover the range between at least 1 and 30 MHz at the X band and between 1 and 200 MHz at the W band. Typical r.f. pulse powers are a few hundred watts at pulse lengths between 1 and 200 μs with repetition times between 1 and 10 ms. In most cases the r.f. coils are terminated by a 50-Ω load. For pulses with 1 kW power and a length of 100 μs at a repetition time of 10 ms, this load must be able to withstand an average power of 10 W. The r.f. source, pulse former, and amplifier are standard devices. Performance of the pulse ENDOR spectrometer depends decisively on the design of the r.f. coils, the resonator, and on matching between the r.f. transmission line and coils.

Simply adding r.f. coils to a standard resonator does not work, since most good m.w. resonator designs feature rather massive metal walls that cannot be penetrated by r.f. The most elegant solution, at least at moderate m.w. frequencies, is to place the r.f. coil inside the resonator as close to the sample tube as possible. This can be done with a Helmholtz coil with rectangular shape (Bietsch and von Schütz 1993). Such a coil has straight wires that are parallel to the cylinder axis of a dielectric resonator. These wires do not cause a decrease in quality factor Q_L, as they do not shortcut the m.w. field.

Alternatively, the m.w. resonator can be made transparent or at least partially transparent to the r.f. field. Such approaches are based on the large difference between the m.w. and r.f. wavelengths. In most cases the m.w. frequency exceeds the r.f. frequency by at least a factor of 680 (ratio of the resonance frequencies of electron spins with $g = g_e$ and protons). Thus, the conducting wall of a bridged loop-gap resonator (Pfenninger et al. 1988) can be made sufficiently thick to be impenetrable to m.w. fields and at the same time sufficiently thin to be transparent for r.f. fields. The r.f. coil can then be placed outside the resonator. Since such a coil has a somewhat larger volume than a coil that encloses only the sample, but not the resonator, more r.f. power is required to generate an r.f. field of the same amplitude, B_2. On the other hand, a more massive coil can be used and thermal effects on the sample or resonator are smaller at a given power. At high frequencies (95 GHz and above) the most successful design places the coil outside a TE_{011} (cylindrical) cavity that is made of ring sections connected by only a single conducting bridge (Burghaus et al. Möbius 1992). Such a cavity is almost transparent to r.f. fields (0.5 dB attenuation at 140 MHz) yet has a quality factor that is only moderately compromised with respect to a massive TE_{011} cavity.

Matching between the r.f. transmission line and the coils is achieved by a matching network consisting of capacitances and, occasionally, resistances. The design of a matching network is aimed either at broadband matching without the requirement for adjustments during an r.f. sweep or at obtaining a resonant enhancement of r.f. field B_2. In the latter case, the matching network has to be tuned to the current r.f. frequency, which can be realized by adjusting a variable capacitance by a stepper motor via a feedback loop (Forrer et al. 1990). At the same incident r.f. power, a tuned circuit achieves higher B_2 fields and thus shorter r.f. π pulses than a broadband matching network. The disadvantages are limitations in the rate of the r.f. sweep and the fact that the power is converted to heat in the coil close to the sample rather than in an external load. At high r.f. frequencies, a tunable resonant matching network requires very short coaxial feeds between the network and the coil, which is hard to realize for low-temperature measurements. Broadband matching networks contain a resistance. Considerations for a network that works well for proton ENDOR at 95 GHz m.w. frequency have been described (Rohrer et al. 1995). Reasonable broadband matching over the whole range of ENDOR frequencies at W band (2 200 MHz) is achieved in a Bruker E680 spectrometer. Matching over such a broad frequency range is not uniform, at least with long coaxial feeds between the network and coil. Combined with the thermal effects due to the proximity of the r.f. coil and the small high-frequency resonator, this leads to baseline artefacts that may be difficult to distinguish from broad ENDOR signals. These artefacts can be converted to uniform noise, and thus averaged, by varying the r.f. stochastically rather than sweeping it linearly (Epel et al. 2003).

3.2. Sensitivity Considerations on Mims and Davies ENDOR

In both established pulsed ENDOR techniques (Mims 1965; Davies 1974) sensitivity varies with the magnitude of the hyperfine coupling, A. In both cases sensitivity approaches zero if A approaches zero. In Mims ENDOR sensitivity as a function of A depends strongly on interpulse delay τ between the first two m.w. pulses and only slightly on pulse length. In contrast, in Davies ENDOR there is no dependence on the interpulse delays but a strong dependence on the length t_{inv} of the first m.w. pulse. Except for the case where only very large couplings $A > 50$ MHz are to be detected, this implies that the first m.w. pulse in Davies ENDOR has to be longer than the pulses in Mims ENDOR. This in turn means that a smaller fraction of the spins takes part in Davies compared to Mims ENDOR experiments. The intrinsic sensitivity of Mims ENDOR should thus be higher. On the other hand, Mims ENDOR features blind spots, i.e., the ENDOR effect is almost totally suppressed at certain frequencies. For ideal (infinitely short) m.w. pulses, Mims ENDOR efficiency is given by

$$F_{\text{ENDOR}} = \frac{1}{2}\sin^2(A\tau) , \tag{10}$$

where A is expressed in angular frequency units. As delay τ must be at least as long as dead time t_d, blind spots cannot be avoided for frequencies larger than $1/(2t_d)$. At the X band with a typical dead time of 80 ns, the first blind spot corresponds to a hyperfine coupling as small as 6.25 MHz. For m.w. pulses with finite length, suppression at the blind spots is not total, as illustrated in Figure 8, but still very pronounced. However, the comparison of Mims ENDOR with $\tau = 400$ ns (dotted line in Figure 8) and Davies ENDOR with $t_{inv} = 100$ ns (solid line) by simulations also shows that the average sensitivity is better for Mims ENDOR up to hyperfine couplings beyond the first blind spot. For very small hyperfine couplings $A < A_1$ (arrow in Figure 8), Mims ENDOR with fixed interpulse delay τ is the most sensitive experiment. For $A > A_1$, the blind spots can be almost averaged by summation of 50 Mims ENDOR experiments with τ values in the range from 100 to 492 ns (dashed line in Figure 8). This experiment is more sensitive than Mims ENDOR with fixed τ for $A > A_1$ and superior to Davies ENDOR for $A < A_2$. For larger couplings, Davies ENDOR is the most sensitive experiment.

There is one caveat about these simulations: they assume that the polarization grating generated by the first two m.w. pulses in the Mims ENDOR experiment (Schweiger and Jeschke 2001) has the same lifetime as the hole burnt by the first m.w. pulse in the Davies ENDOR experiment. As the polarization grating is a finer structure in the frequency domain, it has a shorter lifetime if spectral diffusion is significant. To decide which experiment is better for a given class of samples, one needs to quantify the influence of spectral diffusion on the stimulated echo. This can be done by comparing the decay of the echo signal as a function of interpulse delay T for the Davies observer sequence $\pi(t_{inv})$–delay(T)–$\pi/2$(16 ns)–delay(τ)–π(32 ns)–delay(τ)–*echo* (upper sequence in Figure 7a) and the Mims observer sequence $\pi/2$(16 ns)–delay(τ)–$\pi/2$(16 ns)–delay(T)–$\pi/2$(16 ns)–delay(τ)–*echo* (upper sequence in Figure 7b). Let $T_0 = \tau + 100$ ns (the shortest delay that does not require phase cycling), while T_{ENDOR} is the delay required for inserting the r.f. π pulse. The influence of spectral diffusion can then be quantified via the echo integrals by

$$R = \frac{V_{\text{Mims}}(T_{\text{ENDOR}})V_{\text{Davies}}(T_0)}{V_{\text{Davies}}(T_{\text{ENDOR}})V_{\text{Mims}}(T_0)} . \tag{11}$$

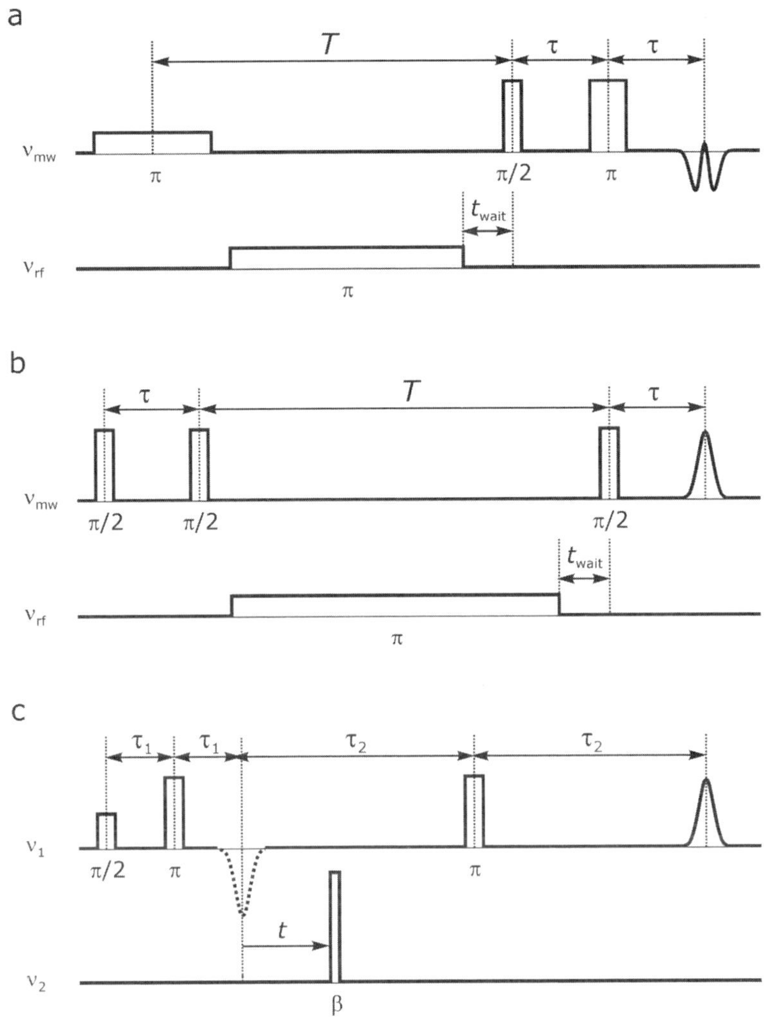

Figure 7. Pulse sequences for double resonance experiments: (a) Davies ENDOR, (b) Mims ENDOR, (c) four-pulse DEER.

A value of $R = 1$ indicates negligible spectral diffusion during an ENDOR experiment and favors Mims ENDOR, while a value of $R << 1$ corresponds to strong spin diffusion and favors Davies ENDOR. In fact, the functions plotted in Figure 8 for Mims ENDOR can be multiplied by R to obtain an estimate of the range of hyperfine couplings for which each experiment is advantageous. Knowing the echo decay as a function of T for the observer sequences is also useful for deciding on the optimum length of the r.f. pulse (see below).

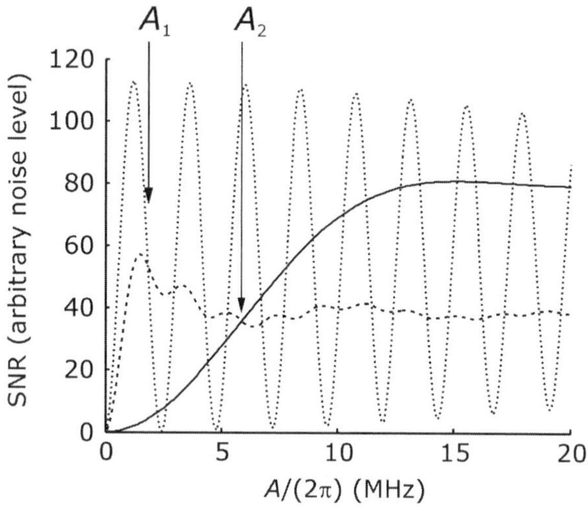

Figure 8. Signal-to-noise ratio as a function of hyperfine coupling for several pulsed ENDOR experiments at arbitrary, but common noise level (simulation, an electron phase memory time T_m of 1 μs was assumed). The same total number of accumulations and optimum integration windows for all experiments was assumed. Mims ENDOR with τ = 400 ns (dotted line) is the best experiment for $A < A_1$. Summation of 50 Mims ENDOR experiments with equidistant τ values between 100 and 492 ns (dashed line) performs best for $A_1 < A < A_2$. Davies ENDOR with a length t_{inv} = 100 ns for the first m.w. pulse (solid line) is best for $A > A_2$.

3.3. Setting up Mims ENDOR

Mims ENDOR sensitivity is optimum for the shortest $\pi/2$ m.w. pulses that are feasible on a certain spectrometer. Depending on available m.w. power and the resonator, typical values are between 8 and 24 ns at X band and between 30 and 200 ns at W band. Pulse power and phase should be adjusted for the Mims ENDOR observer sequence (upper sequence Figure 7b) rather than for a $\pi/2$–τ–π two-pulse echo. The length of the integration window should match the full width at half height of the echo. For detecting couplings between 2 and 5 MHz, the minimum feasible τ value should be used, which is typically 100–120 ns at X-band frequencies. If couplings larger than 5 MHz are expected, averaging over τ (see Figure 8) is strongly recommended. The maximum τ value in such averaging determines the sensitivity toward small couplings. A value of 500 ns should be amply sufficient in most practical cases. Notable exceptions are distance measurements by intramolecular ENDOR or other measurements of very small hyperfine couplings of rare spins. This case is discussed in §3.5.

The r.f. pulse should be a π pulse. The r.f. π pulse length for a given spectrometer, probehead, and r.f. frequency can be determined for a standard sample by a transient nuclear nutation experiment (Schweiger and Jeschke 2001). In this experiment the r.f. frequency is set to an ENDOR line of the standard sample and kept fixed, while the length of the r.f. pulse is incremented. With a good digital oscilloscope, such as a Bruker's SpectJet, it is also possible to preselect the r.f. pulse length and adjust the power while observing changes in the echo amplitude online. Starting at a very low value, the length (or power) should be incremented until the first minimum of the echo intensity is observed. These adjustments work best with single crystals as standard samples. For measurements between room temperature and 80 K, γ-irradiated crystals of α-glucopyranoside or malonic acid are well suited. For measurements at temperatures between 4 and 20 K, glycine doped with 1% $CuCl_2$ is a good standard sample. When using powder samples or frozen solutions, one often observes only a saturation of the ENDOR effect with increasing pulse length or power rather than an oscillatory behavior. A typical length of an r.f. π pulse with broadband matching is 10 μs at proton frequencies in both X-band and W-band ENDOR. With a resonant (tuned) circuit, pulse lengths as short as 2 μs can be obtained. However, short r.f. pulses cause power broadening of the ENDOR lines and may thus decrease resolution. To obtain the resolution of 100 kHz that is usually feasible in proton ENDOR, the r.f. π pulse should not be shorter than 10 μs. For rare spins, such as ^{31}P or ^{14}N, intrinsic line widths can be significantly lower than 100 kHz. An r.f. pulse length of 40 μs is then recommended unless this leads to strong sensitivity loss due to spectral diffusion (see §3.2).

The rising edge of the r.f. pulse must be timed well after the second m.w. pulse of the observer sequence (see Figure 7b). An overlap is recognized easily by an echo suppression that occurs even if the r.f. is off resonance. After the falling edge of the r.f. pulse, a time $t_{wait} > 1$ μs should pass before the final m.w. observer pulse is applied (Figure 7b). Ringing of the r.f. pulse, in particular at low frequencies, may otherwise distort the baseline.

High-power r.f. amplifiers usually generate 3rd and 5th harmonics of the amplified frequency when they are operated at maximum output power. Indeed, the matching network may generate such harmonics even if they are not present in the r.f. that comes from the amplifier. In ENDOR spectra, harmonics manifest themselves by ghost signals at 1/3 or 1/5 of the true ENDOR frequencies. In particular, proton ghost signals are frequently encountered. For heteronuclear ENDOR, it is therefore advisable to use a low-pass filter with a cutoff frequency of 10 MHz for X-band work and 100 MHz for W-band work. The filter must be connected directly to the probe head, even if this somewhat compromises the efficiency of the matching network.

3.4. Setting up Davies ENDOR

The fraction of spins taking part in a Davies ENDOR experiment is determined by the length t_{inv} of the first observer pulse (inversion pulse). This length should be 100 ns or shorter; if a longer inversion pulse is needed to see the hyperfine couplings desired, it would be best to measure Mims ENDOR. The spins taking part in the experiment should be detected as completely as possible, i.e., the observer echo should be formed with the shortest $\pi/2$ and π pulses that can be generated. The observer echo shape then looks as in Figure 9: a narrow positive echo superimposed by a broad inverted echo. The narrow positive echo does not exhibit an ENDOR effect. Therefore, maximum Davies ENDOR sensitivity is obtained for an integration window that matches the length t_{inv} of the inversion pulse and thus the width of the inverted echo (see Figure 9). Simulations show that this is true irrespective of the magnitude of the hyperfine coupling.

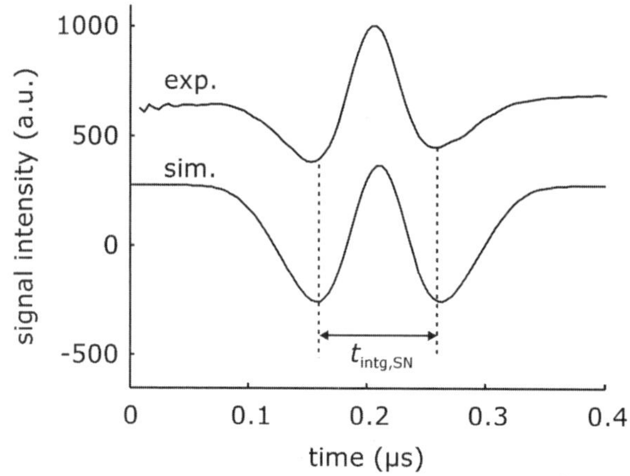

Figure 9. Experimental (upper trace, coffee) and simulated (lower trace, Gaussian line with 7-G full width at half height) signal for a Davies-ENDOR observer sequence β(100 ns) – delay (12 μs) – $\pi/2$ (16 ns) – delay (400 ns) – π (32 ns) – delay (200 ns) and optimum width t_{intg} of the integration window.

Setup starts with optimizing the power and phase for the final two m.w. pulses by maximizing the narrow positive echo. In this step the inversion pulse channel is at maximum attenuation (minimum power). Then the power of the inversion pulse is adjusted by searching for a maximum negative broad echo. Note that the phase of the inversion pulse does not matter.

Considerations on the power and duration of the r.f. pulse as well as on its timing are the same as for Mims ENDOR. Note that adjustment of r.f. power by online echo observation is done more easily and more precisely by Mims ENDOR.

3.5. Distance Measurements by ENDOR

Intermolecular hyperfine couplings are usually strongly dominated by the dipole–dipole interaction between electron and nuclear spins. Relevant exceptions are protons of neighboring molecules that are hydrogen-bonded to the paramagnetic molecule and ^{19}F nuclei, which have substantial isotropic hyperfine couplings for very small spin densities. If the dipole–dipole interaction dominates, distance measurements in supramolecular assemblies such as biomembranes are feasible, as this interaction is inversely proportional to the cube of the distance between the spins. The distance range depends critically on the ability to measure very small hyperfine couplings. According to Eq. (10), the measurement of small couplings requires a long interpulse delay τ in Mims ENDOR. However, the intensity of the observer echo decreases with increasing τ due to electron spin relaxation. It can be shown that in the limit of small couplings, maximum sensitivity is obtained at $\tau = T_m$, where T_m is the phase memory time of the electron spins (Zänker et al. 2004). More precisely, in that context T_m is the decay time of a stimulated echo with respect to τ. This decay time may depend on interpulse delay T between the second and third pulses, as spectral diffusion during that time contributes to the decay.

Due to this dependence on spectral diffusion, T_m strongly varies with the matrix. Typical values for $T = 12$ μs are between 800 and 1000 ns at a temperature of 80 K, but in some membrane protein preparations we have found values as short as 400 ns. As T_m decreases with increasing delay T, the sensitivity of the experiment may decrease significantly when the r.f. π pulse is prolonged. For electron spin ^{31}P distances of up to 1 nm in phospholipids in solution, a pulse length of 40 μs was found to be optimal (Zänker et al. 2004). Measurements with this pulse length are also feasible in phospholipid bilayers both at X- and W-band frequencies. The technique can be used to determine the distance of closest approach of a spin label to the layer of phospholipid head groups.

4. PULSED ELDOR (DEER)

4.1. Pulsed ELDOR Spectrometer

Extending a pulsed ESR spectrometer to a pulsed ELDOR spectrometer is less elaborate than extending a CW ESR spectrometer to ELDOR operation or a pulsed ESR spectrometer to ENDOR operation. For pulsed ELDOR, the same probe heads and the same high-power amplifier can be used as for basic pulse ESR. If a spare pulse channel is available, the same pulse-forming unit can be used as well. The m.w. power from a second source with adjustable frequency can then be fed in as input into one channel of the m.w. pulse-forming unit (see Figure 4). The same is true for high-frequency spectrometers with a design in which pulses are formed at a relatively low microwave frequency and the high frequency is obtained by several steps of frequency doubling or by mixing with continuous-wave high frequency (upconversion). In these designs, only a second source for the low frequency has to be added. Such instrumentation is sufficient for high-quality distance measurements by pulsed ELDOR techniques, if the four-pulse double electron electron resonance (DEER) sequence (Pannier et al. 2000) is applied. With a more elaborate setup, using a bimodal resonator and separate high-power amplifiers for the two m.w. frequencies (Milov et al. 1998), the more simple three-pulse sequence (Milov et al. 1984) can also provide essentially dead-time-free high-quality data. Note, however, that in a setup with a bimodal resonator the difference of the two m.w. frequencies is essentially fixed and cannot be adapted to the problem at hand. For experiments such as two-frequency soft ESEEM, a phase lock between the two frequencies has to be implemented (Schweiger et al. 1990). However, for pulsed ELDOR distance measurements a phase lock is detrimental, as echoes between the pump pulse and observer pulses would then add up during signal accumulation. They cancel if the sources are incoherent.

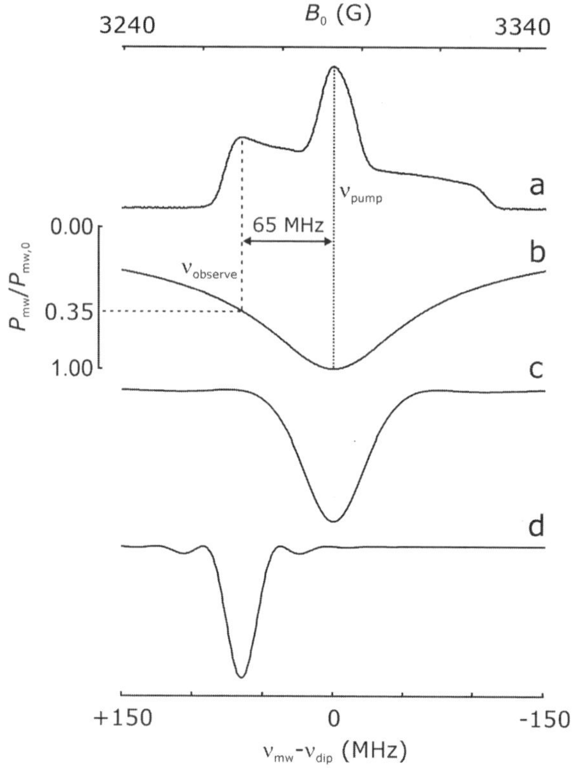

Figure 10. Optimum observer and pump pulse settings in the four-pulse DEER experiment. (a) Echo-detected field-swept ESR spectrum of a nitroxide biradical with a spin-to-spin distance of 5 nm (field ordinate). (b) Simulated idealized m.w. mode assuming $Q_L = 100$ (frequency ordinate power abscissa). (c) Simulated excitation profile of a pump pulse with a length of 12 ns and rise and fall times of 4 ns. (d) Simulated excitation profile of an observer π pulse with a length of 32 ns and rise and fall times of 4 ns.

We now consider what excitation bandwidth of the pulses is required to optimize the sensitivity of pulsed ELDOR distance measurements. This sensitivity depends on the total amplitude of the observed echo and on the fraction of pumped spins, which determines the modulation depth. In the absence of phase noise, the signal-to-noise ratio in the wanted signal is proportional to the product of amplitude and depth. Phase noise is proportional to the echo amplitude; hence, in its presence increasing the modulation depth helps, but increasing echo amplitude does not. Similarly, the ratio of electron–electron modulation to (unwanted) nuclear modulation depends on modulation depth, but not on total echo amplitude. Hence, increasing the modulation depth is more advantageous than increasing the observer echo amplitude. It follows that the pump pulse at frequency ν_2 (for the pulse sequence, see Figure 7c) should be applied at the maximum of the absorption spectrum and should have the maximum excitation bandwidth that is possible without causing a strong overlap with the excitation band of the observer pulses (Figure 10). We found that a pump pulse with a length of 12 ns (for the excitation profile, see Figure 10c) and observer pulses with a pulse length of 32 ns (Figure 10d) can be applied if the frequency difference is 60 MHz or larger (see Eq. (9)). These conditions are well suited for nitroxides at the X band, as the local maximum at the low-field edge of the absorption spectrum and the global maximum at the center are 60–70 MHz apart (Figure 10a). A good pulse ELDOR setup for nitroxide work should thus allow for π pulses as short as 12 ns. For work on transition metal centers, even shorter pump pulses (and larger frequency differences) would be favorable. With a standard TWT amplifier (1-kW power), 12-ns π pulses can be created, for instance, in a fully overcoupled ($Q_L \approx 100$) Bruker 3 mm split-ring resonator. Loop-gap resonators that fit 0.9 mm tubes or TPX™ (poly(4-methylpentene-1)) capillaries allow for 8-ns π pulses even in the wings of the mode, but are more difficult to handle at low temperatures and may be more difficult to overcouple to very low Q_L.

Strong overcoupling is needed at X-band frequencies to sufficiently broaden the m.w. mode to fit both frequencies (Figure 10b). At W-band or higher frequencies, a critically coupled mode may have a sufficient bandwidth (see Eq. (4)), in particular, since the excitation bandwidths at full m.w. power tend to be lower, so that a smaller difference between pump and observer frequency is feasible.

Any ELDOR experiment that can be performed with two frequencies can in principle also be performed with field-step techniques (Schweiger and Jeschke 2001). In this case all pulses can be applied at the center of the m.w. mode. Our own experience suggests that field-step setups are more difficult to handle in practice. There are two problems. First, to apply a field step, electron coherence has to be stored as polarization and later be reconverted to coherence. During this process inevitably half of the magnetization is lost. Second, to avoid signal distortions the additional field has to decay to a very low and very reproducible value before detection. This introduces an additional waiting time, during which spectral diffusion and spin diffusion may alter the polarization pattern and thus lead to further signal decay or even to spurious frequencies. These problems tend to overcompensate the advantage of complete excitation and detection at the center of the mode.

4.2. Setting up Four-Pulse DEER

To extract reliable distance distributions and to obtain reproducible modulation depths, the four-pulse DEER experiment has to be set up very precisely. Set up starts with measurement of an echo-detected field-swept ESR spectrum (see §2.3) and selection of the magnetic field at which the pump pulses should be applied (global maximum of the absorption spectrum as in Figure 10a, unless there is a particular reason to pump elsewhere). The pump frequency ν_2 is then set to coincide with the spectrometer frequency ν_1. Next the flip angle of the pump pulse is adjusted, using the inverted-echo sequence $\beta(t_{pump},\nu_2)$–T–$\pi/2(t_{observe}/2,\nu_1)$–$\tau$ – $\pi(t_{observe},\nu_1)$–τ *echo*. The inverted-echo sequence is used for two reasons. First, the pump frequency is not coherent with the reference frequency of the mixer; hence, a usual echo at the pump frequency would have a different phase in each shot and average to zero. Second, the pump pulse indeed has to be optimized for maximum inversion over its whole excitation bandwidth rather than for echo formation. For broad lines, the required flip angle β differs from π and depends on the line shape. For an infinitely broad line, simulations indicate $\beta \approx 3\pi/4$, while for an infinitely narrow line $\beta = \pi$.

For nitroxides, setup of the pump pulse at the X band works well with t_{pump} = 12 ns, $t_{observe}$ = 32 ns, T = 400 ns, and τ = 200 ns. The power of the pump channel is slowly increased until the echo exhibits maximum inversion. Usually this step is done with the pump (and spectrometer) frequencies coinciding with the center of the m.w. mode.

After adjusting pump pulse power, the pump frequency ν_2 is kept fixed. The spectrometer frequency ν_1 is changed by $\Delta\nu = -2.8$ MHz/G · g/g_e · ΔB_0, where ΔB_0 is the difference between the observer and pump position in the field-swept ESR spectrum. For nitroxides, it is sufficiently precise to approximate $g \approx g_e$. We use $\Delta B_0 \approx -23$ G, so that the frequency has to be *increased* by $\Delta\nu \approx 65$ MHz. Then the power and phase of the observer π pulse ($+\langle x\rangle$ channel) are adjusted with a sequence $\pi/2(t_{observe}/2,\nu_1)$–$\tau$–$\pi(t_{observe},\nu_1)$–$\tau$ *echo*, using the channel attenuator for the power adjustment but the main signal phase shifter for phase adjustment. The observer $\pi/2$ pulse is phase cycled in DEER to eliminate receiver offsets. Hence, the $+x$ and $-x$ channels have to be set using the sequence $\pi/2(t_{observe},\nu_1)$–τ–$\pi(2*t_{observe},\nu_1)$–τ–*echo* and the *channel* phase shifters. After that, a standing DEER experiment $\pi/2(t_{observe},\nu_1)$–τ_1–$\pi(t_{observe},\nu_1)$–τ_1–$\beta(t_{pump},\nu_2)$–τ_2–$\pi(t_{observe},\nu_1)$–τ_2–*echo* (as in Figure 7c, but with t = 0) is used to readjust the *signal* phase (main phase shifter). This adjustment is initially best done with τ_1 = 200 ns and τ_2 = 800 ns. If the solvent or the spin label is deuterated, τ_1 = 400 ns is a better choice. Selection of the τ_2 value for the actual DEER measurement depends on the expected distances. A detailed discussion and a description of variable-time DEER as a technique for sensitivity enhancement are given by Jeschke et al. (2004a).

4.3. Relations between Setup and Data Analysis

Software for data analysis is available at http://www.mpip-mainz.mpg.de/~jeschke/distance.html and is also included with the accompanying CD-ROM of this book. The software package includes a manual. Using this software, distance distributions $P(r)$ can be extracted from pulsed ELDOR data. This corresponds to a moderately ill-posed problem (Jeschke et al. 2002) that can be considerably stabilized by implementing the constraint $P(r) > 0$ (Jeschke et al. 2004b). Yet to maintain stability it is of utmost importance that as few parameters as possible need to be fitted, that systematic differences between the theoretically expected and experimentally observed dipolar evolution function are as small as possible, and that the signal-to-noise ratio is as high as possible. Minimization of the number of fit parameters requires that the signal can be normalized. For this the maximum at dipolar evolution time zero and the zero level at infinite dipolar evolution time have to be known. To know the maximum the experiment must be free of dead time. The zero level is not exactly known as the detector may have a small offset. To eliminate this receiver offset completely, the observer $\pi/2$ pulse is phase cycled $+x/-x$ and the two signals subtracted.

Systematic differences between the theoretically expected and experimentally observed dipolar evolution functions are mainly due to nuclear modulation and orientation selection. The depth of the dipolar modulation can be optimized with respect to the depth of nuclear modulations by using the largest feasible excitation bandwidth for the pump pulses and by adjusting pump pulse power to complete inversion. Nuclear modulations can be partially averaged by varying τ_1 in 8 steps of 8 ns and adding the signals. Orientation selection cannot generally be avoided. As it is less dramatic at lower fields, distance measurements should be performed at high fields only if there is a good reason to do so or if orientation selection is intended. To check experimentally for orientation selection effects, pump and observer positions have to be varied. The signal-to-noise ratio is optimized by careful adjustment of the phase and power of all channels, and in many cases by applying the variable-time DEER experiment (Jeschke et al. 2004a).

Another frequently encountered problem during long measurements is the phase drift of the signal. It is therefore advisable to adjust the phase carefully with the standing DEER experiment before starting the measurement and to use quadrature detection. That way phase drifts can be recognized and small drifts can be compensated by a phase correction of the whole signal trace after the measurement. A large phase drift (>10°) points to stability problems of the spectrometer that should be solved.

With proper setup procedures, modulation depths for the same sample should be reproducible to within less then 2% of the total signal. Modulation depth scaled signals of the same sample should differ only by white noise.

5. ACKNOWLEDGMENTS

Many ideas in this chapter are based on discussions with Arthur Schweiger and Jörg Forrer during my graduate studies in Zürich and later discussions with Peter Höfer and Antoine Wolff from Bruker and Christian Bauer in Mainz. I thank Yevhen Polyhach for proofreading the manuscript and making many suggestions to improve its clarity.

6. PROBLEMS

P0. Sample Preparation

The two experimental problems are based on standard samples for measurements at room temperature that can be prepared easily: ground coffee. Strongly roasted varieties give a somewhat stronger signal. Answers to the problems were prepared with Cafét Siesta (100% Arabica, available from the German supermarket chain Penny Market). Line width, radical concentration, and relaxation times may differ for other brands of coffee. Coffee is a sample with a signal of comparable strength to spin-labeled membrane proteins. The same experiments can be performed with less averaging, i.e., in a shorter time with a coal sample. For instance, the coal standard sample that comes with a Bruker pulsed ESR spectrometer is suitable as well.

P1. Optimization of CW ESR Conditions

Find the maximum m.w. power (minimum attenuation) and maximum modulation depth that can be used for measuring an undistorted CW ESR line shape for the coffee sample.

P2. Echo Integration in Field-Swept ESR

Measure the echo-detected field-swept ESR spectrum of coffee with a sequence $\pi/2$(16 ns)–delay(200 ns)–π(32 ns)–delay–*echo* (a) with an integration window length $t_{\text{intg}} = 200$ ns, and (b) with $t_{\text{intg}} = 32$ ns. Compare signal-to-noise ratio and line shape.

P3. Choice of Resonator

Assume that you want to make a pulsed ELDOR distance measurement at the X band (9.6 GHz) between an organic radical and a metal center ($S = 1/2$, g = 2.1) for which you expect to need a resonator bandwidth of 500 MHz and pump pulses with a flip angle of $3\pi/4$ and a length of 8 ns. Allowing for losses in the transmission line and circulator, the incident m.w. power at your resonator is 200 W. Estimate the effective volume of a resonator that would allow for such an experiment.

7. ANSWERS

A1. Optimization of CW ESR Conditions

For Cofét Siesta, signal intensity increases with less than the square root of the m.w. power for $P_{\text{mw}} >$ mW, i.e., 40-dB attenuation on a Bruker Elexsys 580 spectrometer (see Figure 11). The line width at 40 dB attenuation, measured with a modulation amplitude of 1 G, is 11 G. Hence, an undistorted line shape can be measured with maximum sensitivity at 40-dB attenuation and 2-G modulation amplitude. Note that values may differ somewhat for other brands of coffee.

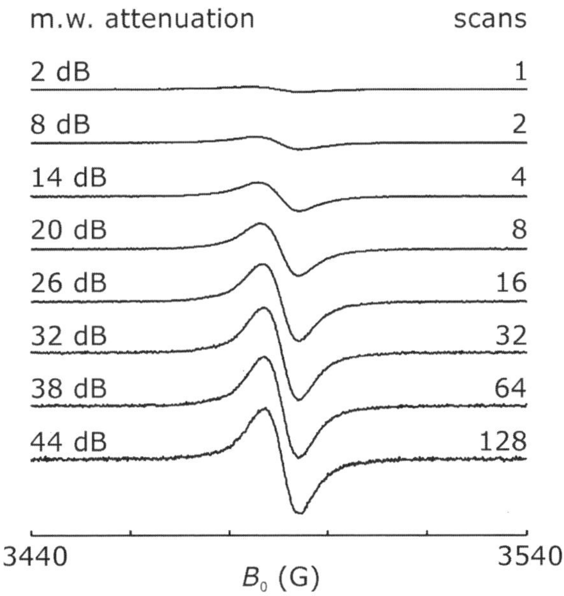

Figure 11. CW ESR measurements with different m.w. power level on a coffee sample. In the absence of saturation, all spectra should have the same amplitude.

A2. Echo Integration in Field-Swept ESR

Typical results are shown in Figure 12. The signal-to-noise ratio is significantly better with the 32-ns window, but the line is also significantly broadened.

A3. Choice of Resonator

$Q_L = \nu_{mw}/\Delta\nu = 19.2$. $B_1 = \beta h/(2\pi g \mu_B t_p) = 1.6 \cdot 10^{-3}$ T. $V_C = 2\mu_0 Q_L P_{mw}/(\nu_{mw} B_1^2) = 392$ mm^3 = 392 μL. Note, however, that the sample volume is much smaller.

8. REFERENCES

Bietsch W, von Schütz JU. 1993. ???full cite: article title???. *Bruker Report* **139**:12.

Borbat PP, Freed JH. 2000. Double quantum ESR and distance measurements. In *Distance measurements in biological systems by EPR*, pp. 383–459. Ed L Berliner, SS Eaton, GR Eaton. New York: Kluwer.

Burghaus O, Rohrer M, Götzinger T, Plato M, Möbius K. 1992. A novel high-field high-frequency EPR and ENDOR spectrometer operating at 3 mm wavelength. *Meas Sci Technol* **3**:765–774.

Davies ER. 1974. New pulse ENDOR technique. *Phys Lett* **A47**:1–2.

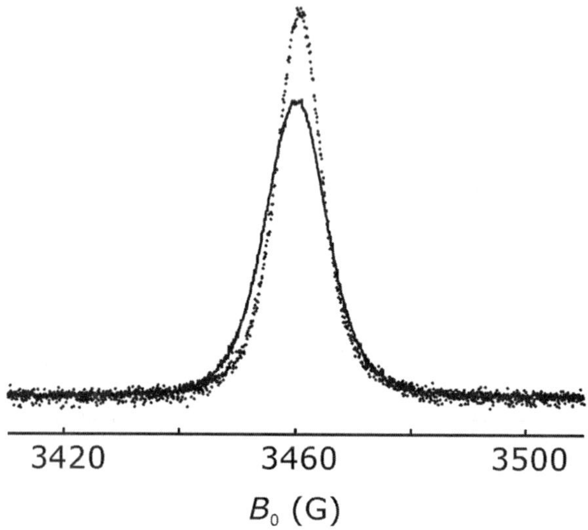

Figure 12. Echo-detected field-swept EPR on a coffee sample with integration gate widths of 32 (solid line) and 200 ns (dots). Amplitudes have been scaled so that the integral of the absorption spectrum is the same for both cases.

Epel B, Arieli D, Baute D, Goldfarb D. 2003. Improving W-band pulsed ENDOR sensitivity: acquisition and pulsed special TRIPLE. *J Magn Reson* **164**:78–83.

Feher G. 1957. Sensitivity considerations in microwave paramagnetic resonance absorption techniques. *Bell Syst Techol J* **36**:449–484.

Forrer J, Pfenninger S, Eisenegger J, Schweiger A. 1990. A pulsed ENDOR probehead with the bridged loop-gap resonator: construction and performance. *Rev Sci Instrum* **61**:3360–3367.

Froncisz, W, Hyde JS. 1982. The loop-gap resonator: a new microwave lumped circuit electron-spin-resonance sample structure. *J Magn Reson* **47**:515–521.

Gemperle C, Schweiger A. 1991. Pulsed electron-nuclear double resonance methodology. *Chem Rev* **91**:1481–1505.

Höfer P, Grupp A, Nebenführ H, Mehring M. 1986. Hyperfine sublevel correlation (HYSCORE) spectroscopy: a 2D electron-spin-resonance investigation of the squaric acid radical. *Chem Phys Lett* **132**:279–282.

Jeschke G, Pannier M, Godt A, Spiess HW. 2000. Dipolar spectroscopy and spin alignment in electron paramagnetic resonance. *Chem Phys Lett* **331**:243–252.

Jeschke G, Koch A, Jonas U, Godt A. 2002. Direct conversion of EPR dipolar time evolution data to distance distributions. *J Magn Reson* **155**:72–82.

Jeschke G, Bender A, Paulsen H, Zimmermann H, Godt A. 2004a. Sensitivity enhancement in pulse EPR distance measurements. *J Magn Reson* **169**:1–12.

Jeschke G, Panek G, Godt A, Bender A, Paulsen H. 2004b. Data analysis procedures for pulse ELDOR measurements of broad distance distributions. *Appl Magn Reson* **26**:223–244.

Milov AD, Ponomarev AB, Tsvetkov YuD. 1984. Electron–electron double resonance in electron spin echo: model biradical systems and the sensitized photolysis of decalin. *Chem Phys Lett* **110**:67–72.

Milov AD, Maryasov AG, Tsvetkov YuD. 1998. Pulsed electron double resonance (PELDOR) and its application in free-radicals research. *Appl Magn Reson* **15**:107–143.

Mims WB. 1965. Pulsed ENDOR experiments. *Proc Roy Soc London* **A283**:452–457.

Pannier M, Veit S, Godt A, Jeschke G, Spiess HW. 2000. Dead-time free measurement of dipole-dipole interactions between electron spins. *J Magn Reson* **142**:331–340.

Pfenninger S, Forrer J, Schweiger A, Weiland T. 1988. Bridged loop gap resonator: a resonant structure for pulsed electron-spin-resonance transparent to high-frequency radiation. *Rev Sci Instrum* **59**:752–760.

Poole ChP. 1997. Electron paramagnetic resonance (a comprehensive treatise on experimental techniques), 2nd ed. New York: Dover.

Rohrer M, Plato M, MacMillan F, Grishin Y, Lubitz W, Möbius K. 1995. Orientation-selected 95 GHz high-field ENDOR spectroscopy of randomly oriented plastoquinone anion-radicals. *J Magn Reson* **A116**:59–66.

Schweiger A, Jeschke G. 2001. Principles of pulse electron paramagnetic resonance. Oxford: Oxford UP.

Schweiger A, Gemperle C, Ernst RR. 1990. Soft pulse electron-spin-echo-envelope modulation spectroscopy (Soft ESEEM). *J Magn Reson* **86**:70–81.

Weil JA, Bolton JR, Wertz JE. 1994. *Electron paramagnetic resonance*. New York: Wiley.

Zänker PP, Jeschke G, Goldfarb D. 2004. Distance measurements between paramagnetic centers and a planar object by matrix Mims ENDOR. *J Chem Phys* **122**, 024515-11.

CHAPTER 3

ADVANCED ESR SPECTROSCOPY IN MEMBRANE BIOPHYSICS

Janez Štrancar

EPR Center, Laboratory of Biophysics,
"Jožef Stefan" Institute, Ljubljana, Slovenia

1. INTRODUCTION

1.1. Biophysical Background

Our current knowledge about **biological membranes** shows that they belong to the most important cell structures. Mass transport and signal transduction obviously appear to be vital to physiological functions of biomembranes as they enable cellular compartmentalization and control over it at the same time. Many experiments and theoretical considerations in the past decades have shown that membranes consist of a laterally heterogeneous lipid bilayer with a large number of different protein molecules embedded in the lipid bilayer. Heterogeneity exists at any level — from the biochemical to the physical level, meaning that different constituents and supramolecular structures in membranes interact via different interactions and exhibit different motional characteristics. This complexity — as the most striking property of any biological system — remains a tough problem also for the up-to-date experimental and theoretical approaches.

As will be shown in this chapter, recent developments in **electron spin resonance spectroscopy (ESR)** bring new possibilities to the further exploration of biosystems.

> The appropriateness of the ESR nanosecond time window and its sensitivity make **spin-label ESR** one of the most informative techniques, able to resolve the puzzle of membrane heterogeneity and intermolecular interactions as well as its role in cell signaling and mass transport.

Recent numerical approaches in combination with spin-label ESR experiments allow characterization of coexisting membrane structures in the cases of specific labeling as well as for nonspecific labeling, resulting in extremely valuable information about living structures in their native environment with only the small perturbation of the addition of spin probes.

Fast spectral simulation algorithms valid for physiological conditions, hybrid evolutionary optimization techniques responsible for efficient inverse-problem solving, and good solution-condensation algorithms have evolved spin-label ESR into an even more powerful technique for studying the dynamics and structure–function relationship of coexisting membrane constituents, like membrane lipid domains, carbohydrate surface aggregates, and conformational freedom of membrane peptides and proteins.

The aim of this chapter is therefore to present the key steps involved in this membrane characterization procedure and that can be applied as more general principles for any form of spectroscopy:

- Motional averaging in a spin-labeled biomembrane
- Strategies for calculating powder spectra
- Solving an inverse problem and condensation of the results

In the characterization procedure all the important properties of the experiment should be built into the simulation model. Furthermore, all the relevant characteristics of the model should enable a valuable characterization of the system within a reasonable time. To make a characterization procedure a useful tool in membrane biophysics as well as in more general cases, the following **three important requirements** need to be considered:

- Simulation should take into account all important characteristics of the ESR experiment and the complexity of the biological system
- Simulation should be fast enough to allow application of a robust scheme for inverse problem solving
- Simulation should be accurate enough to be sensitive to physiological changes

1.2. Mathematical Background

To clarify the notation of the various mathematical symbols and operations, they are introduced here briefly.

A **vector quantity** will be denoted by a one-sided arrow, e.g., magnetic field $\vec{B}$. In case of a **unit vector** with a length of 1, it will be stressed by a double-sided arrow, e.g., the unit vector of a coordinate system $\overleftrightarrow{x}$. A **tensor** quantity will be indicated by an underline, e.g., $\underline{A}$, a **unit tensor** (with diagonal elements 1 and off-diagonal elements 0) by $\underline{1}$. **Scalar products** between vectors, between a vector and matrices, and between matrices will be shown without a dot:

$$\vec{B}\ \vec{S} = \begin{pmatrix} B_x & B_y & B_z \end{pmatrix} \begin{pmatrix} S_x \\ S_y \\ S_z \end{pmatrix} = B_x S_x + B_y S_y + B_z S_z,$$

$$\vec{B}\ \underline{g} = \begin{pmatrix} g_{xx} & g_{xy} & g_{xz} \\ g_{yx} & g_{yy} & g_{yz} \\ g_{zx} & g_{zy} & g_{zz} \end{pmatrix} \begin{pmatrix} S_x \\ S_y \\ S_z \end{pmatrix} = \begin{pmatrix} g_{xx}S_x + g_{xy}S_y + g_{xz}S_z \\ g_{yx}S_x + g_{yy}S_y + g_{yz}S_z \\ g_{zx}S_x + g_{zy}S_y + g_{zz}S_z \end{pmatrix},$$

$$(1)$$

$$\underline{R}\ \underline{P} = \begin{pmatrix} R_{xx} & R_{xy} & R_{xz} \\ R_{yx} & R_{yy} & R_{yz} \\ R_{zx} & R_{zy} & R_{zz} \end{pmatrix} \begin{pmatrix} P_{xx} & P_{xy} & P_{xz} \\ P_{yx} & P_{yy} & P_{yz} \\ P_{zx} & P_{zy} & P_{zz} \end{pmatrix}$$

$$= \begin{pmatrix} R_{xx}P_{xx} + R_{xy}P_{yx} + R_{xz}P_{zx} & R_{xx}P_{xy} + R_{xy}P_{yy} + R_{xz}P_{zy} & R_{xx}P_{xz} + R_{xy}P_{yz} + R_{xz}P_{zz} \\ R_{yx}P_{xx} + R_{yy}P_{yx} + R_{yz}P_{zx} & R_{yx}P_{xy} + R_{yy}P_{yy} + R_{yz}P_{zy} & R_{yx}P_{xz} + R_{yy}P_{yz} + R_{yz}P_{zz} \\ R_{zx}P_{xx} + R_{zy}P_{yx} + R_{zz}P_{zx} & R_{zx}P_{xy} + R_{zy}P_{yy} + R_{zz}P_{zy} & R_{zx}P_{xz} + R_{zy}P_{yz} + R_{zz}P_{zz} \end{pmatrix}.$$

Operators will be indicated with a circumflex accent sign, e.g., scalar operator $\hat{H}$ and vector operator $\hat{\vec{S}}$.

Time and ensemble average are denoted as $\overline{f}$ and $\langle f \rangle$, respectively.

Angles will be denoted by Greek letters.

Indication of the coordinate system (l), where a vector or a tensor quantity X is well defined, it will be explicitly expressed by the left-upper index, e.g., ${}^{(l)}X$. When applied, it should remind the reader of a base of this coordinate system, defined through a set of unit vectors: ${}^{(l)}\mathsf{B} = \left\{ {}^{(l)}\vec{i}, {}^{(l)}\vec{j}, {}^{(l)}\vec{k} \right\}$.

In membrane spectroscopy four important coordinate systems are relevant:

- Nitroxide molecular coordinate system (n)
- Membrane coordinate system (m)
- Laboratory coordinate system (l)
- Coordinate system of local magnetic field (lm)

The introduction of these coordinate systems originates in the fact that various physical quantities can be well defined only in one of these coordinate systems. This is indicated in the following list of quantities and coordinate systems in which these quantities are well defined.

According to the properties above, the bases of the coordinate systems can be defined as indicated in Table 2 and Figure 1.

Coordinate systems *(a)* and *(b)* can be connected by means of **unitary transformation** R, denoted by $R \equiv R^{(a)\to(b)} : {}^{(a)}\mathsf{B} \to {}^{(b)}\mathsf{B}$. The inverse transformation is denoted by a complex conjugate (real transposed) $\overline{R^{(a)\to(b)}}$. Generally, a unitary transformation $R^{(a)\to(b)}$ can be expressed by three consequent rotations $R_{z''}^{(a'')\to(b)}(\varphi)\, R_{y'}^{(a')\to(a'')}(\vartheta)\, R_{z}^{(a)\to(a')}(\psi)$ about a temporary z-, y'-, or z''-axis (see Eq. (2)).

Table 1. Coordinate Systems with Well-Defined Quantities

Quantity Q	Coordinate system, in which Q is well defined
External magnetic field B	Laboratory coordinate system *(l)*
Local magnetic field B^l	Local magnetic field coordinate system *(lm)*
Electron spin S	Local magnetic field coordinate system *(lm)*
Nuclear spin I	Local magnetic field coordinate system *(lm)*
Zeeman tensor g	Nitroxide (molecular) coordinate system *(n)*
Hyperfine tensor A	Nitroxide (molecular) coordinate system *(n)*
Membrane normal n	Membrane coordinate system *(m)*

Table 2. Definition of Coordinate Systems

Coordinate system (cs)	Definition of base
Nitroxide (molecular) cs *(n)*	x-axis parallel to NO bond, z-axis parallel to π orbital
Membrane cs *(m)*	z-axis parallel to membrane normal
Laboratory cs *(l)*	z-axis parallel to external magnetic field
Local magnetic field cs *(lm)*	z-axis parallel to local magnetic field ${}^{(lm)}\vec{B} = {}^{(n)}\underline{g}\,{}^{(l)}\vec{B}$

$$R^{(a)\to(b)} = \begin{pmatrix} \cos\varphi & \sin\varphi & 0 \\ -\sin\varphi & \cos\varphi & 0 \\ 0 & 0 & 1 \end{pmatrix} \begin{pmatrix} \cos\vartheta & 0 & \sin\vartheta \\ 0 & 1 & 0 \\ -\sin\vartheta & 0 & \cos\vartheta \end{pmatrix} \begin{pmatrix} \cos\psi & \sin\psi & 0 \\ -\sin\psi & \cos\psi & 0 \\ 0 & 0 & 1 \end{pmatrix}. \quad (2)$$

In case of the coordinate systems mentioned previously, the unitary transformations are presented in Figure 2. **Transformation matrices** are derived by application of Euler rotation matrices as indicated above:

- For transformation from the laboratory to the membrane coordinate system (the first rotation around the z-axis in the laboratory coordinate system is omitted, since the Hamiltonian is invariant to a rotation around an external magnetic field axis):

$$R^{(l)\to(m)} = \begin{pmatrix} \cos\Phi & \sin\Phi & 0 \\ -\sin\Phi & \cos\Phi & 0 \\ 0 & 0 & 1 \end{pmatrix} \begin{pmatrix} \cos\Theta & 0 & \sin\Theta \\ 0 & 1 & 0 \\ -\sin\Theta & 0 & \cos\Theta \end{pmatrix}. \quad (3)$$

- For transformation from the membrane to the nitroxide coordinate system:

$$R^{(m)\to(n)} = \begin{pmatrix} \cos\varphi & \sin\varphi & 0 \\ -\sin\varphi & \cos\varphi & 0 \\ 0 & 0 & 1 \end{pmatrix} \begin{pmatrix} \cos\vartheta & 0 & \sin\vartheta \\ 0 & 1 & 0 \\ -\sin\vartheta & 0 & \cos\vartheta \end{pmatrix} \begin{pmatrix} \cos\psi & \sin\psi & 0 \\ -\sin\psi & \cos\psi & 0 \\ 0 & 0 & 1 \end{pmatrix}. \quad (4)$$

- For transformation from the laboratory to the local magnetic field coordinate system and vice versa

$$R^{(l)\to(lm)} = \frac{\overline{R^{(l)\to(m)}}\ \overline{R^{(m)\to(n)}}\ g\ R^{(m)\to(n)}\ R^{(l)\to(m)}}{\left|\overline{R^{(l)\to(m)}}\ \overline{R^{(m)\to(n)}}\ g\ R^{(m)\to(n)}\ R^{(l)\to(m)}\ {}^{(l)}\vec{k}\right|}, \quad (5)$$

$$R^{(lm)\to(l)} = \left|\overline{R^{(l)\to(m)}}\ \overline{R^{(m)\to(n)}}\ g\ R^{(m)\to(n)}\ R^{(l)\to(m)}\ {}^{(l)}\vec{k}\right| \overline{R^{(l)\to(m)}}\ \overline{R^{(m)\to(n)}}\ g^{-1}\ R^{(m)\to(n)}\ R^{(l)\to(m)}. \quad (6)$$

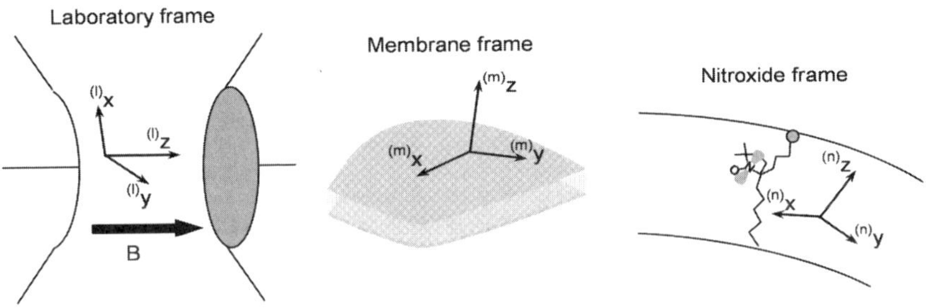

Figure 1. Schematic presentation of the coordinate systems.

1.3. Quantum-Physical Background

To follow this chapter, the reader should be familiar with basic operations of quantum physics, i.e., working with operators and calculating eigenvalues of the spin operators as well as of the Hamiltonian operator. It is not the aim of this chapter to teach the basics of quantum physics; however, for the sake of clarity we briefly review the basic principles.

If the system state is described by a single set of quantum numbers $\{\xi\}$, or shortly in Dirac bracket notation $|\xi\rangle$, then its eigenvalue (eigenenergy) E_ξ can be calculated by application of a Hamiltonian operator function in the following way:

$$\hat{H}|\xi\rangle = E_\xi|\xi\rangle \to E_\xi = \frac{\langle\xi|\hat{H}|\xi\rangle}{\langle\xi|\xi\rangle} \xrightarrow{\langle\xi|\xi\rangle=1} E_\xi = \langle\xi|\hat{H}|\xi\rangle. \quad (1)$$

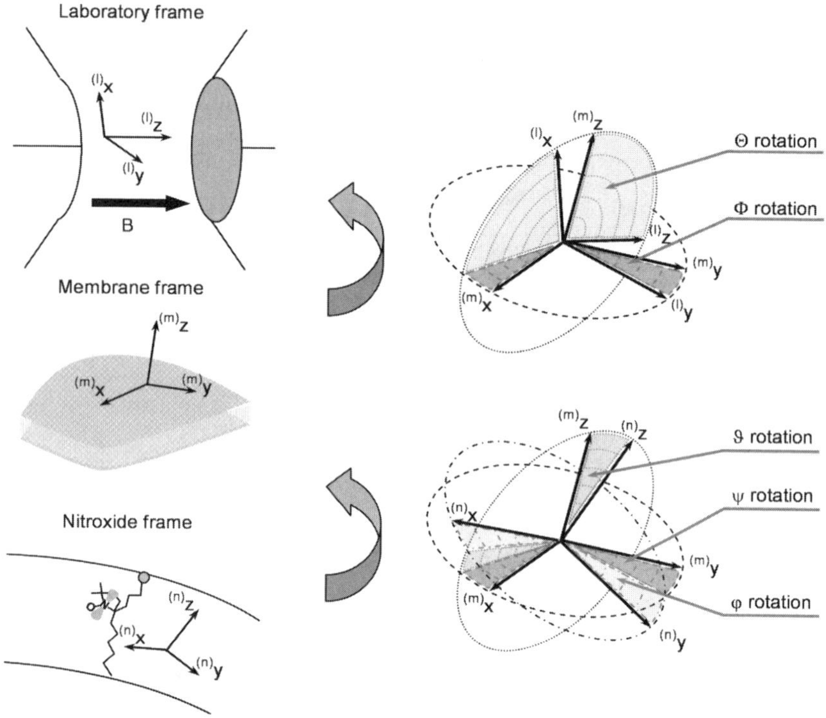

Figure 2. Schematic presentation of the transformations between coordinate systems using Euler rotations.

However, usually there are N states $|\xi_i\rangle$ allowed in a real system. In such a case the eigenvalues (eigenenergies) should be calculated as the principal values of the Hamiltonian matrix defined as

$$\underline{E} = \begin{pmatrix} E_{11} & \cdots & E_{1N} \\ \vdots & \ddots & \vdots \\ E_{N1} & \cdots & E_{NN} \end{pmatrix} \leftarrow E_{ij} = \langle \xi_i | \hat{H} | \xi_j \rangle . \tag{2}$$

Since ESR spectroscopy involves the transition of spin states, the Hamiltonian operator includes the spin operators. Consequently, one should be familiar with the basics of spin algebra. In case of nitroxide spin labels, the spin states of the electron spin, $S = ½$, and nitrogen nuclear spin of $I = 1$ are involved, implying the usage of the following set of base functions:

$$\{ |M_S, M_I\rangle \} = \{ |\tfrac{1}{2},1\rangle, |-\tfrac{1}{2},1\rangle, |\tfrac{1}{2},0\rangle, |-\tfrac{1}{2},0\rangle, |\tfrac{1}{2},-1\rangle, |-\tfrac{1}{2},-1\rangle \}. \tag{3}$$

Here, the short notation of base functions $|M_S, M_I\rangle$ is used instead of $|S, M_S, I, M_I\rangle$. The basic rules of spin algebra and the eigenvalues of the various spin operators are

$$\begin{aligned}
&\hat{S}_z|S,M_S,I,M_I\rangle = M_S|S,M_S,I,M_I\rangle,\\
&\hat{S}_\pm|S,M_S,I,M_I\rangle = \sqrt{(S\mp M_S)(S\pm M_S+1)}\,|S,M_S\pm 1,I,M_I\rangle,\\
&\hat{I}_z|S,M_S,I,M_I\rangle = M_I|S,M_S,I,M_I\rangle,\\
&\hat{I}_\pm|S,M_S,I,M_I\rangle = \sqrt{(I\mp M_I)(I\pm M_I+1)}\,|S,M_S,I,M_I\pm 1\rangle,\\
&\langle S,M_S,I,M_I|S,M_S',I,M_I'\rangle = \delta_{M_S,M_S'}\delta_{M_I,M_I'}.
\end{aligned} \tag{10}$$

Expansion of the vector spin operator in the basis of the coordinate system of the local magnetic field can be written in the following form:

$$\begin{aligned}
\hat{\vec{S}} &= \hat{S}_x\ \vec{i} + \hat{S}_y\ \vec{j} + \hat{S}_z\ \vec{k}\\
&= \tfrac{1}{2}\left(\hat{S}^+ + \hat{S}^-\right)\ \vec{i} + \tfrac{1}{2i}\left(\hat{S}^+ - \hat{S}^-\right)\ \vec{j} + \hat{S}_z\ \vec{k}.
\end{aligned} \tag{11}$$

2. MOTIONAL AVERAGING IN A SPIN-LABELED BIOMEMBRANE

2.1. Spin Labeling Experiment

To be able to detect an ESR spectrum, a stable unpaired-electron system should be part of the system or introduced into the system.

> A spin labeling experiment involves implementation of **nitroxide-based spin probes.**

Usually, introduction of a spin probe is a **nonspecific addition of spin-labeled analogs of common biomolecules** that can be found in biomembranes (or other biosystems). The nitroxide group, which contains a stable free radical, can be attached at different positions of a molecule. If a nitroxide moiety can be justified as a small perturbation to a (larger) biomolecule or supramolecular structure, then the partitioning properties of a spin-labeled molecule are very similar to the partitioning properties of the original molecules — a property that can be used for subsystem identification.

On the other hand, a spin label can be introduced into a biosystem also in a **specific** manner, using a biochemical **targeting** procedure. In this case, its position is well defined, and the complexity that is determined from spectral simulation directly points to local conformations of the targeted biomolecule. Since the presented approach can be used also to detect conformations at nonspecific sites, special care should be taken to identify all possible binding sites and to include the reaction rates of nitroxide binding to these sites.

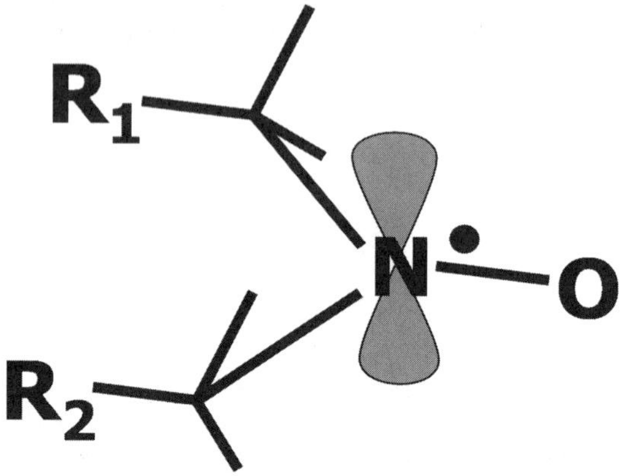

Figure 3. Nitroxide group.

If physiologically relevant information should be detected, the **temperature** of a spin-label experiment should be chosen in quite a narrow interval. Note that "physiological temperature" depends on the biological species, e.g., 0–50°C for various bacteria, 37°C for mammalians, etc. At such temperatures organisms react on a millisecond timescale, and local biosignaling processes on biomembranes occur on the ten microseconds timescale. The local dynamics of the membrane constituents should be fast enough to enable lipid or protein supramolecular structures to find their favored state within this timeframe. Global conformational changes that are involved in biosignaling should evolve through reconfiguration over several thousand of conformational states. That can only be achieved with fast local molecular motions with correlation times at the sub-nanosecond timescale. This fact is confirmed by many molecular dynamics simulations and strongly influences the modeling scheme of the ESR spectral simulations for physiological temperatures.

> Physiological temperatures involve rotational motion on the nanosecond timescale, which is fast on the ESR timescale. This implies partial averaging of the nitroxide magnetic properties over the allowed local rotational configurations. Additional slower time evolution of labeled subsystems will be determined as coexisting motional/polarity patterns.

However, some ESR experiments cannot be performed at physiological temperatures. When a much lower temperature has to be used, slow rotational motions come into play. To simulate such experiments one has to implement slow-motion schemes. However, one has to be aware of the fact that the resulting anisotropy of the system and rate of rotational motion in the slow-motion ESR regime create a numerically ill-posed problem, i.e., in complex membrane systems and cell suspensions, the anisotropy and rate of rotational motion cannot be distinguished well enough.

2.2. The Hamiltonian for Spin Labels

The spin Hamiltonian of a nitroxide spin label can be approximated by a combination of a **Zeeman term** (interaction between electron spin S and external magnetic field B) and a **hyperfine term** (interaction between electron spin S and nitrogen nuclear spin I):

$$\hat{H} = \mu_B \vec{B}\ \underline{g}\ \hat{\vec{S}} + \hat{\vec{I}}\,\underline{A}\ \hat{\vec{S}}\,. \tag{12}$$

The superhyperfine interaction between the electron spin and neighboring proton spins can be effectively taken into account in the linewidth calculation by an additional broadening. The nuclear Zeeman interaction between the nuclear spins and the magnetic field are neglected due to (latter) application of selection rules, which imply transitions to occur between electron spin states and not between nuclear spin states.

> Different quantities in Hamiltonian equation (12) are well defined only in their own coordinate system (see Tables 1 and 2).

To explicitly express the coordinate systems in which the quantities are well defined, we can use the notation defined in the §1 and identify the coordinate systems according to their indices. Remember that the magnetic properties are well defined in the nitroxide molecular system *(n)*, the external magnetic field in the laboratory system *(l)*, and the spin operators in the coordinate system of the local magnetic field *(lm)*. Note that the usage of simple spin algebra is limited to the coordinate system where the direction of quantization is known, which is the case only for the coordinate system of the local magnetic field *(lm)*:

$$\hat{H} = \mu_B\, {}^{(l)}\vec{B}\ {}^{(n)}\underline{g}\ {}^{(lm)}\hat{\vec{S}} + {}^{(lm)}\hat{\vec{I}}\ {}^{(n)}\underline{A}\ {}^{(lm)}\hat{\vec{S}}\,. \tag{13}$$

Although the Hamiltonian expression does not depend on the choice of coordinate system, one has to express all the quantities in the same coordinate system to solve the eigensystem equation. Therefore one should introduce transformation matrices between the quantities defined in different coordination systems. Consequently, the Hamiltonian for a single spin probe molecule should be rewritten as

$$\hat{H} = \mu_B \vec{B}\ \overline{R^{(l)\to(n)}}\ \underline{g}\ R^{(lm)\to(n)}\ \hat{\vec{S}} + \hat{\vec{I}}\ \overline{R^{(lm)\to(n)}}\ \underline{A}\ R^{(lm)\to(n)}\hat{\vec{S}}\,. \tag{14}$$

Since the ESR experiment involves several spin probe molecules, actually more than 10^{13}, which is the sensitivity limit for a nitroxide spin-label experiment on the current ESR spectrometers, Eq. (14) actually becomes

$$\hat{H} = \sum_i \left[\mu_B \vec{B}\ \overline{R_i^{(l)\to(n)}}\ \underline{g}\ R_i^{(lm)\to(n)}\ \hat{\vec{S}} + \hat{\vec{I}}\ \overline{R_i^{(lm)\to(n)}}\ \underline{A}\ R_i^{(lm)\to(n)} \hat{\vec{S}} \right]. \qquad (15)$$

Note that summation over the nitroxide molecules includes summation over their different orientations; however, the electron and nuclear spin states are equivalent for all spin probes. The same is valid for the magnetic properties of the spin probes. Orientational dependencies are buried within the transformation matrices.

Introducing the transformation to and from the coordinate system of the local magnetic field as given by Eqs. (5) and (6), the **Zeeman term** of Eq. (14) becomes

$$\hat{H}_{\text{Zeeman}} = \mu_B B \left| \overline{R^{(l)\to(m)}}\ \overline{R^{(m)\to(n)}}\ \underline{g}\ R^{(m)\to(n)}\ R^{(l)\to(m)}\ \vec{k} \right| \vec{k}\ \hat{\vec{S}}. \qquad (16)$$

By expanding the vector spin operator in the basis of the local magnetic field coordinate system as shown in Eq. (11), the Zeeman term can be simplified to

$$\hat{H}_{\text{Zeeman}} = \mu_B B\ g^{\text{eff}}\left(\vec{\Omega}\right)\ \hat{S}_z, \qquad (17)$$

where effective Zeeman coupling is defined as

$$g^{\text{eff}}\left(\vec{\Omega}\right) = \sqrt{\vec{k}\ \overline{R^{(l)\to(m)}}\ \overline{R^{(m)\to(n)}}\ \underline{g}\ \ \underline{g}\ R^{(m)\to(n)}\ R^{(l)\to(m)}\ \vec{k}}\ . \qquad (18)$$

Note that the effective Zeeman coupling, $g^{\text{eff}}\left(\vec{\Omega}\right) = g^{\text{eff}}\left(\Theta, \Phi; \psi, \vartheta, \varphi\right)$, includes the "powder-like" membrane-normal orientation dependence $\{\Theta, \Phi\}$ as well as the orientation dependence of fast changing rotational conformations defined with Euler angles $\{\psi, \vartheta, \varphi\}$. To take this into account while partial averaging due to fast rotational motion, we introduce

$$\underline{\tilde{G}}\left(\psi, \vartheta, \varphi\right) = \overline{R^{(m)\to(n)}}\ \underline{g}\ \underline{g}\ R^{(m)\to(n)}. \qquad (19)$$

Similarly, we can transform the **hyperfine term** of Eq. (14):

$$\hat{H}_{\text{hyperfine}} = \left(\overline{R^{(lm)\to(l)}}\ \hat{\vec{I}} \right) \overline{R^{(l)\to(m)}}\ \overline{R^{(m)\to(n)}}\ \underline{A}\ R^{(m)\to(n)}\ R^{(l)\to(m)}\ R^{(lm)\to(l)}\ \hat{\vec{S}}. \qquad (20)$$

Using the transformation from/to the coordinate system of the local magnetic field. the hyperfine term can be rewritten as

$$\begin{aligned} \hat{H}_{\text{hyperfine}} &= \left(g^{\text{eff}}\left(\vec{\Omega}\right)\right)^2 \hat{\vec{I}}\, \overline{R^{(l)\to(m)}\, R^{(m)\to(n)}}\, \underline{g}^{-1}\, \underline{A}\, \underline{g}^{-1}\ {}^{(m)\to(n)}\ R^{(l)\to(m)}\, \hat{\vec{S}} \\ &= \hat{\vec{I}}\ \underline{A}^{\text{eff}}\left(\vec{\Omega}\right)\ \hat{\vec{S}} \end{aligned}, \qquad (21)$$

where the effective hyperfine coupling (tensor) is defined as

$$\underline{A}^{\text{eff}}\left(\bar{\Omega}\right)=\left(g^{\text{eff}}\left(\bar{\Omega}\right)\right)^{2}\ \overline{R^{(l)\to(m)}}\ \overline{R^{(m)\to(n)}}\ \underline{g}^{-1}\underline{A}\ \underline{g}^{-1}\ R^{(m)\to(n)}\ R^{(l)\to(m)}. \tag{22}$$

Similarly to the effective Zeeman term, the effective hyperfine coupling, $\underline{A}^{\text{eff}}\left(\bar{\Omega}\right)=\underline{A}^{\text{eff}}\left(\Theta,\Phi;\psi,\vartheta,\varphi\right)$, includes the "powder-like" membrane-normal orientation dependence $\{\Theta,\ \Phi\}$ and the orientation dependence of fast changing rotational conformations defined with Euler angles $\{\psi,\ \vartheta,\ \varphi\}$. Note that both dependencies are involved also in the g^{eff} term, making the expression for A^{eff} even more complicated. In the same way as $\underline{\tilde{G}}$, we introduce

$$\underline{\tilde{A}}\left(\psi,\vartheta,\varphi\right)=\overline{R^{(m)\to(n)}}\ \underline{g}^{-1}\ \underline{A}\ \underline{g}^{-1}\ R^{(m)\to(n)}. \tag{23}$$

Using the effective Zeeman and hyperfine term, the Hamiltonian can be written in the following compact form:

$$\hat{H}=\mu_{B}B\ g^{\text{eff}}\left(\bar{\Omega}\right)\ \hat{S}_{z}+\ \hat{\vec{I}}\ A^{\text{eff}}\left(\bar{\Omega}\right)\ \hat{\vec{S}}. \tag{24}$$

Further simplification of the effective hyperfine term depends on the approximation one takes into account when solving the eigensystem equation. However, before solving the eigensystem equation, an important first step is taking into account, e.g., fast rotational motions in the membrane coordinate system, which implies partial averaging in the $\underline{\tilde{G}}\left(\psi,\vartheta,\varphi\right)$ and $\underline{\tilde{A}}\left(\psi,\vartheta,\varphi\right)$ terms.

2.3. Introducing Fast Motion of the Spin Probe

> When the spin probe wobbles, the effective Zeeman and hyperfine couplings change with orientation (they involve dependencies on $\{\Theta,\ \Phi\}$ as well as $\{\psi,\ \vartheta,\ \varphi\}$).

Suppose a wobbling spin probe in a membrane in a situation where the spin probe also is able to laterally diffuse. We already define that the wobbling — changing of $\{\psi,\ \vartheta,\ \varphi\}$ in time — is fast on the ESR timescale. On the other hand, the mean displacement of the spin probe molecule in the membrane is at largest a few lipid molecules per ESR timescale. Since the curvature of the membrane is usually not extremely high, we can say that the time variation of $\{\Theta,\ \Phi\}$ is much slower as compared to the time variation of $\{\psi,\ \vartheta,\ \varphi\}$ — actually we will assume $\{\Theta,\ \Phi\}$ to be constant on the ESR timescale. In such a case, the orientation dependence can be separated into two major contributions:

- Orientation of the molecule (or nitroxide group) relative to molecule average orientation in the membrane, that is, membrane normal ($\{\psi,\ \vartheta,\ \varphi\}$ dependences)

- Orientation of the local membrane normal relative to the external magnetic field ($\{\Theta, \Phi\}$ dependences)

However, sometimes the translational diffusion could be higher and the local curvature could be larger, such as in the case of membrane defects. In such cases one has to be careful in separating the two orientational dependencies.

> Fast rotational motion of a spin-labeled molecule on the ESR timescale averages the magnetic properties over all the possible orientations relative to the membrane normal.

Consequently, time averaging should be taken into account over the terms with $\{\psi, \vartheta, \varphi\}$ dependences in Eqs. (18) and (22). The $\underline{\tilde{G}}(\psi,\vartheta,\varphi)$ and $\underline{\tilde{A}}(\psi,\vartheta,\varphi)$ terms are transformed as follows:

$$\begin{aligned} \underline{\tilde{G}}(\psi,\vartheta,\varphi) &\mapsto \underline{\tilde{G}}^{\text{avg}} = \overline{\underline{\tilde{G}}(\psi(t),\vartheta(t),\varphi(t))}\,, \\ \underline{\tilde{A}}(\psi,\vartheta,\varphi) &\mapsto \underline{\tilde{A}}^{\text{avg}} = \overline{\underline{\tilde{A}}(\psi(t),\vartheta(t),\varphi(t))}\,. \end{aligned} \tag{25}$$

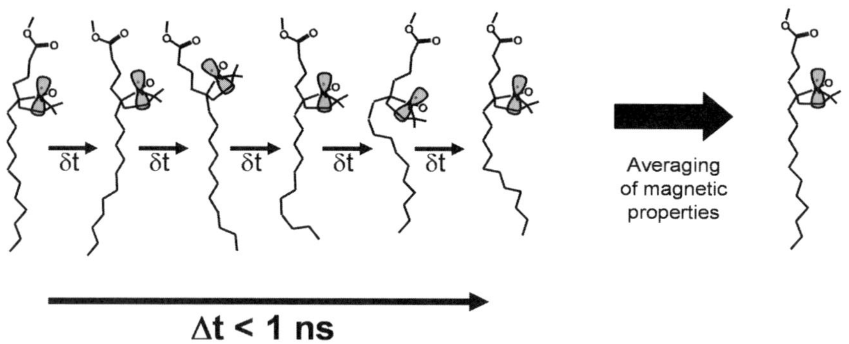

Figure 4. Averaging of the magnetic properties (Zeeman and hyperfine couplings).

The elements of $\underline{\tilde{G}}$ and $\underline{\tilde{A}}$ involve complicated expressions of trigonometric functions that should be averaged according to the motional restriction

$$\overline{f(\Omega)} = \int f(\Omega)\frac{dP}{d\Omega}d\Omega\,. \tag{26}$$

The definition intervals are $[-\pi/2, \pi/2]$ for φ, $[0, \pi/2]$ for ϑ, and $[-\pi/2, \pi/2]$ for ψ, as indicated in Eq. (25). The full expressions for the averaged effective Zeeman coupling components and hyperfine coupling are given in Appendix 1 of this chapter (Eqs. (A.1)–(A.12)). Note that the terms involving sin(φ), cos(φ), sin(2φ), sin(ψ), cos(ψ), and sin(2ψ) vanish after averaging due to symmetry.

The averages in Eq. (25) can be determined in the following ways:

- Deduction from molecular dynamics simulations,
- Phenomenological expression with order parameters,
- Calculation from conformational probabilities, which can be derived from reorienting potentials.

The last two approaches are discussed in the following text.

> If probability functions are not directly known, the averages in Eq. (25) may be expressed in terms of **order parameters**.

$$\overline{\cos^2(\vartheta_{zz})} = \overline{\cos^2(\vartheta)} = \frac{2S_z + 1}{3}, \tag{27}$$

$$\overline{\cos^2(\vartheta_{zy})} = \overline{\sin^2(\vartheta)}\ \overline{\sin^2(\varphi)} = \frac{2S_y + 1}{3} = \frac{1}{3}(1 - S_z)(1 - \sigma)\ . \tag{28}$$

The most common order parameter, S_z, represents ordering along the long molecular axes in the case of a membrane spin label. The two extreme situations are: a completely isotropic case (a solution with no preferred orientation) and a completely ordered case. Note that S_z counts for an averaged projection of a tumbling molecule on its averaged direction. Typically, the value of S_z of a fatty acid spin probe in disordered phases of membranes is below 0.3, while it usually does not exceed 0.7 in the ordered phase.

The other order parameter, σ counts for the asymmetry/restrictions of the rotational motions around the long axes (averaged molecule position), with definition intervals being the same as for S_z. Partial averaging due to rotational motions around the long axes strongly depends on the size and conformational freedom of the spin-labeled molecule as well as the properties of the local environment — its sterical restrictions and dynamics. For example, the rotation of a spin-labeled lipid molecule around its long axes is more restricted than for a spin-labeled fatty acid molecule, resulting in higher σ. The same is true for a spin probe being close to rigid molecules like cholesterol. Note that σ rarely exceeds 0.7.

To be able to calculate all averages directly, one needs to know or define the probability functions.

In a **wobble or cone model**, the probability functions are defined as step functions:

$$\left\{\begin{array}{l} 0 < |\xi| < \xi_0 \Rightarrow \dfrac{dP}{d\xi} = \text{const} \\ \xi_0 < |\xi| < \dfrac{\pi}{2} \Rightarrow \dfrac{dP}{d\xi} = 0 \end{array}\right\} \quad \text{for } \xi = \{\psi, \vartheta, \varphi\}\ . \tag{29}$$

This leads to the following expressions:

$$\overline{\cos(\vartheta)^2} = \tfrac{1}{3}\left(\cos^2(\vartheta_0) + \cos(\vartheta_0) + 1\right), \tag{30}$$

$$\overline{\sin(\varphi)^2} = \frac{\varphi_0 - \frac{1}{2}\sin(2\varphi_0)}{2\varphi_0} = \frac{1}{2}\left(1 - \frac{\sin(2\varphi_0)}{2\varphi_0}\right), \tag{31}$$

$$\overline{\sin(\psi)^2} = \frac{\psi_0 - \frac{1}{2}\sin(2\psi_0)}{2\psi_0} = \frac{1}{2}\left(1 - \frac{\sin(2\psi_0)}{2\psi_0}\right), \tag{32}$$

$$\overline{\sin(2\vartheta)} = \tfrac{2}{3}\left(\cos(\vartheta_0) + 1\right)\sin(\vartheta_0), \tag{33}$$

$$\overline{\cos(\psi)} = \frac{\sin(\psi_0)}{\psi_0}. \tag{34}$$

Using appropriate probability functions, the averages can be expressed in terms of maximal wobbling angles. Note that in general one could have three maximal wobbling angles $\{\psi_0, \vartheta_0, \varphi_0\}$; however, two of them, ψ_0 and φ_0, are quite similar in this model. In fact, in the two limiting cases, close to no restrictions and close to complete restriction, ψ_0 and φ_0 are identical. Moreover, when ϑ_0 is small, ψ_0 and φ_0 are identical again. Therefore, we can often assume that ψ_0 is equal to φ_0. A second reason for using only one restriction parameter is purely numerical, since application of all three angles would make their determination ill-posed in an inverse-problem-solving case.

Alternatively, the probability function can be defined through a **Boltzmann distribution**:

$$\frac{dP}{d\Omega} = e^{-\frac{U(\Omega)}{kT}}, \tag{35}$$

and **reorienting potential** $U(\Omega)$ in generalized form of a Maier-Saupe potential:

$$U(\Omega) = U(\vartheta,\varphi) = A\ \cos^2(\vartheta) + \left(B + C\ \sin(\vartheta)^2\right)\cos^2(\varphi). \tag{36}$$

Using probability function Eq. (35) and reorienting the potential of Eq. (36), the averages of the trigonometric function can be calculated numerically in terms of potential constants *A*, *B,* and *C*. Below is only an example of how an average is calculated in a more general way:

$$\overline{\cos(\vartheta)^2} = \frac{\int \cos(\vartheta)^2 e^{\frac{A}{kT}\cos(\vartheta)^2 + \left(\frac{B}{kT} + \frac{C}{kT}\sin(\vartheta)^2\right)\cos(\varphi)^2} d\left(\cos(\vartheta)\right) d\varphi}{\int e^{\frac{A}{kT}\cos(\vartheta)^2 + \left(\frac{B}{kT} + \frac{C}{kT}\sin(\vartheta)^2\right)\cos(\varphi)^2} d\left(\cos(\vartheta)\right) d\varphi}. \tag{37}$$

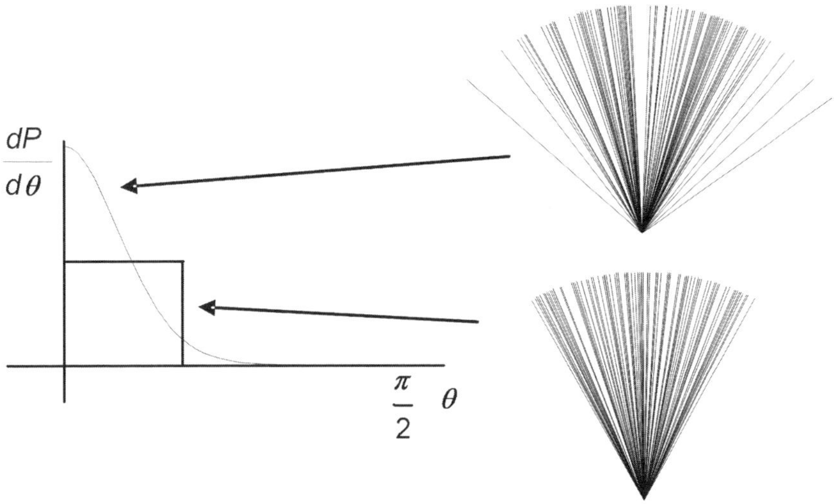

Figure 5. Schematic presentation of two different directional probability functions (on the left) and the corresponding distributions that yield the same $\overline{\cos(\vartheta)^2}$ and consequently the same order parameter S_z; box-like distribution function and lower distribution pattern (lower right corner) correspond to the Cone (wobble) model, whereas the smooth distribution and the upper distribution pattern (upper right corner) correspond to a Boltzmann distribution with a simple symmetric Maier-Saupe reorienting potential (see (36)); for clarity, only one angle (ϑ) is presented.

2.4. Derivation of Resonant Fields

After motional averaging, the total spin Hamiltonian of Eq. (15) can be rewritten as

$$\begin{aligned}\hat{H} &= \sum_i \Big[\mu_B \vec{B}\ \overline{R_i^{(l)\to(m)}}\ \underline{\tilde{G}}^{\text{avg}}\ R_i^{(l)\to(m)}\ \hat{\vec{S}} \\ &\quad + \left| \vec{k}\ \overline{R_i^{(l)\to(m)}}\ \underline{\tilde{G}}^{\text{avg}}\ R_i^{(l)\to(m)}\ \vec{k} \right|^2 \hat{\vec{I}}\ \overline{R_i^{(l)\to(m)}}\ \underline{\tilde{A}}^{\text{avg}}\ R_i^{(l)\to(m)}\ \hat{\vec{S}} \Big] \\ &= \sum_i \Big[\mu_B B\ g^{\text{eff}}(\Theta_i,\Phi_i)\ \hat{S}_z + \hat{\vec{I}}\ \underline{A}^{\text{eff}}(\Theta_i,\Phi_i)\ \hat{\vec{S}} \Big].\end{aligned} \tag{38}$$

Once again we want to stress that motional averaging cannot average out the orientation dependence of the local membrane normal relative to the magnetic field direction, unless lateral diffusion is high or membrane curvature is strong.

To find the eigenvalues of spin Hamiltonian Eq. (38), the all-spin operators in the hyperfine term should be expanded as shown for the electron spin operators in Eq. (11):

$$
\begin{aligned}
\hat{H}_i = [\mu_B B\ & g^{\text{eff}}(\Theta_i,\Phi_i) + A^{\text{eff}}_{zz}(\Theta_i,\Phi_i)\ \hat{I}_z \\
& + \overline{A^{\text{eff}}_{xzyz}}(\Theta_i,\Phi_i)\ \hat{I}_+ + A^{\text{eff}}_{xzyz}(\Theta_i,\Phi_i)\ \hat{I}_-]\hat{S}_z \\
& + \left[\overline{A^{\text{eff}}_{xzyz}}(\Theta_i,\Phi_i)\ \hat{I}_z + \overline{A^{\text{eff}}_{xyxy}}(\Theta_i,\Phi_i)\ \hat{I}_+ + A^{\text{eff}}_{xxyy}(\Theta_i,\Phi_i)\ \hat{I}_-\right]\ \hat{S}_+ \\
& + \left[A^{\text{eff}}_{xzyz}(\Theta_i,\Phi_i)\ \hat{I}_z + A^{\text{eff}}_{xxyy}(\Theta_i,\Phi_i)\ \hat{I}_+ + A^{\text{eff}}_{xyxy}(\Theta_i,\Phi_i)\ \hat{I}_-\right]\ \hat{S}_- \ .
\end{aligned}
\tag{39}
$$

The full expressions for effective tensor couplings g^{eff}, $A_{zz}{}^{\text{eff}}$, $A_{xzyz}{}^{\text{eff}}$, $A_{xyxy}{}^{\text{eff}}$, and $A_{xxyy}{}^{\text{eff}}$ are given in Appendix A (Eqs. (A.13)–)A.17)). Note that the overbar in Eq. (39) indicates the complex conjugate of the corresponding quantities.

The **Hamiltonian matrix** (see §1 for an explanation) now becomes

$$
\langle \hat{H} \rangle = \left(\begin{array}{cc|cc|cc}
+\frac{1}{2}\left(G^{\text{eff}} + A^{\text{eff}}_{zz}\right) & \overline{A^{\text{eff}}_{xzyz}} & +\frac{1}{2}\overline{A^{\text{eff}}_{xzyz}} & \overline{A^{\text{eff}}_{xyxy}} & 0 & 0 \\
A^{\text{eff}}_{xzyz} & -\frac{1}{2}\left(G^{\text{eff}} + A^{\text{eff}}_{zz}\right) & A^{\text{eff}}_{xxyy} & -\frac{1}{2}\overline{A^{\text{eff}}_{xzyz}} & 0 & 0 \\
\hline
+\frac{1}{2}A^{\text{eff}}_{xzyz} & A^{\text{eff}}_{xxyy} & +\frac{1}{2}\left(G^{\text{eff}}\right) & 0 & +\frac{1}{2}\overline{A^{\text{eff}}_{xzyz}} & \overline{A^{\text{eff}}_{xyxy}} \\
A^{\text{eff}}_{xyxy} & -\frac{1}{2}A^{\text{eff}}_{xzyz} & 0 & -\frac{1}{2}\left(G^{\text{eff}}\right) & A^{\text{eff}}_{xxyy} & -\frac{1}{2}\overline{A^{\text{eff}}_{xzyz}} \\
\hline
0 & 0 & +\frac{1}{2}A^{\text{eff}}_{xzyz} & A^{\text{eff}}_{xxyy} & +\frac{1}{2}\left(g^{\text{eff}} - A^{\text{eff}}_{zz}\right) & -\overline{A^{\text{eff}}_{xzyz}} \\
0 & 0 & A^{\text{eff}}_{xyxy} & -\frac{1}{2}A^{\text{eff}}_{xzyz} & -A^{\text{eff}}_{xzyz} & -\frac{1}{2}\left(g^{\text{eff}} - A^{\text{eff}}_{zz}\right)
\end{array}\right)
\tag{40}
$$

where G^{eff} corresponds to $\mu_B \cdot B \cdot g^{\text{eff}}$.

The Hamiltonian matrix is a block-tridiagonal Hermitian (self-adjoint) matrix. Although the Hamiltonian matrix has complex elements, please note that it is a Hermitian matrix and therefore has real eigenvalues.

In general the matrix of Eq. (40) cannot be solved analytically. Although there are some efficient numerical methods to find the eigenvalues of Eq. (40), some approximations can be made without reducing the accuracy significantly. The matrix below shows the order of magnitudes of the individual matrix elements in field units of mT for in the case of X-band ESR:

$$
\underline{E} = \langle \hat{H} \rangle = \left(\begin{array}{cc|cc|cc}
101 & 1-i & 1-i & 1-0.1i & \emptyset & \emptyset \\
1+i & -101 & 1 & -1+i & \emptyset & \emptyset \\
\hline
1+i & 1 & 100 & \emptyset & 1-i & 1-0.1i \\
1+0.1i & -1-i & \emptyset & -100 & 1 & -1+i \\
\hline
\emptyset & \emptyset & 1+i & 1 & 99 & -1+i \\
\emptyset & \emptyset & 1+0.1i & -1-i & -1-i & -99
\end{array}\right).
\tag{41}
$$

Note that in Eq. (41) i is the imaginary unit. The exact solution of the matrix above would give us the following eigenvalues:

$$
\{E_i\} = \{102.26,\ -102.25,\ 100.02,\ -100.01,\ 97.77,\ -97.77\}.
\tag{42}
$$

By neglecting different non-diagonal terms, the solutions can be found also analytically; however different approximations yield solutions with different relative accuracy:

$S_z I_z$ terms only (relative accuracy 1%, absolute accuracy 1 mT)

$$\{E_i\} = \{101,\ -101,\ 100,\ -100,\ 99,\ -99\} \tag{43}$$

$(S_z, S_+, S_-)\, I_z$ terms only (relative accuracy 1%, absolute accuracy 1 mT)

$$\{E_i\} = \{101.01,\ -101.01,\ 100,\ -100,\ 99.01,\ -99.01\} \tag{44}$$

$S_z\,(I_z, I_+, I_-)$ terms only (relative accuracy 0.01%, absolute accuracy 0.01 mT)

$$\{E_i\} = \{102.24,\ -102.24,\ 100,\ -100,\ 97.76,\ -97.76\} \tag{45}$$

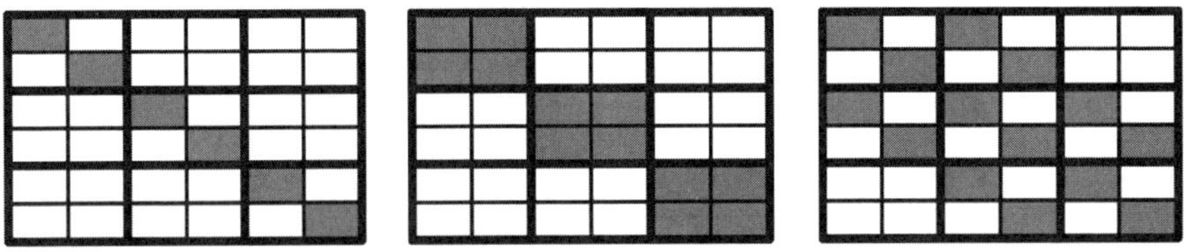

Figure 6. Schematic presentation of different approximations to Hamiltonian matrices: "$S_z\, I_z$ terms only" (left), "$(S_z, S_+, S_-)I_z$ terms only" (center), "$S_z\,(I_z, I_+, I_-)$ terms only" (right).

> Among the solutions that can be found analytically, the highest accuracy is achieved by including the diagonals of the super-block-diagonal and sub-block diagonal.

For the approximation "*$S_z\,(I_z, I_+, I_-)$ terms only*" the following general expressions for the eigenvalues can be derived:

$$E_{\left|\pm\frac{1}{2}, M_I\right\rangle} = \pm\frac{1}{2}\left(\mu_B B\ g^{\mathrm{eff}}(\Theta_i,\Phi_i) + M_I\sqrt{\left(A_{zz}^{\mathrm{eff}}(\Theta_i,\Phi_i)\right)^2 + 2\left|A_{xzyz}^{\mathrm{eff}}(\Theta_i,\Phi_i)\right|^2}\right), \tag{46}$$

yielding the energies of the ESR transitions:

$$\Delta E_{M_I} = \mu_B B\ g^{\mathrm{eff}}(\Theta_i,\Phi_i) + M_I\sqrt{\left(A_{zz}^{\mathrm{eff}}(\Theta_i,\Phi_i)\right)^2 + 2\left|A_{xzyz}^{\mathrm{eff}}(\Theta_i,\Phi_i)\right|^2}\,. \tag{47}$$

Finally, the resonant fields are calculated from resonant equation

$$\Delta E_{M_I} = h\nu \rightarrow B = \frac{h\nu - M_I\sqrt{\left(A_{zz}^{\mathrm{eff}}(\Theta_i,\Phi_i)\right)^2 + 2\left|A_{xzyz}^{\mathrm{eff}}(\Theta_i,\Phi_i)\right|^2}}{\mu_B\ g^{\mathrm{eff}}(\Theta_i,\Phi_i)}\,. \tag{48}$$

Note that the typical absolute error of a resonant field value of about 0.01 mT for X-band ESR in this approximation is much less than the typical linewidths of a nitroxide spin probe. An accurate spectral simulation tool can therefore apply up to 100 points-per-mT sweep without inducing spectral distortions.

In case of higher ESR frequencies, the absolute error decreases almost linearly with frequency and the relative error almost quadratically.

Note that when one replaces ^{14}N for ^{15}N, which has a nuclear spin of ½, the resonant frequencies can be calculated more directly. In this case the Hamiltonian matrix looks like

$$\left\langle \hat{H} \right\rangle = \left(\begin{array}{cc|cc} +\frac{1}{2}\left(G^{\text{eff}} + \frac{1}{2}A_{zz}^{\text{eff}}\right) & \frac{1}{2}\overline{A_{xzyz}^{\text{eff}}} & +\frac{1}{2}\overline{A_{xzyz}^{\text{eff}}} & \overline{A_{xyxy}^{\text{eff}}} \\ \frac{1}{2}A_{xzyz}^{\text{eff}} & -\frac{1}{2}\left(G^{\text{eff}} + \frac{1}{2}A_{zz}^{\text{eff}}\right) & A_{xxyy}^{\text{eff}} & -\frac{1}{2}\overline{A_{xzyz}^{\text{eff}}} \\ \hline +\frac{1}{2}A_{xzyz}^{\text{eff}} & A_{xxyy}^{\text{eff}} & +\frac{1}{2}\left(g^{\text{eff}} - \frac{1}{2}A_{zz}^{\text{eff}}\right) & -\frac{1}{2}\overline{A_{xzyz}^{\text{eff}}} \\ A_{xyxy}^{\text{eff}} & -\frac{1}{2}A_{xzyz}^{\text{eff}} & -\frac{1}{2}A_{xzyz}^{\text{eff}} & -\frac{1}{2}\left(g^{\text{eff}} - \frac{1}{2}A_{zz}^{\text{eff}}\right) \end{array} \right), \tag{49}$$

where G^{eff} corresponds to $\mu_B \cdot B \cdot g^{\text{eff}}$. In the approximation "*S_z (I_z, I_+, I_-) terms only*" the above expression can be transformed to

$$\left\langle \hat{H} \right\rangle = \left(\begin{array}{cc|cc} +\frac{1}{2}\left(G^{\text{eff}} + \frac{1}{2}A_{zz}^{\text{eff}}\right) & 0 & +\frac{1}{2}\overline{A_{xzyz}^{\text{eff}}} & 0 \\ 0 & -\frac{1}{2}\left(G^{\text{eff}} + \frac{1}{2}A_{zz}^{\text{eff}}\right) & 0 & -\frac{1}{2}\overline{A_{xzyz}^{\text{eff}}} \\ \hline +\frac{1}{2}A_{xzyz}^{\text{eff}} & 0 & +\frac{1}{2}\left(g^{\text{eff}} - \frac{1}{2}A_{zz}^{\text{eff}}\right) & 0 \\ 0 & -\frac{1}{2}A_{xzyz}^{\text{eff}} & 0 & -\frac{1}{2}\left(g^{\text{eff}} - \frac{1}{2}A_{zz}^{\text{eff}}\right) \end{array} \right), \tag{50}$$

which can be solved analytically. After solving the eigensystem, a general expression for resonant fields becomes

$$\Delta E_{M_I} = h\nu \rightarrow B = \frac{h\nu - M_I\sqrt{\left(A_{zz}^{\text{eff}}(\Theta_i,\Phi_i)\right)^2 + 4\left|A_{xzyz}^{\text{eff}}(\Theta_i,\Phi_i)\right|^2}}{\mu_B\ g^{\text{eff}}(\Theta_i,\Phi_i)}. \tag{51}$$

3. STRATEGIES FOR CALCULATING POWDER SPECTRA

3.1. From Resonant Fields Values to Stick Spectra

A simple ESR spectrum of a nitroxide spin probe with a particular orientation is represented by a Lorentzian-shaped triplet, actually acquired in its first-derivative form due to field modulation.

$$\overline{P}(B) \propto \frac{1}{1+\left(\frac{B-B_r(M_I)}{\Delta B(M_I)}\right)^2} \xrightarrow{\text{Modulation}} \frac{d\overline{P}(B)}{dB} \propto \frac{\left(B_r(M_I)-B\right)}{\left(1+\left(\frac{B-B_r(M_I)}{\Delta B(M_I)}\right)^2\right)^2}. \tag{52}$$

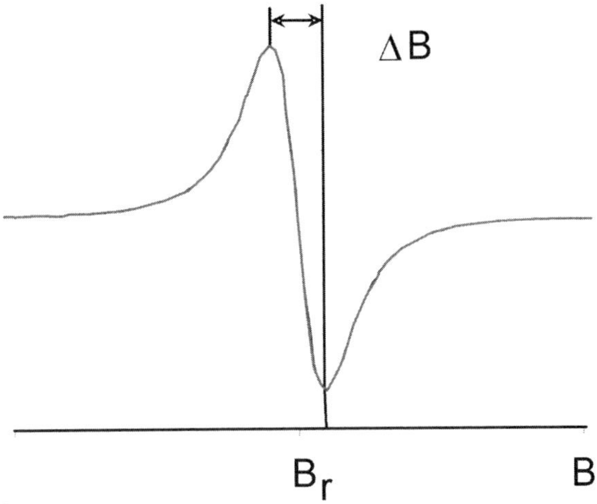

Figure 7. Typical ESR lineshape: — derivative of a Lorentzian function.

The resonant-field positions $B_r(M_I)$ are calculated according to resonance equation Eq. (48) at the corresponding orientation. Linewidths $\Delta B(M_I)$ are determined by a linewidth model, for example, by the Redfield theory in the case of a fast motion approximation.

The Lorentzian lineshape arises from the exponential decay of the magnetization in a rotating frame of reference.

At physiological temperatures and non-saturating ESR conditions, spin dephasing (decay of coherent magnetization) represents the most common relaxation mechanisms due to the motion of spin probe molecules.

Since a common spin-labeled sample involves many spin probe molecules oriented in different directions, the ESR spectrum should be calculated as a convolution of the resonant field distribution $d\rho/dB_r$, taking into account the Lorentzian lineshape for all triplet lines:

$$\frac{d\bar{P}(B)}{dB}(M_I) \propto \int_{B_r} \frac{d\rho(B_r(M_I))}{dB_r} \frac{d\bar{P}(B;B_r(M_I))}{dB} dB_r \,. \tag{53}$$

Calculation of the convolution in Eq. (53) represents the most time-consuming step in the calculation procedure for ESR spectra. Therefore, a lot of effort is required to reduce its numerical cost in any implementation algorithm.

Every such convolution calculation costs around 1 MFLO, or around a millisecond on 1-GFLOPS processor.

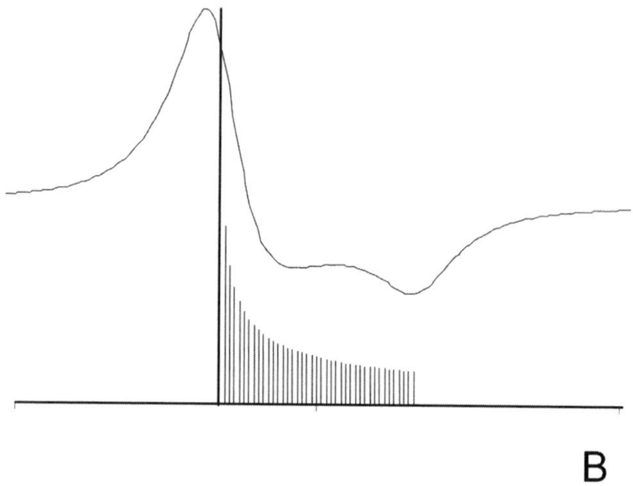

Figure 8. Typical numerical convolution in an ESR lineshape calculation: the resonant field probability function (histogram-like "stick spectra"), which is convoluted with the Lorentzian lineshape from Figure 6, is represented together with the resulting spectral lineshape.

3.2. Summation Schemes and Resolution: Reducing the Computation Time

> Generally, the convolution integral (or sum) can be done with a **Fast Fourier Transform algorithm** (FFT) or **direct summation scheme**.

To choose a fast algorithm consider the following points:

- The costs of direct convolution and the FFT algorithm increase with spectral size N like N^2 and $N \cdot \mathrm{Log}_2(N)$, respectively, making the FFT algorithm more efficient than a direct scheme in convoluting large spectra (several thousands points);
- "Zero" intervals of a resonant field distribution with known resonant field extremes enable a significant acceleration of the direct scheme, making it more efficient than FFT in convoluting small spectra (up to thousand points);
- To avoid FFT wiggles on the spectral wings, zero padding or equivalent approaches are needed, giving rise to additional costs.

According to the maximal hyperfine splitting of 7 mT in the case of a nitroxide and resolution requirements of around 0.01 mT, the spectral size can be reduced to 1024 calculated points. For this case an accelerated direct scheme can be more efficient and accurate than an FFT scheme. However, for higher-resolution requirements and related higher spectral sizes, FFT will become more efficient at the same accuracy.

3.3. Matching Resolution and Accuracy

Before performing a convolution integral (or sum), resonant field distribution $d\rho/dB_r$ has to be constructed for all three resonant lines. Remember that this distribution results in different motional-averaged orientations of the spin probe molecules, which actually represent the orientation distribution of the membrane normal vector $d\rho/d\Omega$:

$$\frac{d\rho(B_r(M_I))}{dB_r} = \frac{d\rho(\Omega)}{d\Omega}\frac{d\Omega(\Omega; M_I)}{dB_r}. \tag{54}$$

The orientation distribution is basically described by the geometry of the spin-labeled sample taking into account the dynamic and static properties on the partitioning as well.

Note that differential $d\Omega = \sin(\Theta)\, d\Theta\, d\Phi$ also contains the phase space volume.

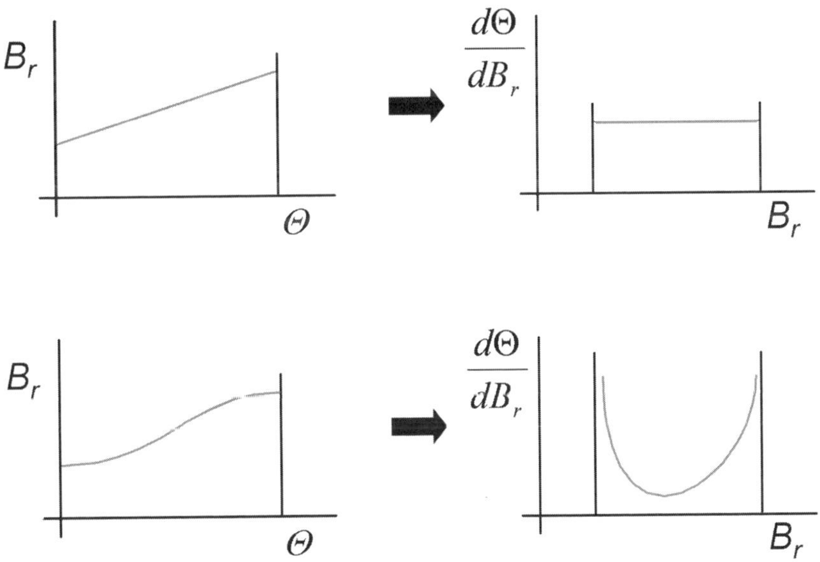

Figure 9. Linear (top) and realistic (bottom) resonant field dependences on the direction and the corresponding derivatives that are defined in Eq. (53) and used in Eq. (54).

Construction of the resonant field distribution involves summation over different directions. To reduce the time cost, the size of this summation ensemble should be optimized according to the resolution requirements and anisotropy of the effective magnetic properties (see Table 3).

The resolution should be equal to the magnetic-field-sweep-step-per-point (which indirectly depends on the linewidths), whereas the anisotropy of the magnetic field properties is determined as the maximal interval of the resonant fields among all three orthogonal directions and all three lines.

Table 3. Resolution Requirements and Anisotropy of the Effective Magnetic Properties

No. of points per sweep	**Resolution requirement**	S_z ordering	σ ordering	Maximal resonant field intervals
$\Delta N/\Delta B$ [points/mT]	**δB [mT]**	S_z [..]	σ [..]	$\Delta B_{r,Mi} \rightarrow$ **Max(ΔB_r) [mT]**
100	**0.01**	0.2	0	0.4, 0.2, 0.7 → **0.7**
100	**0.01**	0.5	0	1.0, 0.4, 1.8 → **1.8**
100	**0.01**	0.5	1	1.8, 1.0, 2.0 → **2.0**
100	**0.01**	0.8	1	2.1, 1.1, 3.2 → **3.2**

In the last example from Table 3 one needs to calculate the resonant field interval of 3.2 mT with a resolution of 0.01 mT. In this case at least, 320 spin probes should be arranged in different directions to cover one quarter of a (phase space) sphere to match the resolution requirement. If the resonant field would change linearly with direction (see Figure 9, top), a linear grid of 400 directions would be sufficient to produce a smooth resonant field distribution. However, this dependence is far from being linear (see Eq. (54) and Figures 9 and 10). To satisfy the resolution requirement everywhere on the resonant field interval and accurately construct a smooth (and correct) resonant field distribution; the number of points should be increased by one order of magnitude.

In order not to exceed the accuracy of the resonant field far beyond the resolution limit and simultaneous dramatically increase the time cost, a two-dimensional linear interpolation of resonant fields can be applied. The simplest two-dimensional linear interpolation grid of 400 can be extended nine times with only a twofold increase in time cost at the same accuracy relative to the resonant field grid used in the convolution summation.

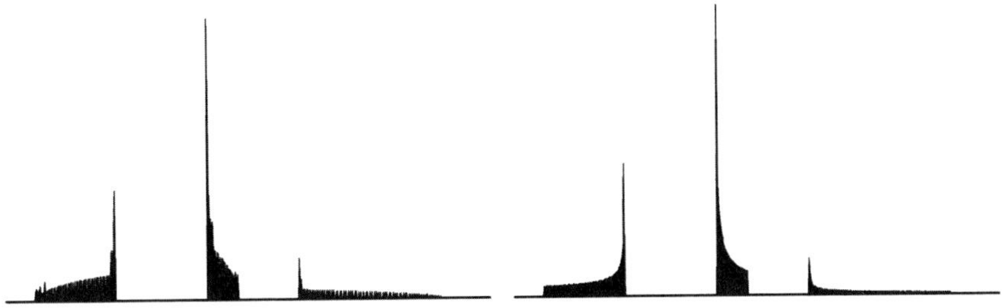

Figure 10. Non-smooth (left) and smooth (right) distributions as a result of rough and fine meshes and highly nonlinear resonant field dependences on the direction (from Figure 9).

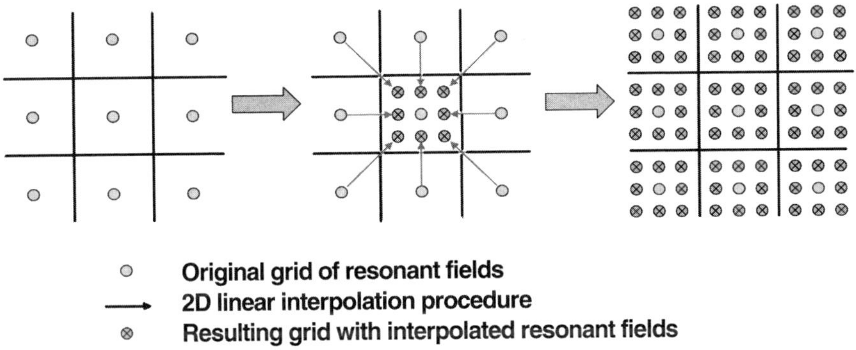

Figure 11. Schematic presentation of a grid expansion by a 2D linear interpolation.

3.4. Implementing Different Geometries and Directors Distributions

The most straightforward calculation of the resonant field distribution implies simple summation over the azimuthal and elevation angles (see Figure 12). In this general notation, any geometry can be implemented within $d\rho/d\Omega$ of Eq. (54).

In many cases, simple summation over the azimuthal and elevation angle leads to a highly non-smooth distribution of directions, resulting in a non-optimized summation scheme (see Figure 9).

In the isotropic case ($d\rho/d\Omega = 1$), many calculation steps are used to cover a small part of a phase space when the elevation angle and volume of the phase space approach zero (see Figure 12). Therefore, other summation schemes should be used to minimize time cost.

> An efficient summation scheme is usually derived with the strategy of equalizing the phase space volume.

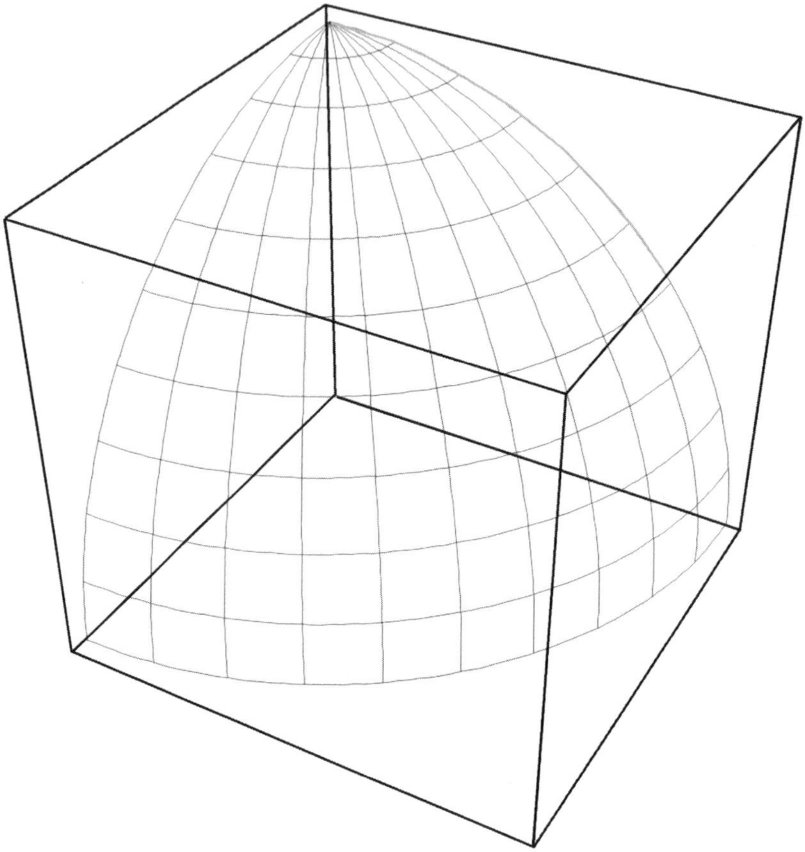

Figure 12. General and simple summation over azimuthal and elevation angles.

One of the most simple and also quite efficient summation schemes involves reduction of the azimuthal grid proportional to the elevation angle (see Figure 13).

Splitting the spherical surface into an ordered mesh of triangles could also be an efficient summation scheme. A typical example is the application of a geodesate, icosahedron, or dodecahedron (meshed projection of a icosahedron or dodecahedron onto a sphere; see Figure 14).

A simple experimental system that can be covered by an "isotropic" model is a model membrane dispersion (dispersion of liposomes in a buffer). Also dispersion of biological cells with other-than-spherical shape can be covered by this model, if the cells do not stack and are equally oriented in all directions.

Other geometries can be implemented by introducing directional probability $d\rho/d\Omega$ into Eq. (54).

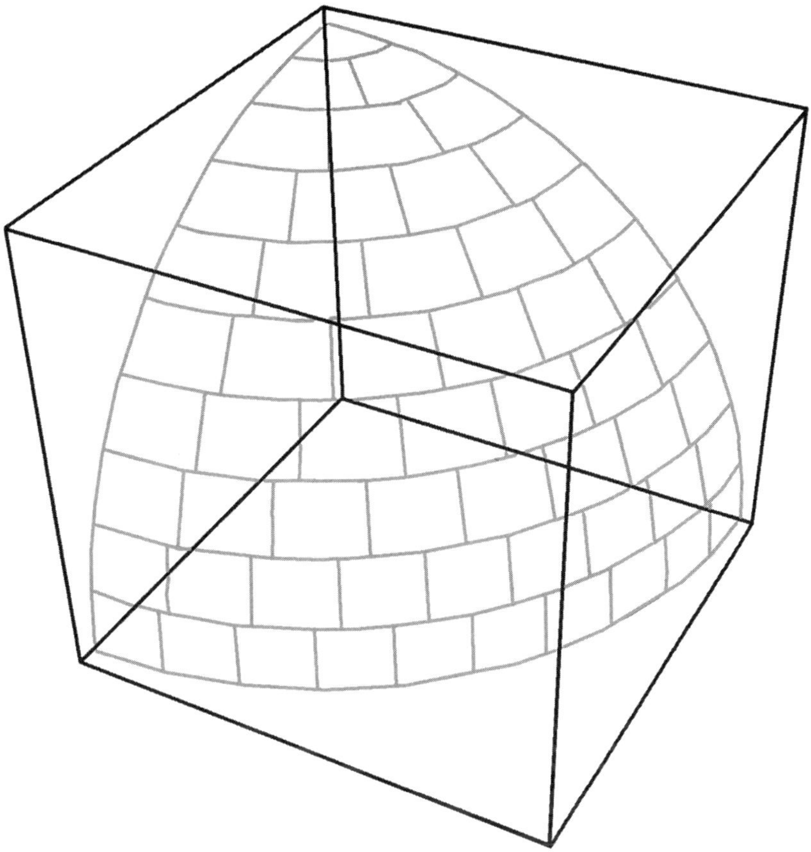

Figure 13. Efficient summation scheme: summation with reduction of azimuthal grid proportional to the elevation angle.

The directional probability can be defined in different ways, through a geometry consideration or a statistical probability distribution. In case of an oriented membrane, for example, one can define the directional probability with a reorienting potential (Maier-Saupe form) and a Boltzmann distribution function:

$$\frac{d\rho}{d\Theta} \tilde{\propto} e^{-\frac{U}{kT}} = e^{\frac{A}{kT}\cos^2(\Theta)} . \tag{55}$$

Since an efficient summation scheme depends on the geometry, any summation scheme should be optimized independently (of the "isotropic" case) to achieve reliable results, fastest computation times and accurate simulated spectra. For example, a summation scheme with a reduction of the azimuthal grid proportional to

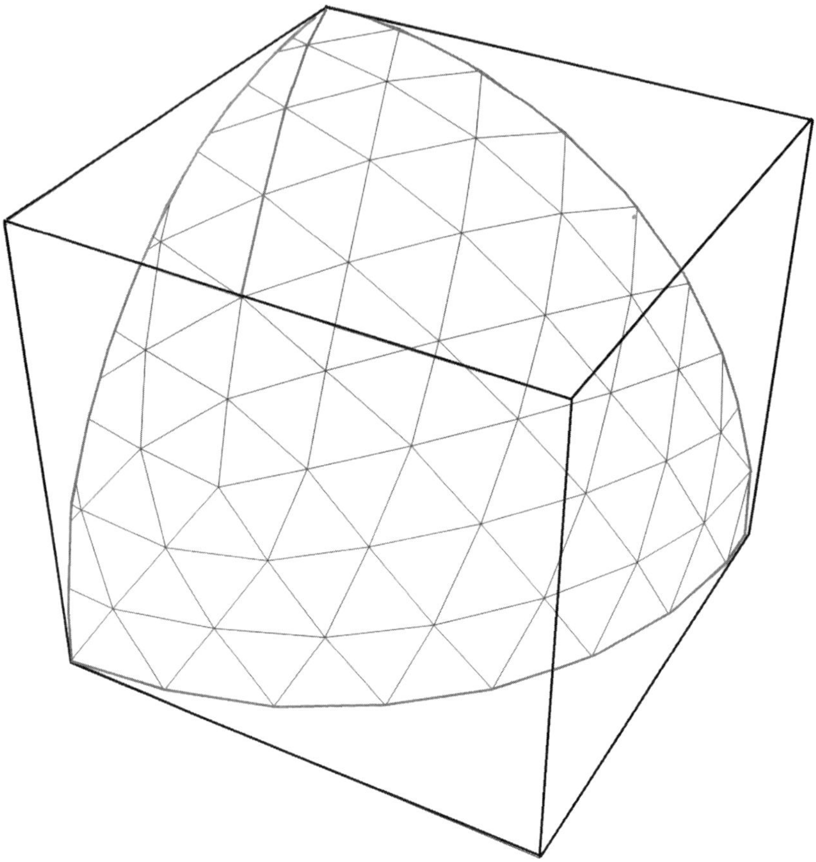

Figure 14. Another example of efficient summation schemes: summation over the geodesate of the icosahedron.

the elevation angle that is optimized for the isotropic case can be quite inadequate in the case of oriented membranes. This is especially true if the orientation of the membrane director is close to $\Theta = 0$. In this case the directional distribution vanishes close to the $\Theta = \pi/2$, where the "volume factor" is maximal.

In the case of oriented membranes, an efficient summation scheme could again be derived with the strategy of equalizing the phase space volumes. Due to the very low orientation probabilities at directions perpendicular to the membrane (acyl chains cannot be oriented in the membrane plane), we can modify the summation scheme from Figure 13 by reducing the grid of elevation angles near $\Theta = \pi/2$ (see Figure 15).

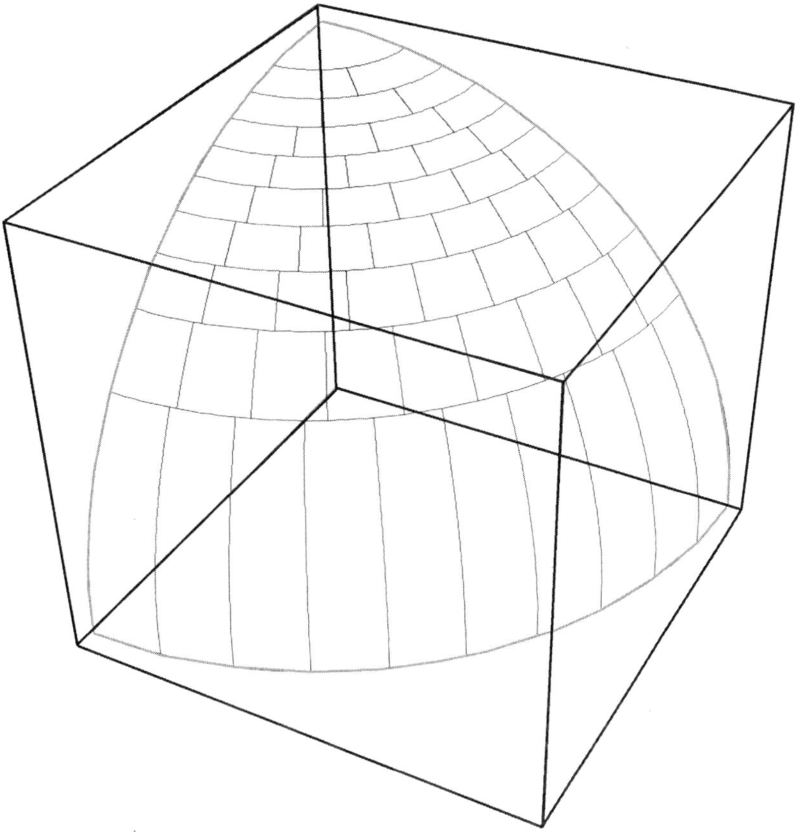

Figure 15. More efficient summation schemes for oriented membranes.

Using the above principles, optimization of summation schemes can be done automatically, making the simulation models more robust and fast at the same time.

In case of various membrane defects like pores and necks, one can define a directional probability using a modified phase space volume.

A pore or neck is a point-like defect type that can occur in a membrane stack. A pore actually represents a point where two monolayers of the same bilayer come close to each other. The whole structure possesses an inverse cylindrical symmetry with the lipid headgroup oriented to the symmetry axis of the structure (see Figure 16, left). Necks are points where two monolayers of the neighboring bilayers are close to each other. The whole structure possesses a cylindrical symmetry with the lipid headgroup oriented outward the symmetry axis of the structure (see Figure 16, right). Both structures possess a cylindrical symmetry, so there is no φ dependence in the orientational distribution. Since the phase space volume (of the isotropic distribution) $\sin(\Theta)$ can be thought of as part of a spherical area, dS, defined within a constant Φ and Θ grid, one can define a similar expression for the pore/neck geometry by $dS \propto 1-\gamma\sin(\Theta)$, where γ represents the ratio between bilayer thickness d and pore outer diameter D. The definition interval for γ is [0,1]. Close-to-zero values correspond to extremely large holes in pores, whereas close-to-one values correspond to pores without holes. The same expression is valid also for neck-type defects by taking $\gamma = ½$.

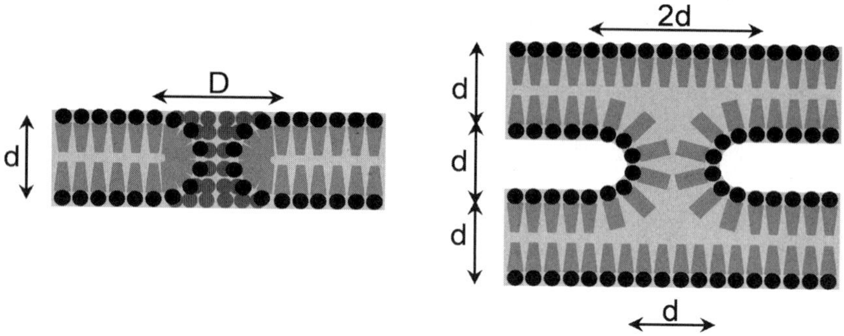

Figure 16. Pore and neck geometry: cross-section of the pore (left) and neck (right).

3.5. Implementing the Transition Probability Correction Factor

ESR transitions are induced by an oscillating microwave magnetic field, perpendicular to the static magnetic field. Using the "Fermi golden rule of quantum mechanics," the transition probability between the two spin states +½ and –½ can be expressed by the off-diagonal element of the Hamiltonian matrix:

$$w \propto \left| \left\langle +\tfrac{1}{2}, M_I \right| \hat{H}_{\text{microwave}} \left| -\tfrac{1}{2}, M_I \right\rangle \right|^2 , \qquad (56)$$

where the microwave-field interaction term can be written in the form of a Zeeman term:

$$\hat{H}_{\text{microwave}} = \mu_B \vec{B}_{\text{microwave}}\ \underline{g}\ \hat{\vec{S}}\,. \qquad (57)$$

Taking into account proper coordinate systems where the quantities in the above equation are well defined as well as the proper direction of the microwave magnetic field (${}^{(l)}x$-axis), the Zeeman term becomes

$$\hat{H}_{\text{microwave}} = \mu_B B_{\text{microwave}}\ {}^{(l)}\vec{i}\ \overline{R^{(l)\to(m)}}\ \overline{R^{(m)\to(n)}}\ \underline{g}\ R^{(m)\to(n)}\ R^{(l)\to(m)}\ R^{(lm)\to(l)}\ \hat{\vec{S}}\,. \qquad (58)$$

Introducing the definitions in Eqs. (5), (6), and (11) and the motional averaged effective Zeeman coupling g^{eff} from Eq. (18), Eq. (63) can be written in a more compact form as

$$\hat{H}_{\text{microwave}} = \mu_B B_{\text{microwave}} \; g^{\text{eff}}(\Theta,\Phi) \; \tfrac{1}{2}\left(\hat{S}^+ + \hat{S}^-\right). \tag{59}$$

The transition probability equation therefore transforms into

$$w \propto \left(g^{\text{eff}}(\Theta,\Phi)\right)^2 \; \left|\left\langle +\tfrac{1}{2}, M_I \left| \tfrac{1}{2}\left(\hat{S}^+ + \hat{S}^-\right) \right| -\tfrac{1}{2}, M_I \right\rangle\right|^2 \propto \left(\tfrac{1}{2} g^{\text{eff}}(\Theta,\Phi)\right)^2. \tag{60}$$

In numerical algorithms one can use the following "unit-free" correction factor instead:

$$w \propto \left(\frac{g^{\text{eff}}(\Theta,\Phi)}{g_0}\right)^2. \tag{61}$$

Note that the relative effect of this correction is only about 0.8%, which corresponds to an error that is easily introduced by model averaging, summation schemes, and convolution integrals. Therefore, the transition probability correction is rarely needed in nitroxide spectra simulation.

4. SOLVING AN INVERSE PROBLEM AND CONDENSATION OF RESULTS

4.1. Phase Space of the Problem: Parameterization

To characterize a system by means of any experimental approach, one should understand the value of the experimental response function and its dependence on various boundary (stress) conditions. In case of ESR spectroscopy, one should therefore be able to simulate the changes in the spectral lineshape due to interactions within the system and due to interaction with external fields.

> Any simulation involves parameterization of the system, i.e., a set of parameters that defines the response function — the ESR spectra in case of ESR spectroscopy.

Some of the parameters (let us denote them *group 0 constants*) involved in a spectral simulation model do not define the exploring system, but are connected directly to the measuring conditions. In the case of ESR spectroscopy these constant parameters are:

- ESR band properties such as microwave power, frequency, and center field

- Scan properties: field sweep interval and scan speed
- RC filtering parameters including modulation conditions
- Magnetic properties of paramagnetic molecules in a reference environment (tensors).

A second group of parameters that affects the response function (denoted *group A parameters*) defines the properties of the system. In the case of ESR spectroscopy these are:

- Motional averaging parameters that measures the anisotropy of motion (e.g., order parameters)
- Motional rate parameters (e.g., rotation correlation times)
- Environmental corrections to magnetic properties (e.g., polarity correction factors)
- Parameters that measure inter-paramagnetic interactions (e.g., spin–spin exchange between molecules of the same or different species)

In addition to the parameters above, any complex-system modeling should define

- Heterogeneity of the system
- Distribution properties (of the paramagnetic species) between different compartments

These parameters are easily introduced by expansion of the basic set of *group A* parameters. Any compartment (domain, aggregate, phase, etc.) therefore possesses a unique set of *group A* parameters in addition to a *proportion* that defines the amount of nitroxide spin probe molecule in this particular compartment. Complex modeling can also involve the definition of the distribution functions as well as the dynamical properties of the system on the system scale (possible rearrangements, non-equilibrium transport, fast exchange between compartments, etc.).

> To provide a more reliable characterization of a complex system, more different spectroscopic and other experiments have to be applied.

4.2. Solving an Inverse Problem

The experimental response of ESR defined by the *group 0 constants* is used to determine the *group A parameters* of a certain biological system. When characterizing a complex biological system, only efficient modeling and a comparison of the experimental and modeled response function enable extraction of biosystem parameters. Resolving the *group A parameters* via system response modeling is usually called **inverse problem solving** (see Figure 17). However, inverse problem solving strongly depends on the complexity of the phase space, which (in a single experiment) relates to the number of *group A* parameters.

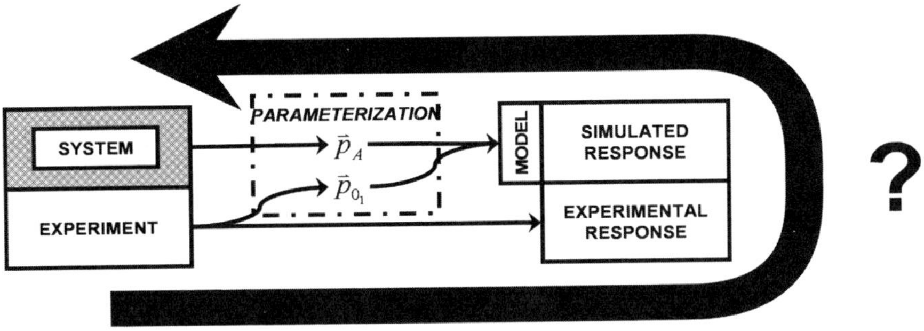

Figure 17. Inverse problem solving.

The simplest model for the ESR spectral shape simulations involves at least $N_A \geq 4$ spectral parameters (*group A*), whereas the response of a more complex system with m compartments can be defined with at least $m \cdot (N_A + 1) - 1$ parameters.

Note that the complexity of the phase space can increase significantly. When modeling the response of a complex system also the resolution limit of ESR should be considered. This is the maximum number of spectral parameters that can be extracted from a single ESR experiment and turns out to be about 30. This number depends on the ESR linewidths and maximal magnetic field anisotropy. Note also that some pairs of parameters can be partially (anti-)correlated, making determination of the parameters even more complicated.

Solving the inverse problem can be a non-trivial task, to be performed successfully.

A typical membrane simulation model can involve, for example, 3 domains with 5 + 1 parameters per domain, resulting in 17 parameters that should be determined from the ESR spectra to characterize the biosystem. To define each of the 17 parameters with an average error of only 10% (10 points per axis), this would result in a phase space volume of 10^{17} points. If each spectrum costs at least 1 ms of calculation time, it would take around 10^6 years to scan the whole phase space.

The enormous size of an ordinary phase space that is involved in modeling of the response function of a complex system calls for robust optimization methods that can be applied to solve the inverse problem. Simple phase space scans will not be efficient at all.

4.3. Characterization Process and Possible Ways of Solving the Inverse Problem

In a typical characterization, the experimentally measured response function has to be compared with a simulated one defined with a set of *group 0* constants and a set of *group A* parameters.

The main task of the characterization procedure is to vary a set of *group A parameters* to find the best match of the corresponding spectral lineshape to the experimentally measured spectral lineshape.

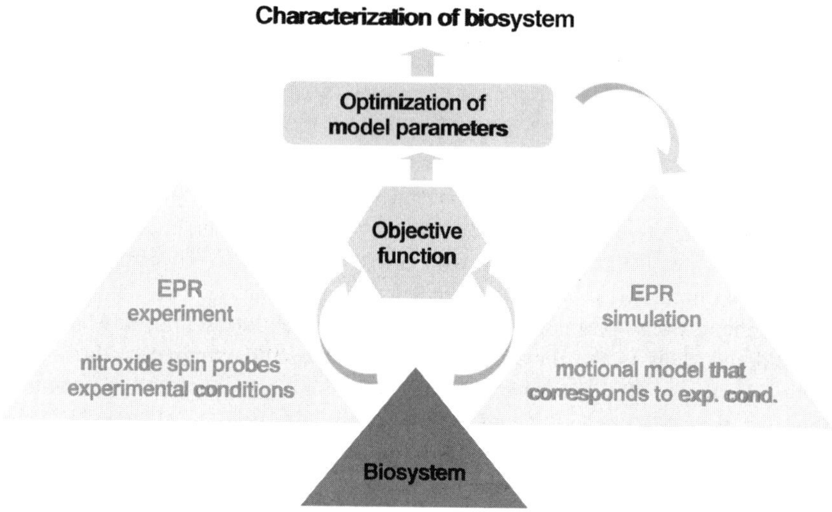

Figure 18. Characterization of a biosystem with ESR spectral simulation in a motional averaging model.

To efficiently navigate the optimization procedure, a fitness function is introduced. It measures the goodness of fit of the simulated spectrum with the experimental one. Most frequently, the measure is chosen to be the reduced χ^2, i.e., the sum of the squared residuals between the experimental and simulated spectra divided by the squared standard deviation of the experimental points σ, and by the number of points in the experimental spectrum N:

$$\chi^2 = \frac{1}{N}\sum_{j=1}^{N}\frac{(y_j^{\text{exp}} - y_j^{\text{sim}})^2}{\sigma^2}. \tag{62}$$

Standard deviation σ is assessed numerically from points in the simulated spectrum regions where the derivatives are close to zero, usually at both ends of the spectrum. Several other fitness functions can be selected as well (like island-weighted χ^2, correlation length functions, maximal difference functions, difference at constant positions, etc.). The convergence of the optimization method can depend on the applied objective function especially due to a different sensitivity to broad lines. In general some numerical transformations may be needed before an objective function can be applied.

In addition, convergence of the optimization routine is affected also by a set of "internal" optimization constants — *group B constants*, which define the scalable properties of the optimization routine. These constants should be optimized against a pool of synthetic spectra (examples) from certain types of problems to guarantee the convergence of the optimization technique for this type of problems. In case of more robust optimization schemes, the tuned *group B* constants are more widely applicable; however, in general *group B constants* should be optimized for each particular class of problems, before the real characterization process is begun.

> Generally, case independence of the convergence quality is one of the important properties of an optimization method.

4.4. Application of Different Optimization Schemes

Various optimization methods can be applied to solve the inverse problem. However, not all are generally efficient in the biosystem characterization process.

Optimization methods can be divided according to two criteria:

- Means of moving through the phase space
- Size of the optimizing set.

Accordingly, four groups of methods can be identified:

- Deterministic – single individual optimization
- Stochastic – single individual optimization
- Deterministic – population optimization
- Stochastic – population optimization

The above method classes are sorted according to their speed, fine-tuning capabilities, and, on the contrary, to the robustness of the method. The fastest methods are deterministic (single individual optimizations), whereas the most robust are stochastic (population optimizations).

The basic scheme of any optimization method includes initialization and the main loop where parameters are changed according to predefined rules. At each evaluation of the main loop the fitness function is checked to fulfill predefined stop criteria. Optimization methods, however, differ in the way each of these steps is performed.

According to the size and partial correlation in phase spaces that can be found in biosystem characterization processes, robustness becomes a much more important issue than speed, although speed can be a limiting factor. To select a robust method, we can search within a class of population-based optimizations that are in general less sensitive to local extremes as well as to model speciations. Optimization methods that do not calculate the spectral derivatives explicitly are especially useful, since the numerical calculation of spectral derivatives is in general a noise-amplifying process.

In our case, evolutionary optimization (i.e., the generational genetic algorithm created by Holland) is taken as a base for inverse problem solving due its robustness and genotype-driven movements in large-phase-space search. This enables a global search independent of detection of local minima. However, due to its inability in terms of fine-tuning, it is efficient to hybridize it with a deterministic method like the Simplex-Downhill "amoeba" of Nelder and Mead — which also does not require a derivative calculation. To get an idea of the implementation, we summarize some important properties below.

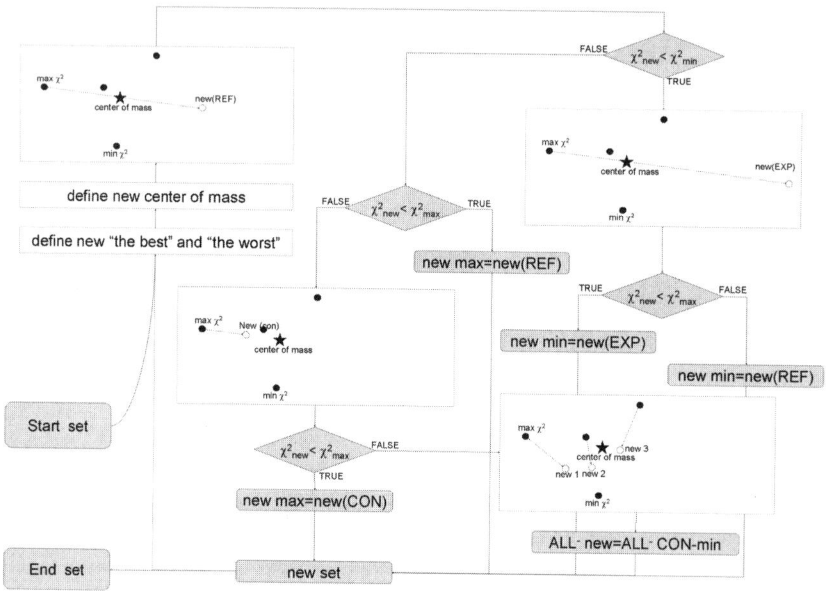

Figure 19. The main loop of SD optimization.

The **Simplex Downhill** (SD) **algorithm** is based on linear transformations of a population of N_p+1 points (each point is a set or vector of all spectral parameters $\{p_i\}$), where N_p is the number of optimizing parameters.

A starting population is usually created from an input vector $\{p_{i,0}\}$ by modifying the *i*th parameter of the *i*th point of the population for a small step δ_i.

The main loop consists of four linear transformations of the worst or all point(s) relative to the center of mass (originally calculated without the worst point; see Figure 19):

- Reflection of the worst point across center of mass
- Reflection of the worst point across center of mass and expansion away
- Contraction of the worst point in the direction of the center of mass
- Contraction of all except the best point toward the best point

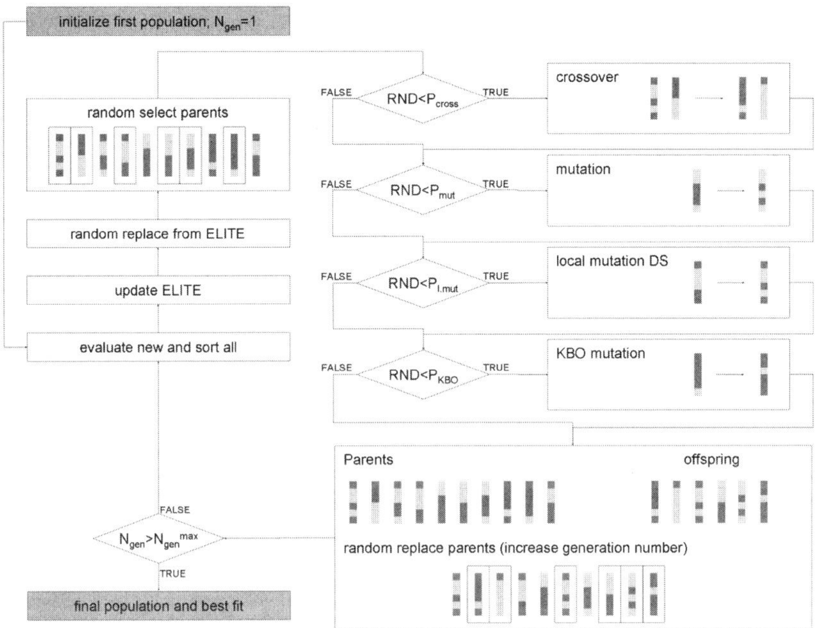

Figure 20. The main loop of **GA (EO)** and **HEO** optimization.

The **generational Genetic (GA) Algorithm** is based on genetic-like transformations of parameters of a large population of M points, where $M \gg N_p$ (each point is a set or vector of spectral parameters $\{p_i\}$), where N_p is the number of optimizing parameters. GA belongs to a family of evolutionary optimization methods; therefore, it is sometimes denoted as EO. When GA is hybridized with a local search operator (like SD) and knowledge-based operators, it is called **Hybrid Evolutionary Optimization** (HEO).

A starting population is usually created randomly within the definition intervals $[p_{i,\min}, p_{i,\max}]$.

The main loop consists of the selection, real coding, and application of genetic operators in the following scheme (see Figure 20):

- Fitness (χ^2) function evaluation for all new individuals

- Sorting
- Elitism implementation
- Fitness-proportional or tournament selection
- Genetic-operations application:
  - Multi-point/uniform crossover
  - Uniform mutation
  - Local mutation (Simplex Downhill in hybrid version)
  - Knowledge-based operators (in hybrid version)
- Random replacement

Table 4. Comparison of Advantages and Disadvantages of SD and GA

Simplex Downhill	Genetic algorithm
Advantages	
Accurate, able to fine-tune	Robust
Fast	Model independent
Robust	High probability for finding global minimum (minima)
Model independent	No need for any starting points provided by the user
Disadvantages	
Low probability for finding global minima	Slow, computational very demanding
Need for reasonably good starting point provided by the user in order to be efficient	Disability for fine-tuning of the solution

Here are some notes on the implementation of HEO:

- To improve convergence, the evolutionary optimization approach can be combined with a more conservative approach, called **elitism**, which keeps track of the best individuals found so far and does not allow replacement of any member of this "elite" with worse individuals.
- Efficient hybridization can be achieved with implementation of **DS as a local mutation operator,** performed with a very low probability to maintain diversity within the population.
- To help the algorithm override the problem of partial correlations between pairs or group of parameters, **knowledge-based mutation operators** (KBOs) can be applied. These operators swap the values within such a group of partially correlated parameters with a very low probability. In this way the algorithm induces special jumps in a phase space to search for better solutions.

4.5. Accuracy and Reproducibility of Solutions

At the end of the characterization procedure, the solutions have to be classified in terms of reproducibility and accuracy.

Reproducibility is related to the effectiveness, robustness, and fine-tuning ability of the optimization routine, and is defined through:

- Means of moving through the phase space
- Starting point(s) determination process
- Ability for global search
- Ability for local search
- Convergence dependence on simulation model

The reproducibility of the method (and model) can be determined via multi-run optimization of spectral parameters using synthetic (simulated) spectra with known parameter sets and contaminated with a low Gaussian noise level. The synthetic spectrum should be produced with the same model as will be used in the optimization. In case of a multi-run stochastic optimization, the determination of reproducibility is straightforward. However, in case of a multi-run deterministic method, special care should be taken to supply a large set of random starting points. The similarity of the synthetic spectra and experimental spectra assures that the reproducibility is approximately the same when optimization is done with a real experimental spectrum or with a synthetic spectrum (as experimental one).

Accuracy defines the most probable range(s) of a real solution (confidence region). It depends on:

- Signal-to-noise ratio of the experimental data
- Partial correlations between different model parameters

Accuracy can be determined in a straightforward manner using multi-run stochastic optimization methods. The area that includes 66% of the solutions around the averaged solution defines the standard confidence region (in the multidimensional space). However, this region also includes the reproducibility range, which has to be subtracted from this confidence region.

In the case when deterministic methods are used, the accuracy determination is not so reliable. The simplest way is to expand $\chi^2(\{p_i\})$ around an optimized value $\{p_{i,0}\}$ and determine the confidence region through a second derivative matrix α (*Hessian matrix*). Note that this method can be used only as a rough approximation only if the optimized value $\{p_{i,0}\}$ is in a real minimum.

The elements of the Hessian matrix $\alpha_{k,l}$ can be calculated as

$$\alpha_{k,l} = \sum_{j=1}^{N} \frac{1}{\sigma^2} \left(\frac{\partial y_j^{\text{sim}}}{\partial p_k} \frac{\partial y_j^{\text{sim}}}{\partial p_l} \right), \tag{63}$$

where N is the number of points in the spectrum, σ is the standard deviation of the experimental points, p_i are the spectral parameters and y_j^{sim} are the values of the simulated spectra.

The inverse of Hessian matrix α is defined as *covariance matrix C*:

$$C \equiv \alpha^{-1}. \tag{64}$$

The diagonal elements can be used to derive a rough approximation to confidence intervals σ_i:

$$\sigma_i = \sqrt{C_{ii}}\,. \tag{64}$$

The off-diagonal elements approximate correlation coefficients $c_{i,j}$:

$$c_{i,j} = \frac{C_{i,j}}{\sqrt{C_{ii}}\sqrt{C_{jj}}}. \tag{66}$$

4.6. Complexity of a Biological System: GHOST Condensation

Before the characterization procedure is applied to determine biosystem complexity, one has to check that the inverse problem is not ill-posed. Frequent divergence or "address violation errors" that are detected within the optimization indicate that the model cannot adequately describe the experimental data or is not sensitive to a certain part of the model phase space (a combination of some parameters in a certain interval). Any of these possibilities calls for immediate model modifications.

However, even if the optimization converges successfully in 100% of cases, there is still a possibility that the model cannot fully describe the experimental system response due to low complexity of the proposed model. This can be recognized by the following features

- Successful convergence with no obvious model errors
- Many equally good solutions provided by the multi-run stochastic optimization routines
- High reproducibility and accurateness of solutions found for the synthetic spectra with the same model and the same optimization routines

In such a case, the appearance of virtual irreproducibility of solutions of real problems results from low **model complexity**.

Often the response of a biosystem is more complex than the simulation model can handle. Consequently, the simulation model cannot lead to a perfect solution, but only to a low-dimensional "projection." Clearly, many good projections exist of the same higher-dimensional object. This fact is implemented in a so-called

GHOST condensation approach based on multiple HEO runs, and to roughly construct a quasi-continuous description of the complex system, even when model complexity is proposed to be finite.

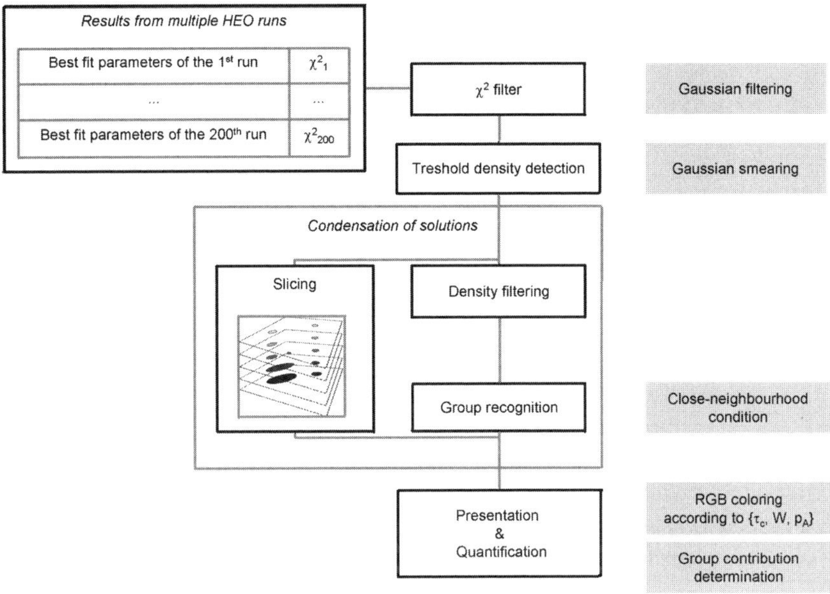

Figure 21. The GHOST approach.

In GHOST condensation, it as assumed that the maximal number of spectral parameters that can be resolved from nitroxide based spectra is between 20 and 30. This corresponds to 4 spectral components. Therefore, a 4-component model is applied in spectral simulation models. The approach is divided into the following four steps (see also Figure 21):

1. **Collection of solutions**: To collect possible projections of the complex system, multiple runs of HEO are performed (up to 200 runs). In an older approach only the best-fit parameter sets of each run were used in the further condensation procedure. However, faster acquisition of solution is obtained by acquiring the best 10 solutions from the population of each run, in combination with new "shaking" operators to maintain solution diversity in a single run (dHEO algorithm).
2. **Filtering of solution**: To assure that only the successful optimization runs were taken into account, 40% of the best-fit sets were passed through the χ^2-filter. In addition, the density of the solutions in the parameter space was calculated to enable density filtering. The latter

is applied to pass through only those solutions that possess a local density higher than the minimum density threshold. With this filter unusual solutions (that are not found in several runs) are thrown away.

3. **Condensation of solutions within a slicing procedure**: Predefined uncertainties of spectral parameters are used to group the solutions according to a neighborhood condition at various density levels. The groups are used to calculate the center of mass of each group of solutions as well as their second moment (spreading of the solution). This is used to discriminate between the discrete and continuous character of a particular group. In addition, slicing by itself can also help to determinate this character: by tracking the center of mass of each group from the highest to the lowest density slice. Discrete groups do not change it center of mass significantly from slice to slice; however, quasi-continuous groups with or without discrete groups superimposed on it can significantly change the center of mass between slices.
4. **GHOST presentation**: The groups of solutions can be presented by two-dimensional cross-section diagrams, such as S–τ_c, S–W, S–p_A, or θ–ϕ, θ–τ_c, θ–W, θ–p_A, depending on the model used (here S, τ_c, W, and p_A represent typical model parameters: order parameter, effective rotational correlation time, additional broadening constant, and polarity correction factor on the hyperfine tensor). In addition, RGB color coding can be applied to enable fast tracking of groups between different GHOST diagrams (see Figure 22).

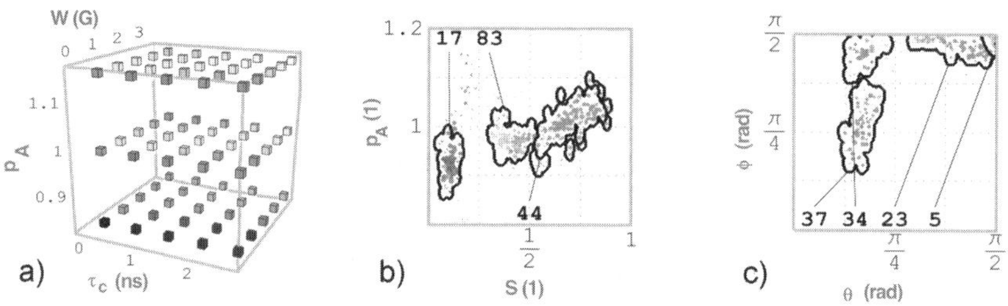

Figure 22.RGB color cube (a) and two examples of GHOST diagrams: MT1 breast cancer cells spin labeled with MeFASL (10,3) growing 5 days at $(10.4 \pm 1.8) \times 10^6$ cells in the culture flask (b), and SDSL of M13 coat protein — cysteine mutant spin labeled at position 34 (c).

As can be seen from the examples in Figure 22, the motional/polarity patterns can be recognized for different biosystems under various conditions. In the case of a nonspecifically spin-labeled membrane (b), some sharply defined solutions are found together with a quasi-continuous distribution of solutions. This indicates that a part of the membrane cannot be modeled well (contrary to healthy cells, which usually possess well-resolved discrete domain structure; data not shown). In case of a site-directed spin labeled cysteine mutant of a membrane protein that was reconstituted in a model membrane (c), coexistence of different motional/polarity patterns is observed. This points to coexisting local conformations of this membrane protein with different restrictions and rates of rotational motion, polarity, and interaction with the oxygen (additional broadening), which together with the polarity information indicate the position of the label relative to the membrane surface, i.e., relative to the water and oxygen partitioning profiles.

Many other models and applications can also be involved in the characterization scheme that is presented in this chapter. According to a recent decrease in the numerical demand of this characterization to roughly one spectrum per 1-GFLOPS processor per day and successful elimination of user interaction from the core of this procedure, this approach is affordable now for any spectroscopic laboratory without a huge investment in computer mainframes or computer clusters.

5. APPENDIX

The explicit expressions for the $\underline{\tilde{G}}$ and $\underline{\tilde{A}}$ terms after averaging are given below. Note that the terms involving sin(φ), cos(φ), sin(2φ), sin(ψ), and sin(2ψ) vanish after averaging due to symmetry reasons:

$$\begin{aligned}\tilde{G}_{xx}^{\mathrm{avg}} &= g_{xx}^2\left(\overline{\cos^2(\varphi)}\ \overline{\cos^2(\vartheta)}\ \overline{\cos^2(\psi)}+\overline{\sin^2(\varphi)}\ \overline{\sin^2(\psi)}\right)\\ &+g_{yy}^2\left(\overline{\sin^2(\varphi)}\ \overline{\cos^2(\vartheta)}\ \overline{\cos^2(\psi)}+\overline{\cos^2(\varphi)}\ \overline{\sin^2(\psi)}\right),\\ &+g_{zz}^2\overline{\sin^2(\vartheta)}\ \overline{\cos^2(\psi)}\end{aligned} \tag{A.1}$$

$$\begin{aligned}\tilde{G}_{yy}^{\mathrm{avg}} &= g_{xx}^2\left(\overline{\cos^2(\varphi)}\ \overline{\cos^2(\vartheta)}\ \overline{\sin^2(\psi)}+\overline{\sin^2(\varphi)}\ \overline{\cos^2(\psi)}\right)\\ &+g_{yy}^2\left(\overline{\sin^2(\varphi)}\ \overline{\cos^2(\vartheta)}\ \overline{\sin^2(\psi)}+\overline{\cos^2(\varphi)}\ \overline{\cos^2(\psi)}\right),\\ &+g_{zz}^2\overline{\sin^2(\vartheta)}\ \overline{\sin^2(\psi)}\end{aligned} \tag{A.2}$$

$$\tilde{G}_{zz}^{\mathrm{avg}} = g_{xx}^2\overline{\cos^2(\varphi)}\ \overline{\sin^2(\vartheta)}+g_{yy}^2\overline{\sin^2(\varphi)}\ \overline{\sin^2(\vartheta)}+g_{zz}^2\overline{\cos^2(\vartheta)}, \tag{A.3}$$

$$\tilde{G}_{xz}^{\mathrm{avg}}=0\,,\ \ \tilde{G}_{xy}^{\mathrm{avg}}=0\,,\ \ \tilde{G}_{yz}^{\mathrm{avg}}=0\,. \tag{A.4–6}$$

Similarly averaged effective hyperfine tensor components are expressed as

$$\begin{aligned}\tilde{A}_{xx}^{\mathrm{avg}} &= \frac{A_{xx}}{g_{xx}^{2}}\left(\overline{\cos^{2}(\varphi)}\ \overline{\cos^{2}(\vartheta)}\ \overline{\cos^{2}(\psi)}+\overline{\sin^{2}(\varphi)}\ \overline{\sin^{2}(\psi)}\right)\\ &+\frac{A_{yy}}{g_{yy}^{2}}\left(\overline{\sin^{2}(\varphi)}\ \overline{\cos^{2}(\vartheta)}\ \overline{\cos^{2}(\psi)}+\overline{\cos^{2}(\varphi)}\ \overline{\sin^{2}(\psi)}\right)\\ &+\frac{A_{zz}}{g_{zz}^{2}}\overline{\sin^{2}(\vartheta)}\ \overline{\cos^{2}(\psi)},\end{aligned} \tag{A.7}$$

$$\begin{aligned}\tilde{A}_{yy}^{\mathrm{avg}} &= \frac{A_{xx}}{g_{xx}^{2}}\left(\overline{\cos^{2}(\varphi)}\ \overline{\cos^{2}(\vartheta)}\ \overline{\sin^{2}(\psi)}+\overline{\sin^{2}(\varphi)}\ \overline{\cos^{2}(\psi)}\right)\\ &+\frac{A_{yy}}{g_{yy}^{2}}\left(\overline{\sin^{2}(\varphi)}\ \overline{\cos^{2}(\vartheta)}\ \overline{\sin^{2}(\psi)}+\overline{\cos^{2}(\varphi)}\ \overline{\cos^{2}(\psi)}\right)\\ &+\frac{A_{zz}}{g_{zz}^{2}}\overline{\sin^{2}(\vartheta)}\ \overline{\sin^{2}(\psi)},\end{aligned} \tag{A.8}$$

$$\tilde{A}_{zz}^{\mathrm{avg}} = \frac{A_{xx}}{g_{xx}^{2}}\overline{\cos^{2}(\varphi)}\ \overline{\sin^{2}(\vartheta)}+\frac{A_{yy}}{g_{yy}^{2}}\overline{\sin^{2}(\varphi)}\ \overline{\sin^{2}(\vartheta)}+\frac{A_{zz}}{g_{zz}^{2}}\overline{\cos^{2}(\vartheta)}, \tag{A.9}$$

$$\tilde{A}_{xz}^{\mathrm{avg}} = 0,\quad \tilde{A}_{xy}^{\mathrm{avg}} = 0,\quad \tilde{A}_{yz}^{\mathrm{avg}} = 0. \tag{A.10–12}$$

With the above averaged expressions one can define the expressions for the effective tensors g^{eff}, $A_{zz}{}^{\mathrm{eff}}$, $A_{xzyz}{}^{\mathrm{eff}}$, $A_{xyxy}{}^{\mathrm{eff}}$, and $A_{xxyy}{}^{\mathrm{eff}}$ of the final spin Hamiltonian of Eq. (39):

$$\left(g^{\mathrm{eff}}(\Theta_i,\Phi_i)\right)^2 = \tilde{G}_{xx}^{\mathrm{avg}}\cos^2(\Phi_i)\sin^2(\Theta_i)+\tilde{G}_{yy}^{\mathrm{avg}}\sin^2(\Phi_i)\sin^2(\Theta_i)+\tilde{G}_{zz}^{\mathrm{avg}}\cos^2(\Theta_i), \tag{A.13}$$

$$\begin{aligned}A_{zz}^{\mathrm{eff}}(\Theta_i,\Phi_i) &= [\tilde{A}_{xx}^{\mathrm{avg}}\cos^2(\Phi_i)\sin^2(\Theta_i)+\tilde{A}_{yy}^{\mathrm{avg}}\sin^2(\Phi_i)\sin^2(\Theta_i)\\ &\quad +\tilde{A}_{zz}^{\mathrm{avg}}\cos^2(\Theta_i)]\left(g^{\mathrm{eff}}(\Theta_i,\Phi_i)\right)^2,\end{aligned} \tag{A.14}$$

$$\begin{aligned}A_{xzyz}^{\mathrm{eff}}(\Theta_i,\Phi_i) &= \left[\frac{1}{2}\left(\tfrac{1}{2}\left(\tilde{A}_{xx}^{\mathrm{avg}}\cos^2(\Phi_i)+\tilde{A}_{yy}^{\mathrm{avg}}\sin^2(\Phi_i)-\tilde{A}_{zz}^{\mathrm{avg}}\right)\sin(2\Theta_i)\right)\right.\\ &\quad \left.+\frac{i}{2}\left(\tfrac{1}{2}\left(\tilde{A}_{xx}^{\mathrm{avg}}-\tilde{A}_{yy}^{\mathrm{avg}}\right)\sin(2\Phi_i)\sin(\Theta_i)\right)\right]\left(g^{\mathrm{eff}}(\Theta_i,\Phi_i)\right)^2,\end{aligned} \tag{A.15}$$

$$\begin{aligned}A_{xyxy}^{\mathrm{eff}}(\Theta_i,\Phi_i) &= \left[\frac{1}{4}\left(\tilde{A}_{xx}^{\mathrm{avg}}\left(\cos^2(\Phi_i)\cos^2(\Theta_i)-\sin^2(\Phi_i)\right)+\tilde{A}_{yy}^{\mathrm{avg}}\left(\sin^2(\Phi_i)\cos^2(\Theta_i)-\cos^2(\Phi_i)\right)\right.\right.\\ &\quad \left.\left.+\tilde{A}_{zz}^{\mathrm{avg}}\sin^2(\Theta_i)\right)+\frac{i}{2}\left(\tfrac{1}{2}\left(\tilde{A}_{xx}^{\mathrm{avg}}-\tilde{A}_{yy}^{\mathrm{avg}}\right)\sin(2\Phi_i)\cos(\Theta_i)\right)\right]\left(g^{\mathrm{eff}}(\Theta_i,\Phi_i)\right)^2,\end{aligned} \tag{A.16}$$

$$\begin{aligned}A_{xxyy}^{\mathrm{eff}}(\Theta_i,\Phi_i) &= \frac{1}{4}\left[\tilde{A}_{xx}^{\mathrm{avg}}\left(1-\cos^2(\Phi_i)\sin^2(\Theta_i)\right)+\tilde{A}_{yy}^{\mathrm{avg}}\left(1-\sin^2(\Phi_i)\sin^2(\Theta_i)\right)+\right.\\ &\quad \left.+\tilde{A}_{zz}^{\mathrm{avg}}\sin^2(\Theta_i)\right]\left(g^{\mathrm{eff}}(\Theta_i,\Phi_i)\right)^2.\end{aligned} \tag{A.17}$$

6. ACKNOWLEDGMENTS

This work is granted by the Slovenian Research Agency (Ministry of Higher Education, Science and Technology, grants P1-0060 and J1-6581). It was also partially supported by the ESF COST D22 action. I wish to thank all my colleagues at the Laboratory of Biophysics at the EPR center, "Jožef Stefan" Institute, in Ljubljana, for many discussions and lots of important notes on this chapter. I am also grateful to Bogdan Filipič, David Stopar, Peter Laggner, and Marcus A. Hemminga, who helped me to evolve this approach to the current situation.

7. INTERESTING REFERENCES AND FURTHER READING

Spin-Labeling ESR Spectroscopy

Arsov Z, Schara MV, Zorko M, Štrancar J. 2004. The membrane lateral domain approach in the studies of lipid–protein interaction of GPI-anchored bovine erythrocyte acetylcholinesterase. *Eur Biophys J* **33**:715–725.

Borbat PP, Costa-Filho AJ, Earle KA, Moscicki JK, Freed JH. 2001. Electron spin resonance in studies of membranes and proteins. *Science* **291**:266–269.

Chiang YW, Shimoyama Y, Feigenson GW, Freed JH. 2004. Dynamic molecular structure of DPPC–DLPC–cholesterol ternary lipid system by spin-label electron spin resonance. *Biophys J* **87**:2483–2496.

Griffith OH, Jost PC. 1976. Lipid spin labels in biological membranes. In *Spin labeling, theory and application*, pp. 453–523. Ed LJ Berliner. New York: Academic Press.

Kurad D, Jeschke G, Marsh D. 2003. Lipid membrane polarity profiles by high-field EPR. *Biophys J* **85**:1025–1033.

Marsh D. 2002. Membrane water-penetration profiles from spin labels. *Eur Biophys J* **31**:559–562.

Rappolt M, Amenitsch H, Štrancar J, Teixeira CV, Kriechbaum M, Pabst G, Majerowicz M, Laggner P. 2004. Phospholipid mesophases at solid interfaces: in-situ X-ray diffraction and spin-label studies. *Adv Colloid Interface Sci* **111**:63–77.

Shin YK, Freed JH. 1989. Dynamic imaging of lateral diffusion by electron spin resonance and study of rotational dynamics in model membranes: effect of cholesterol. *Biophys J* **55**:537–550.

Simulation of ESR Spectra

Della Lunga G, Pezzato M, Baratto MC, Pogni R, Basosi R. 2003. A new program based on stochastic Liouville equation for the analysis of superhyperfine interaction in CW-ESR spectroscopy. *J Magn Reson* **164**:71–77.

Eviatar H, van der Heide UA, Levine YK. 1995. Computer simulations of the electron spin resonance spectra of steroid and fatty acid nitroxide probes in bilayer systems. *J Chem Phys* **102**:3135–3145.

Hemminga MA. 1983. Interpretation of ESR and saturation transfer ESR spectra of spin labeled lipids and membranes. *Chem Phys Lipids* **32**:323–383.

Marsh D. 1981. Electron spin resonance: spin labels. In *Membrane spectroscopy*, pp. 51–142. Ed E Grell. Berlin: Springer.

Schneider DJ, Freed JH. 1989. Calculating slow motional magnetic resonance spectra: a user's guide. In *Biological magnetic resonance: spin labeling, theory and applications*, pp. 1–76. Ed LJ Berliner, J Reuben. New York: Plenum.

Steinhoff HJ, Hubbell WL. 1996. Calculation of electron paramagnetic resonance spectra from Brownian dynamics trajectories: application to nitroxide side chains in proteins. *Biophys J* **71**:2201–2212.

Štrancar J, Sentjurc M, Schara MV. 2000. Fast and accurate characterization of biological membranes by EPR spectra. *J Magn Reson* **142**:254–265.

Van SP, Birrell GB, Griffith OH. 1974. Rapid anisotropic motion of spin labels: models for motion averaging of the ESR parameters. *J Magn Reson* **15**:444–459.

Optimization Methods of ESR Spectral Parameters

Budil DE, Lee S, Saxena S, Freed JH. 1996. Nonlinear-least-squares analysis of slow-motion EPR spectra in one and two dimensions using a modified Levenberg-Marquardt algorithm. *J Magn Reson A* **120**:155–189.

Della Lunga G, Pogni R, Basosi R. 1998. Global versus local minimization procedures for the determination of spin Hamiltonian parameters from electron spin resonance spectra. *Mol Phys* **95**:1275–1281.

Filipič B, Štrancar J. 2003. Evolutionary computational support for the characterization of biological systems. In *Evolutionary computation in bioinformatics*, pp. 279–294. Ed GB Fogel, D Corne. Amsterdam: Elsevier Science.

Filipič B, Štrancar J. 2001. Tuning EPR spectral parameters with a genetic algorithm. *Appl Soft Comp* **1**:83–90.

Kavalenka AA, Filipič B, Hemminga MA, and Štrancar J. 2005. Speeding up a genetic algorithm for EPR-based spin label characterization of biosystem complexity. *J Chem Inf Model* **45**:1628–1635.

Moens P, De Volder P, Hoogewijs R, Callens F, Verbeeck R. 1993. Maximum-likelihood common-factor analysis as a powerful tool in decomposing multicomponent EPR powder spectra. *J Magn Reson A* **101**:1–15.

Evolutionary Optimizations

Eiben AE, Smith JE. 2003. *Introduction to evolutionary computing*. Berlin: Springer.

Goldberg DE. 1989. Genetic algorithms in search, optimization and machine learning. Reading, MA: Addison-Wesley.

Complexity Determination

Stopar D, Štrancar J, Spruijt RB, Hemminga MA. 2005. Exploring the local conformational space of a membrane protein by site-directed spin labeling. *J Chem Inf Model* **45**:1621–1627.

Štrancar J, Koklič T, Arsov Z. 2003. Soft picture of lateral heterogeneity in biomembranes. *J Membr Biol* **196**:135–146.

Štrancar J, Schara MV, Pečar S. 2003. New EPR method for cellular surface characterization. *J Membr Biol* **193**:15–22.

Štrancar J, Koklič T, Arsov Z, Filipič B, Stopar D, Hemminga MA. 2005. Spin label EPR-based characterization of biosystem complexity. *J Chem Inf Model* **45**:394–406.

Author's webpage

Laboratory: http://www.ijs.si/ijs/dept/epr/index.html
Personal: http://www.ijs.si/ijs/dept/epr/Janez.html

CHAPTER 4

PRACTICAL PULSED DIPOLAR ESR (DEER)

Piotr G. Fajer[1], Louise Brown[2] and Likai Song[1]
[1]*Florida State University and National High Magnetic Field Laboratory, Tallahassee, Florida, USA, and* [2]*Macquarie University, Sydney, Australia*

The recent resurgence of ESR (electron spin resonance) in structural biology is in large part due to the development of distance measurements. This application was made possible by targeting of specific sites/domains by cysteine mutagenesis. Four techniques were developed to cover various distance ranges: *exchange* ESR for the very short distances (4–8 Å) (Miick et al. 1992); *static* and *dynamic dipolar cw* ESR (continuous wave ESR) for the 8–25 Å distance range (McHaourab et al. 1997; Rabenstein and Shin 1995); and two pulsed methods: *DEER* (double electron electron resonance) (Milov et al. 1981) and DQC (double quantum coherence) (Borbat et al. 2002) for distances between 17 and 80 Å. The first two methods rely on line shape broadening and are thus limited to strong interactions (short distances). The pulsed methods extract weaker dipolar interactions from spin coherence and are thus sensitive to longer interspin distances. In addition, distances can be measured between the nitroxides and paramagnetic metals that enhance the relaxation of nitroxides (Budker et al. 1995; Voss et al. 1995). All these techniques have been extensively tested and verified across the full distance range of sensitivity on model systems including organic biradicals (Jeschke et al. 2000), proteins (Sale et al. 2005) and DNA (Borbat et al. 2004; Schiemann et al. 2004). The use of extrinsic spin labels for distance measurements allows targeting of specific sites by side-directed spin labeling, which is of great advantage. However, the observed distances are between unpaired electrons at the end of long tethers (spin labels) that are attached to the molecular surface. To interpret these distances in terms of molecular structure, one needs to account for label length and conformation. Currently, the accuracy of the various ESR methods is of the order of 3 Å when label conformation is taken into account (Sale et al. 2005). This chapter will focus entirely on one of the pulsed dipolar methods, DEER (double electron electron resonance). DEER, otherwise known as pulsed electron–electron resonance (PELDOR), was developed in Russia during the 1980s by Milov and collaborators (Milov et al. 1981, 1984). Twenty years later Jeschke proposed a deadtime-free variant of the method, 4-pulse DEER, which, combined with commercially available instrumentation and advances in site-directed labeling, resulted in its resurgence in structural biology (Pannier et al. 2000). An extensive review of the applications is beyond the scope of this chapter, but the interested reader is directed to the many available reviews (Eaton and Eaton 2000; Fajer 2005; Hubbell et al. 2000; Jeschke 2002; Jeschke et al. 2000; Steinhoff 2004). Here we present a simple introduction to DEER, with emphasis on instrumental aspects of data collection and data analysis, and a few examples of recent applications with membrane proteins and aqueous motor proteins.

1. DEER SIGNAL

The DEER signal is a modulation of the echo amplitude of one spin population when another spin population is excited with intense microwave radiation. If the two populations are near each other and the first population "sees" the dipolar field generated by the second population, the echo of the former will be modulated as we change the timing of the second population excitation. In general, the DEER signal is composed of two parts: the decay due to the intermolecular interactions of unpaired spins, and the periodic oscillations generated by the intramolecular interactions of paired spins:

$$I(t) = I_{\text{intra}}(t) \times I_{\text{inter}}(t)\,. \tag{1.1}$$

Modulation of the echo arises from the observed spins experiencing a local magnetic field occurring from the dipolar interactions of the nearby spins. Since the direction of this field is a function of the spin state of the coupled spin, changing the latter selectively identifies the contribution of the dipolar field to the total magnetic field. Imagine that we are observing spins A, excited with a microwave of frequency ν_o, and the $\pi/2$ pulse along the y-axis tips the spins into the xy plane, where they precess with angular rate of ω_A (Figure 1). The presence of field inhomogeneities and different resonance fields results in slightly different angular rates, which dephase the spins. For example, the dipolar field due to nearby spins B adds or subtracts from angular rate $\pm\frac{1}{2}\omega_{AB}$ depending on the spin state of spins B. A π pulse applied to spins A reverses the dephasing and refocuses spin A along the x-axis, creating a spin echo. This $\pi/2$–π pulse sequence, is known as a Hahn echo and is the mainstay of most pulse ESR experiments. Spins A continue to precess in the xy plane, and subsequent π pulses refocus the echo, albeit with smaller amplitude as the phase coherence is lost due to spin–spin relaxation. To extract the dipolar contribution from the total field experienced by spins A, a π pulse is applied to spins B. This "pumping" or "ELDOR" pulse (the name stems from use of a second frequency to excite spins B, ν_p) flips the B spins and thus reverses the dipolar contribution experienced by the A spins from $+\frac{1}{2}\omega_{AB}$ to $-\frac{1}{2}\omega_{AB}$, and vice versa. The change of the dipolar contribution to angular rate of spins A in between the π pulses means that spins A accrue a phase lag and are no longer refocused along the x-axis; hence the intensity of the echo decreases. The accrued phase lag is a function of the dipolar interaction strength, ω_{AB}, and the timing of the ELDOR pulse, which determines how long spins A experience dipolar field $+\frac{1}{2}\omega_{AB}$ versus

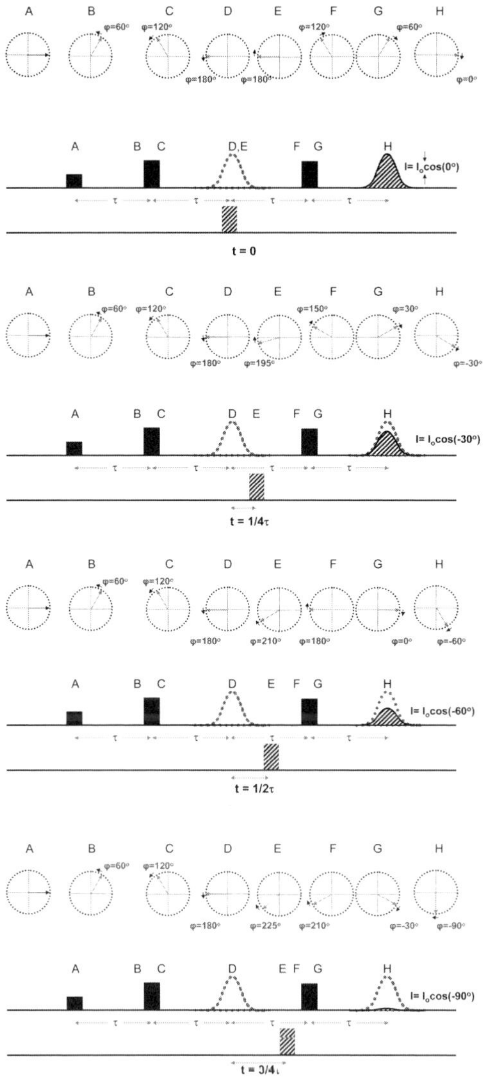

Figure 1. Vector diagram of the DEER experiment. (**A**) $\pi/2$ pulse tips magnetization into the *xy*-plane. Spins *A* precess with ω_A. If the reference frame is set to rotate with the same rate ω_A, spins *A* appear stationary. (**B**) Spins *A* experience local magnetic field due to spin *B* that adds to their angular rate, $\pm\frac{1}{2}\omega_{AB}$. In this example the strength of this dipolar field results in a 60° phase within time *t*. Only the spins experiencing $+\frac{1}{2}\omega_{AB}$ are shown for clarity. (**C**) π pulse inverts the spins in the *xy*-plane. (**D**) After time τ echo is formed and the phase lag due to dipolar interactions is nil. (**E**) Pumping pulse flips the *B* spins, inverting the additional field experienced by spins *A*. (**F**) Spins *A* accumulate negative phase due to inverted field $\pm\omega_{AB}$. (**G**) π pulse inverts the spins in the *xy*-plane. (**H**) Echo is formed but with accumulated phase $-\omega_{AB}t$. For spin experiencing $-\frac{1}{2}\omega_{AB}$ the phase is $+\omega_{AB}t$. In this example the accumulated phase due to dipolar interactions is 0 as $t = 0$.

$-\frac{1}{2}\omega_{AB}$. Figure 1 details this process, where, for simplicity, the rotating frame moves at ω_A to emphasize dephasing due to dipolar interactions, and only the *A* spins experiencing initial $+\frac{1}{2}\omega_{AB}$ are depicted. (The spins initially experiencing $-\frac{1}{2}\omega_{AB}$ are omitted, but their behavior is symmetric across the *x*-axis.) From the vector representation of Figure 1, the total accrued phase difference is $\omega_{AB}(\tau - t)$, where the τ is the interval between the $\pi/2$ and π pulses, and t is the timing of the pumping pulse with respect to the initial echo (2τ after the first $\pi/2$ pulse). In the above example, the strength of the dipolar interaction was chosen to result in the 60° phase before the π pulse. As the diagram illustrates, when the pumping pulse is applied at $t = 0$, the time of the *primary* echo (in *"gray"*), the accrued phase is 0° with respect to the frame rotating at ω_A, resulting in the echo at maximum intensity. When the pumping pulse is moved ¼, ½, and ¾ of an interval between the primary echo and second refocusing π pulse, the phase lag is 30, 60, and 90°, respectively (Figure 1). The echo intensity is reduced to the cosine of the phase lag. The DEER signal is the modulation of the echo intensity as the pumping pulse is swept between the π pulses.

The echo intensity oscillates with

$$I_{\text{intra}}(t) = I_0 \cos(\omega_{AB}(\tau - t)), \tag{1.2}$$

where I_0 is the echo in the absence of dipolar interactions, τ is the time between the $\pi/2$ and π pulses, and t is the timing of the pumping pulse (Milov et al. 1981, 1984).

Dipolar splitting, ω_{AB}, between the two spins at the distance r is given by

$$\omega_{AB} = \frac{\mu_{\text{Bohr}}^2 g_A g_B}{2}(3\cos^2\theta - 1), \tag{1.3}$$

in which the interspin vector subtends an angle θ with an external magnetic field, and where μ_{Bohr} is the Bohr magneton, and g_A and g_B are the **g** tensor of the coupled spins. For the samples randomly distributed with respect to the field, integration of ω_{AB} over angle θ gives a symmetric Pake pattern in which singularities ($\theta =$ 90°) are separated by $2\frac{\mu_{\text{Bohr}}^2 g_A g_B}{\hbar}\frac{1}{r^3}$.

For a homogenous distribution of spins, such as in glass solutions (Milov et al. 1981, 1984),

$$I_{\text{inter}}(t) = \exp(-kCF_B|\tau - t|), \tag{1.4}$$

where C is the concentration of spins A that interact with each other via intermolecular interactions, F_B is the fraction of spins B excited by ν_p, and k is given by

$$k = \frac{8\pi^2\mu_{\text{Bohr}}^2 g_A g_B}{9\sqrt{3}\hbar}. \tag{1.5}$$

1.1. Analysis

Like all other spectroscopic signals, analysis of the DEER or CW-dipolar spectra falls into two categories: (a) deconvolution methods and (b) fitting methods.

The example of the former is a DEER analysis program developed by G. Jeschke (http://www.mpip-mainz.mpg.de/~jeschke/distance.html (Jeschke 2004)) and programs developed in the laboratory of Y.K. Shin for cw spectra (Rabenstein et al. 1995). These programs are widely available and have been discussed in previous publications. The fitting approach first assumes a physical model for distance distribution and then simulates corresponding spectra and tests them against the experimental spectrum. The testing is coupled to optimization procedures like *Simplex*, *Marquardt-Levenberg,* and *NL2SNO* or can be completely random (*Monte Carlo*). The advantage of a fitting approach is the ability to estimate the errors for the assumed distance models and to check for uniqueness for the determined distance distribution. The plots of "goodness of fit" between simulated and experimental spectra, as a function of average distance and their distribution, easily identify other possible solutions (see Figure 2). Uniqueness is a serious problem in the analysis of DEER spectra, as the problem is mathematically ill-defined (Jeschke et al. 2002). The fitting approach stabilizes the solution while assuming a model, but the assumption of a model is also its disadvantage. This is a more philosophical than practical drawback, as one can always create complex distance distributions from multiple components provided there is enough information in the experimental signal to support more complex models. The change in the "goodness of fit" parameter as the number of components is increased is a good guide as to how many Gaussians are justified by the experimental spectra. More rigorous determination is made by applying the *F*-test as a function of the number of Gaussians. More often than not, the signal-to-noise ratio limits the number of parameters that can be extracted from an experimental spectrum. Thus the ability to choose a simple physical model with a limited number of parameters in the fitting approach turns into an advantage. The fitting program DEFit is available at the authors' website (http://fajerpc.magnet.fsu.edu).

The DEER signal for a Gaussian distribution of distances centered at r_0 and with a half width of Δr, $\Gamma(r) = A\ \exp\left(-0.693\frac{(r-r_0)^2}{\Delta r^2}\right)$, is simulated by integrating (1.3):

$$I_{\text{intra}}(t,r) = \int_r \Gamma(r) \int_0^{\pi/2} I_o \sin\theta \cos\left(\frac{\mu_{\text{Bohr}}^2 g_A g_B}{\hbar r^3}(3\cos^2\theta - 1)(\tau - t)\right) d\theta dr\,. \tag{1.6}$$

A similar approach is taken for extracting the dipolar broadening from the cw dipolar spectra. The Pake pattern, Pake(r), is calculated by integration of the field splitting, $D(r)$, over angle θ of the dipole with respect to magnetic field **H**. The integral is convoluted with a resonance absorption line (width $\Delta\Gamma = 3$ G for frozen protonated nitroxide) at the resonance field corresponding to angle θ:

$$\text{Pake}(H,r) = \int_0^{\pi/2} \sin(\theta)\exp\left(-0.693\frac{(H-D(r))^2}{\Delta\Gamma^2}\right)d\theta\,, \tag{1.7}$$

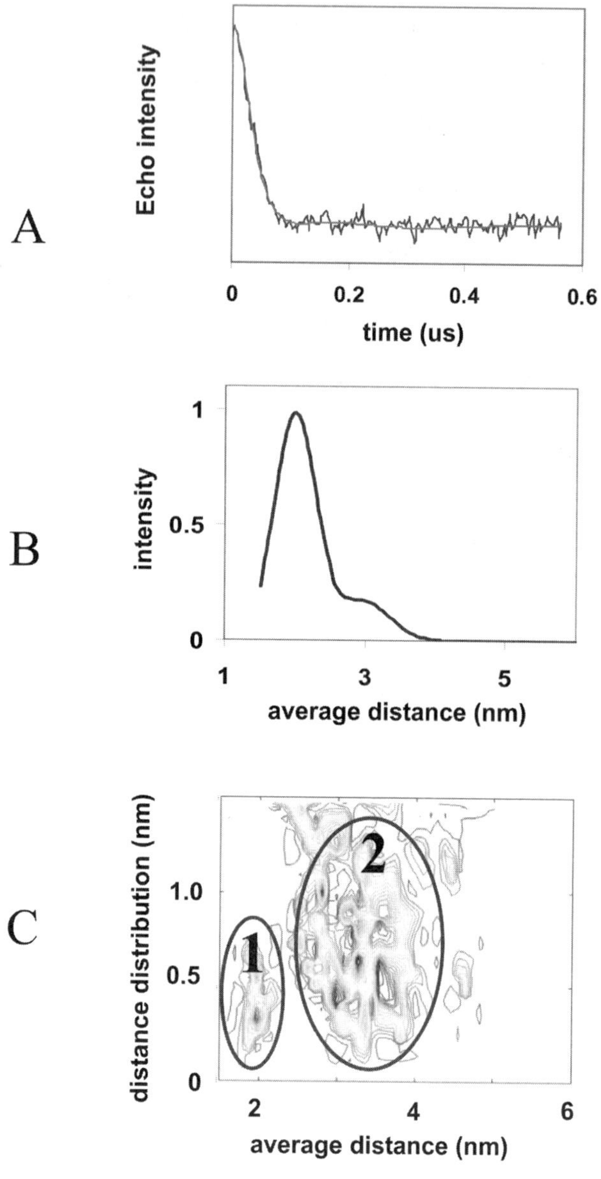

Figure 2. Gaussian fitting of DEER signal (smooth muscle myosin double mutant 282C/595C, MTSSL). (**A**) dipolar evolution function: DEER signal and the best fit; (**B**) Gaussian distribution of spin–spin distances corresponding to the fit; (**C**) the error surface of Gaussian fitting; solutions corresponding to shorter distances are marked as (1), longer distances (2). The contours reflect the χ^2 value for each fit; the lower the contour, the better the fit.

where

$$D(r) = \pm \frac{3g\beta(3\cos^2\theta - 1)}{4r^3}. \quad (1.8)$$

For a Gaussian distribution of distances, instead of having Pake(r) as a broadening function, we have function $B(H,r)$ given by

$$B(H,r) = \int_r \exp\left(-0.693\frac{(r - r_o)^2}{\Delta r^2}\right) \mathrm{Pake}(H,r)dr. \quad (1.9)$$

The spectrum of a singly labeled macromolecule, V_S, is convolved with a broadening function $B(H,r)$ with a Gaussian distribution of distance r:

$$V_{\mathrm{double}} = V_{\mathrm{single}} \otimes B(r). \quad (1.10)$$

The simulated broadened spectrum is then compared to the experimental spectrum and the distance parameters varied (*Simplex, Monte Carlo*) until a satisfactory fit is obtained.

1.2. Sensitivity to Distance and Distance Distribution

The DEER signals are sensitive to both the average interspin distance and to distribution of the distance about the average value. Figure 3 illustrates both points: the curves are simulated for the Gaussian distributions with r varying from 2 to 6 nm. The general trend is to extend the time it takes the modulation to decrease to nil: 200 ns for an interspin distance of 2 nm, 500 ns for 4 nm, and 1400 ns for $r = 6$ nm. In practice, longer evolution times are required in order to reliably remove background arising from intermolecular interactions. The rate of the initial echo decay is the easiest way to estimate the average distance. In principle, one could detect distances that are much longer than 6–8 nm as long as we can extend the evolution period, the time between the π pulses. The maximum evolution time is limited by the phase memory time or spin–spin relaxation time of the nitroxide labels. For protonated nitroxides in H_2O we are generally limited to 2 ms. This sets a practical limit of about 6–8 nm for the maximum observable interspin distance.

The width of the distance distribution defines the oscillations of the DEER curves before their decay to baseline. Narrow distributions, characteristic of structurally homogenous, rigid molecules oscillate few times around zero ($\Delta r = 0.4$ nm for "*circles*" in Figure 3). As the distribution becomes wider these oscillations are dampened. For wide distributions ($\Delta r = 2$ nm) the DEER curves barely drop below baseline (Figure 3).

There is this qualitative difference: the rate of decay and the amplitude of oscillations are advantages of DEER. As a first approximation one can directly read off the average distance as well as the distance distribution from the experimental curves.

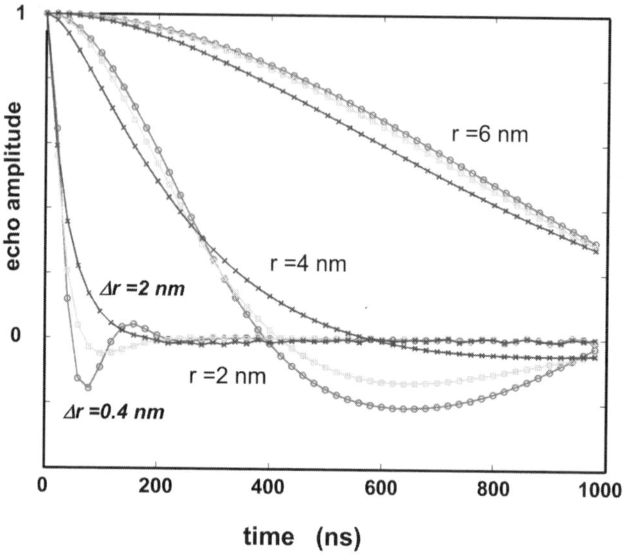

Figure 3. Sensitivity of DEER spectra to average interspin distance and to the width of Gaussian distribution of distances. Each of the set of three curves corresponding to $r = 2$, 4, and 6 nm is simulated for three different Gaussian distributions of distances: 0.4, 1, and 2 nm full width.

1.3. Interpretation in Molecular Terms

The DEER-derived distances are interspin distances. The size and flexibility of the spin labels prevent direct interpretation of distances in terms of the protein backbone. The problem is especially acute when measuring short distances (~10 to 15 Å), where the probe tether length (8–10 Å) grossly influences the interpretation. Figure 4A shows the correlation between the C_β–C_β distances from the atomic models and the same distances determined by ESR. The short (8–20 Å) distances are from the KcsA channel using conventional line-broadening spectroscopy, while the 20–50 Å distances in troponin were measured with pulsed ESR: DEER and metal-nitroxide relaxation between spin labels and Gd^{3+}, a paramagnetic metal. Although there is a strong correlation between the protein and experimental distances, clustering in the 10–20 Å range illustrates the difficulty in interpreting interspin distances (Figure 4A). In this range, the 10 Å C_β–C_β distances are observed as interspin distances anywhere between 8 and 20 Å. Interpretation of such distances in molecular terms is futile.

One approach to account for the spin label length and conformation is to use Monte Carlo conformational searching and Molecular Dynamics (Sale et al. 2005). Monte Carlo searches for the lowest-energy spin label conformers by varying each torsional angle within the spin label and calculating the energy of the spin label and its interaction with the immediate environment. Short (1-ns) molecular dynamics (MD) simulations at 300–500K are then used to search the local conformational space and the dynamics of the lowest energy spin label conformers. For computational efficiency, simulations can be performed using the CHARMM19 extended atom force field with a distance-dependent dielectric constant. Such simulations can take only a few hours on a desktop PC. The details of these simulations including sample programs are given in the Appendix and on the accompanying CD.

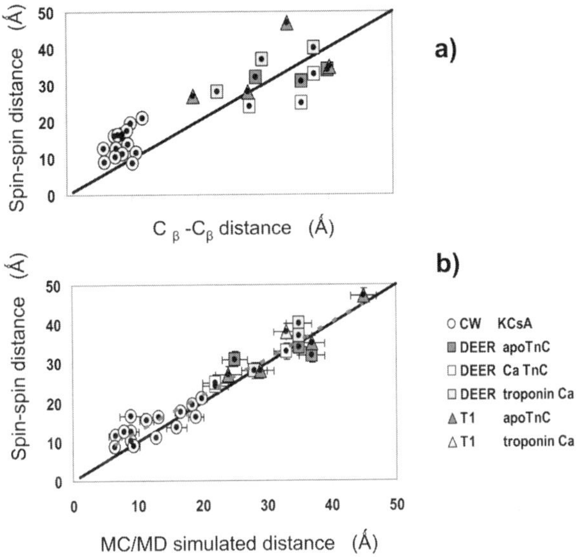

Figure 4. (**A**) Comparison of ESR spin–spin distances to those measured between C_β carbons. Distances measured between β carbons have a scatter of 6 Å when compared to the measured values. (**B**) Correlation between ESR distances and distances between spin labels modeled into the crystal structure using Monte Carlo and molecular dynamics methods; the scatter decreases to 3 Å and the short distances are linearized. Reprinted with permission from Sale *et al.* (2005). Copyright © 2005, American Chemical Society.

When the above strategy was applied to distances measured in three different proteins labeled with a variety of spin labels (iodoacetemide spin label [IASL], maleimide spin label [MSL], and methanethiosulfonate spin label [MTSSL]), the agreement improved dramatically (Figure 4B). The correlation increases by a factor of 4 over the shorter distance range, and the average mean error decreases to 3 Å. An additional benefit of modeling is an estimate of the heterogeneity in label geometry and its contribution to the heterogeneity of the observed distances. The molecular dynamics trajectory of the distance between the labels gives a credible estimate of the distance distribution originating from the dynamics of the probes with respect to protein. Significantly wider distributions of observed distances represent the structural heterogeneity of the protein's backbone.

2. PRACTICAL DEER

DEER experiments can be time-consuming, ranging from minutes to hours depending on the sample concentration, the measured distance, and the distance distribution. Optimization of the experimental variables is thus of importance. Three independent aspects need to be considered: (a) the size of the spin echo; (b) the depth and length of the echo modulation; and (c) the competition of other weak interactions that affect spin coherence. The first of these is an obvious target of optimization: the larger the echo, the smaller the electronic noise, thereby requiring less averaging time. A better signal-to-noise ratio allows longer evolution times, which in turn facilitate measurement of longer distances. The second, echo modulation, contains the dipolar evolution function and is the signal from which distances are extracted. The modulation is dampened by the intermolecular interactions, which should be minimized, and is the function of the fraction of the pumped spins. Finally, the other weak interactions — e.g., hyperfine coupling to the protons — also result in echo modulation. In this case the modulation has a well-defined frequency that can be shifted by using deuterated matrices.

Below we discuss various aspects to be considered when optimizing the DEER experiment. Often, it is a matter of making compromises between opposing requirements.

2.1. Resonators

The resonator of choice for DEER is more critical than for cw ESR. The signal is proportional to the quality factor (Q) of the resonator, but so is the ringing. When the power is turned off, the microwaves oscillate in the resonating chamber before decaying to zero. The better the resonating structure (the higher Q), the longer these oscillations persist. The kilowatt-level microwave pulses need to dissipate before the microwatt-level echo can be observed. The choice is a tradeoff between Q, the filling factor, and the total volume of the sample that the resonator can hold. Split-ring resonators (Bruker MS series) have a higher sensitivity (higher filling factor, lower Q, better power-to-field conversion) than the overcoupled dielectric resonators (MD series). Nevertheless, the inner size of dielectric resonators is larger than that of the split-ring resonators, allowing for more sample to be placed in the sensitive region: 100 μL in MD-5 vs. 15 μL in MS-3. A direct comparison of the echo intensity of a 100 μM free spin label sample in the Bruker ER 4118X-MS3 and ER 4118X-MD5 resonators reveals a twofold stronger signal intensity in the MS-3 resonator (15 μL sample volume) as compared to the MD-5 resonator (100 μL active volume). Thus, given a choice, the split-ring resonator is recommended for pulsed ESR measurements of spin-labeled samples.

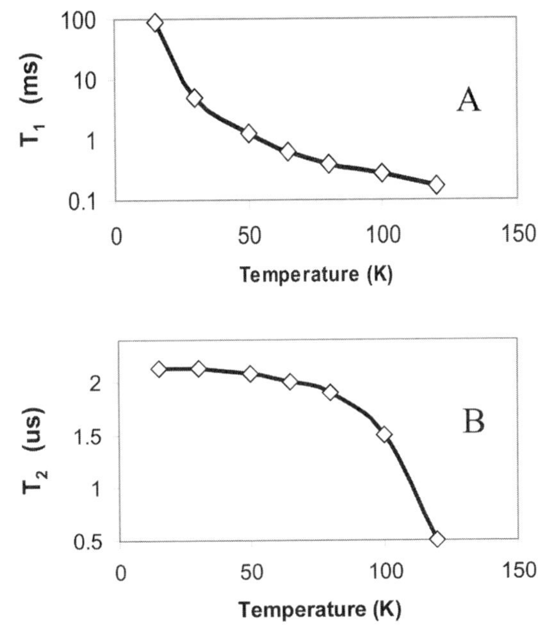

Figure 5. (A) Spin–lattice T_1 and (B) transverse T_2 relaxation times dependence on temperature. T_1 as measured by inversion recovery sequence (π–200 ns–$\pi/2$–t–π); T_2 was measured by $\pi/2$–t–π pulse sequence. Sample was an MTSSL labeled double mutant (55C/83C) of troponin C in 40% glycerol.

2.2. Temperature

Temperature affects the DEER signal in three ways; a decreasing temperature increases the T_1 and T_2 relaxation times and increases the difference in the Boltzmann population (Figure 5). All three define the quality of the DEER signal. T_1, the spin–lattice relaxation time, is the time constant of the magnetization along the z-axis. The spin–spin relaxation time is the time constant of spin coherence in the x,y plane.

The longer T_2, the longer the spin coherence, allowing longer evolution times between the two π pulses. As discussed earlier, longer evolution times facilitate measurement of weaker dipolar interactions and longer distances. Low temperature also increases the intensity of an echo, as net magnetization increases (Figure 6). Temperatures below 20 K would have been preferred if not for the increase in spin–lattice relaxation time. Longer relaxation rates decrease the speed of data averaging: after each pulse one needs to wait for the magnetization to recover to equilibrium. Thus, the optimum temperature is a compromise between echo strength and pulse repetition rate. At a constant averaging rate, the signal-to-noise ratio has a broad maximum between 50 and 80 K (Figures 6c and 7).

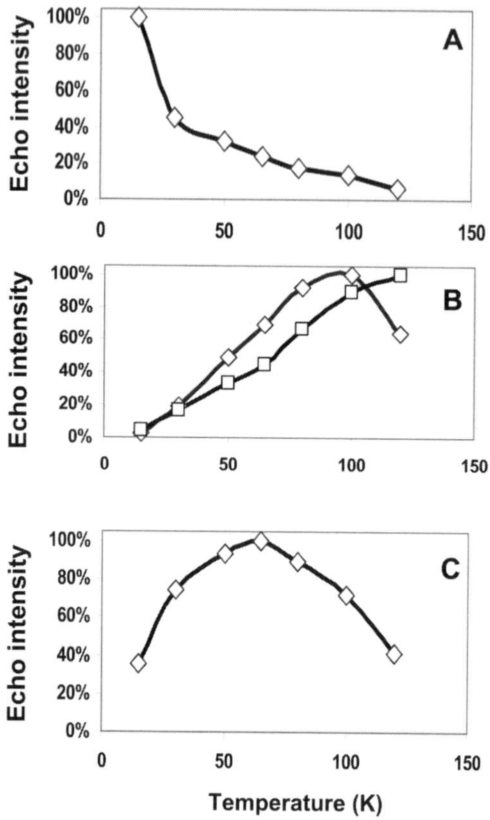

Figure 6. The intensity of the Hahn echo as a function of sample temperature: (**A**) single pulse sequence without data averaging; (**B**) echo averaged for 10 s with the SRT = 1.2 × T_1 and the associated noise; (**C**) the signal-to-noise ratio of the averaged echo. The echo was measured with $\pi/2$–t–π sequence. The sample was troponin C.

2.3. Signal Averaging Rate: Shot Repetition Times (SRTs)

In the four-pulse DEER experiment, the intensity of the refocused echo was collected at constant time after the final π pulse (Figure 1), and thus the quality of the DEER signal was proportional to the S/N of the single shot echo and the amount of averaging. The shot repetition time (SRT) is the interval between the DEER pulse trains and defines the rate of data averaging. The shorter SRT, the more DEER spectra are averaged in a given time period, and S/N increases. However, if the time between the pulse trains is too short to allow magnetization to return to equilibrium, the echo intensity will decrease due to partial saturation. The

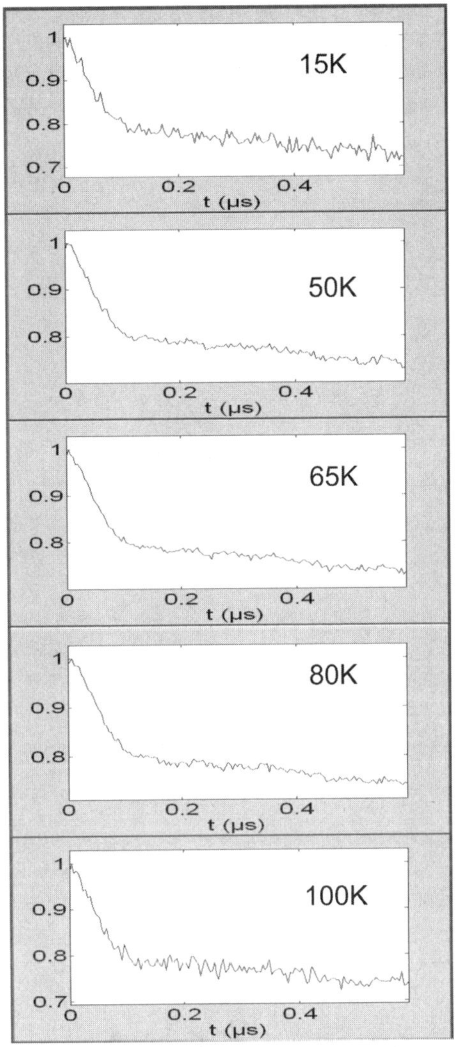

Figure 7. Signal-to-noise ratio of 4-pulse DEER collected at different temperatures. Signal was averaged for 15 min using $\pi/2$–t–π–t'–ELDOR–τ'–π–τ sequence with 16-ns $\pi/2$ pulse and 32-ns π ELDOR pulse at 70-MHz frequency offset. The sample was doubly labeled troponin, as in Figure 6.

optimal SRT is a compromise between T_1 relaxation and the number of averages, and it needs to be determined experimentally. In principle, the magnetization reaches equilibrium after 3 to 5 × T_1 , the spin–lattice relaxation time.

The signal from a four-pulse DEER sequence is determined by both T_1 and T_2 relaxation:

$$S_0 = kC(1 - e^{-SR/T_1})e^{-T_e/T_2}\ . \tag{1.11}$$

where C is the spin concentration, T_e is the time between the π/2 pulse and echo detection, and k is an instrumental constant. The signal intensity, S_{sum}, after total collection time T is

$$S_{sum} = S_0 * (T / SRT)\ . \tag{1.12}$$

The predicted signal intensity in Figure 8 suggests that the shorter SRT, the higher the total intensity; however, the signal-to-noise ratio is of more interest:

$$S / N = S_0 * \sqrt{T / SRT}\ . \tag{1.13}$$

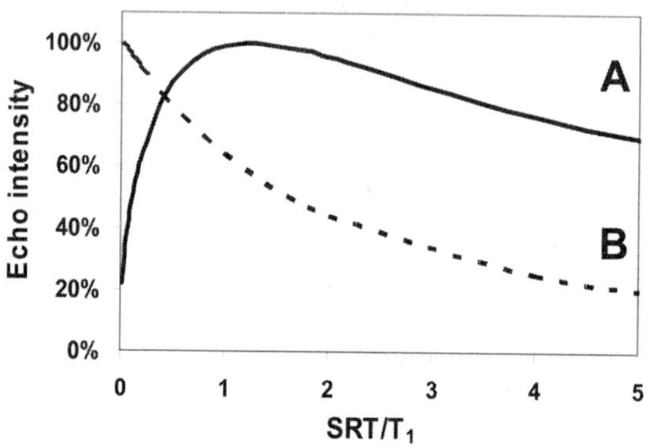

Figure 8. Calculated relative DEER echo intensity (**A**) and S/N ratio (**b**) at different rates of averaging, SRT.

The DEER dipolar evolution signals in Figure 9 suggest that an acceptable compromise is achieved at a SRT of 1.2 × T_1. Although the net magnetization has not recovered to its equilibrium (one needs 4.6 × T_1 to recover 99% of magnetization) and some of the spins are saturated (30%), the increased number of averages, 4.6/1.2 = 3.8, results in twofold improvement of S/N.

2.4. Pumping and Observing Positions

The pumping and observed pulse should be separated enough so that the width of the pumping pulse does not excite the *observed* spins directly. At the same time, the pumping and observed pulses should be non-selective, i.e., excite all the spins irrespective of their orientation, thereby avoiding orientational selection and

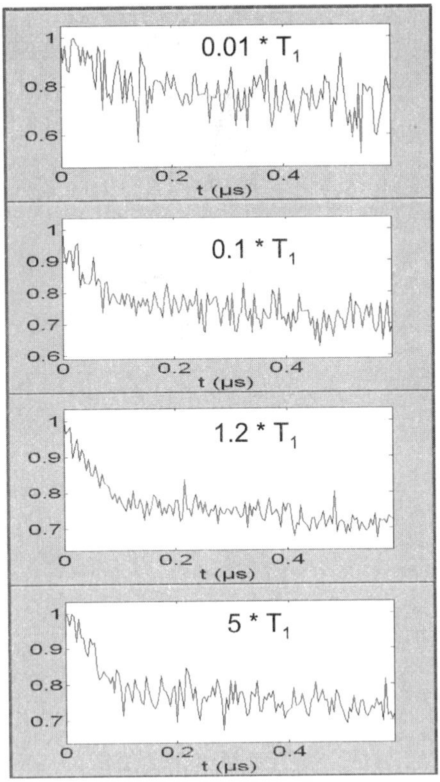

Figure 9. DEER signal of different SRT at 15 K. Conditions as in Figure 7, except for 15-min averaging time.

maximizing the fraction of coupled spins. Although the orientational selectivity cannot be fully avoided for the nitroxides, the optimal conditions are met when the pumping frequency (ν_p) is set to the central manifold of the ^{14}N nitroxide signal and the observed pulse frequency (ν_o) set to low field manifold. Since the magnetic field is kept constant during the experiment, this is achieved by setting ν_o = 60–70 MHz above ν_p (Figure 10). Note that the above arrangement sacrifices the echo amplitude (which is smaller for low field than for center field); however, the modulation depth is larger due to the larger fraction of coupled spins.

The power, duration, and pulse frequency are a function of the available hardware. The cavity bandwidth (reflected in the *Q*-factor) limits the frequency separation of the observed and pumping frequency and determines how much of the pulse power is available at the sample. The filling factor, *Q*, power-to-field conversion of the resonator, and the available power for the observed and pumping pulses all have to be considered when deciding on the microwave frequency vis-à-vis the resonator frequency. For the Bruker MD-5 resonator the *Q* factor is ~400 when the

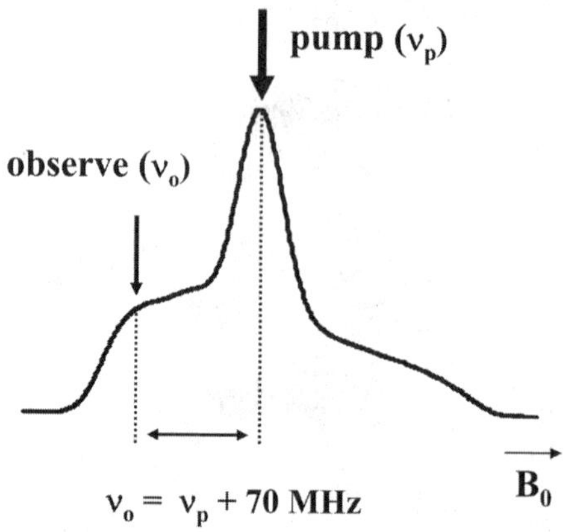

Figure 10. Relationship between the observing and pumping pulses in the DEER experiment.

resonator is over-coupled. With this Q value the optimum DEER signal arises when the observed frequency is matched to the resonator frequency and the pump frequency is set 60–70 MHz lower. For resonators with a lower Q (split-ring series or loop-gap resonators) or for systems with limited pumping power, it might be advantageous to set the pumping pulse to the resonator dip and increase the observed frequency. This configuration favors the strength of the pumping pulse over the observed pulse; the echo might be smaller but the fraction of pumped spins will be bigger. The practitioners of DEER are advised to find the best compromise for their hardware by optimizing the echo height (by adjusting the observed arm attenuation and frequency) and the echo modulation (pumping pulse power and frequency). For details of setting up the DEER experiment, see also Chapter 2 by Jeschke in this volume.

2.5. Sample Preparation

As in any experiments, the results are only as good as the sample: the more concentrated the sample, the better. As long as the sample is homogenously distributed, concentrations up to 500 μM can be tolerated before intermolecular distances are of a similar magnitude as the intramolecular distances (< 70 Å). For protein samples one is usually experimentally limited to concentrations of 50–200 μM so the high concentrations are rarely an issue. Homogenous distribution is a bigger problem than high concentrations; freezing of aqueous samples often tends to exclude macromolecules from the ice crystals, resulting in high local spin concentrations or molecular clusters. To prevent ice formation, DEER is performed in glass solutions of ~40% sucrose or ~30% glycerol. A visual check for crystal formation is the transparency of the frozen sample; ice crystals will scatter light and the sample will appear opaque. More rigorously, one can measure the T_2 of the sample; free nitroxide frozen in 40% glycerol has a T_2 of ~2 μs (at 15–80 K), while in frozen aqueous solutions in which nitroxide clusters are formed T_2 is shortened to 0.5–1.5 μs.

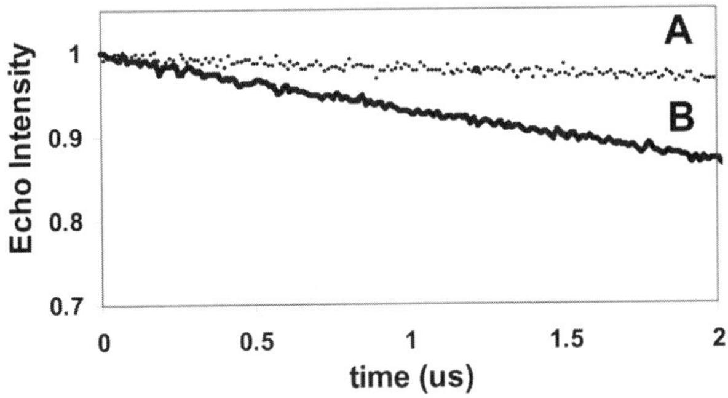

Figure 11. Concentration dependence of DEER spectra: (**A**) 100 μM, (**B**) 500 μM MTSSL in 40% glycerol.

Even with all these precautions, there still are the effects of intermolecular distances. Figure 11 shows a decay of the echo in a 500-μM solution of a free spin label that is faster then the decay in a 100-μM solution. The analysis program of G. Jeschke (http://www.mpip-mainz.mpg.de/~jeschke/distance.html) takes into account local concentration effects by fitting a 3D shell model of homogenously distributed spins to a DEER baseline. A note of caution when using glycerol to form glass samples: glycerol is known to perturb protein–protein interfaces, which may also result in dissociation of macromolecular complexes. An independent check for complex formation (e.g., sedimentation) in the presence of glycerol is recommended.

2.6. Choice of Spin Labels

There are no hard and fast rules with regard to the label (Figure 12) and/or labeling site. The distance distribution recovered from DEER signals consists of the conformational heterogeneity of the macromolecule convoluted with the rotamer heterogeneity of the spin label itself. Generally there is a correlation between the flexibility of the spin label on a given site and the distribution of distances. Figure 13 shows the mobility of the MTSL spin label in various positions on a membrane-bound protein, Band 3, and the corresponding distance distributions recovered from the DEER signals. Not surprisingly, the larger the motional freedom seen with increased spectral averaging results in a larger distribution of the distances.

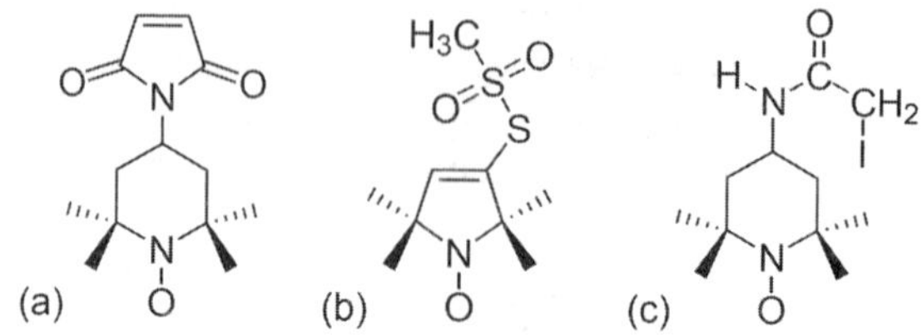

Figure 12. Molecular structures of various spin labels: (**a**) MSL, (**b**) MTSSL, (**c**) IASL.

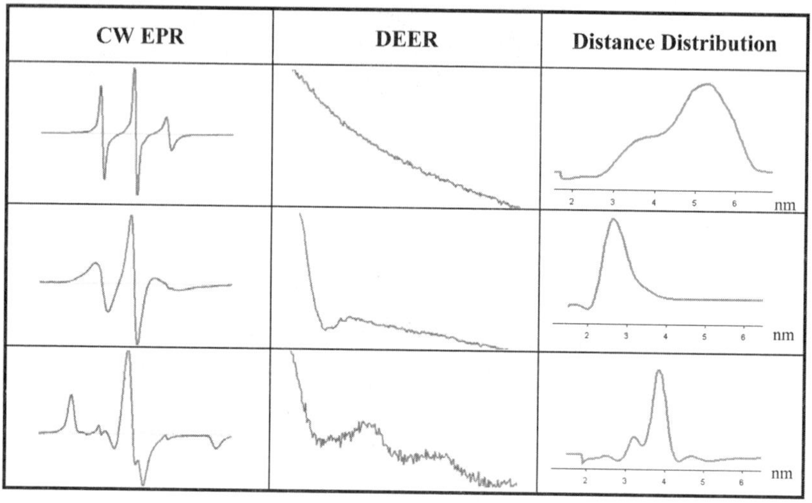

Figure 13. Comparison of spin label mobility (CW ESR), DEER spectra, and distance distribution. Decreasing mobility of the spin labels results in a large modulation depth of the dipolar evolution (DEER) and a narrower distance distribution. CW ESR spectra of MTSSL labeled Band 3 were measured at room temperature; DEER measurements were in frozen glass at 65 K .

This does not always hold. It is difficult to predict how the motional energy landscape at room temperature translates into the conformational landscape when the sample is frozen. The freezing rate (tens of milliseconds to seconds) is invariably slower than the motion (nanoseconds to microseconds); hence the conformational space will be defined by the local minima and energy barriers at low temperatures rather than the motional restrictions at room temperature. If the energy profile is smooth, it is conceivable that the cooling will result in a single rotamer even if the mobility was large to start with. The reverse is also true: the absence of mobility at room temperature should not be taken as evidence of the single-label rotamer state. It is just as likely that the spin label, although immobilized, has many rotamers. To distinguish between label and molecular heterogeneity, it is prudent to utilize more than one label on any given site. Figure 14 shows that the DEER spectra and the distance distribution of a mutant of smooth muscle myosin are largely independent of the size of the label used: the large MSL spin label or the smaller MTSSL spin label. We can conclude that the distance distribution is most likely defined by the backbone heterogeneity. In other cases, e.g., in a DNA repressor protein, different spin labels result in different distance distributions. Clearly, in this case the large distance distribution can be attributed to the label distribution and not the heterogeneity in the protein backbone.

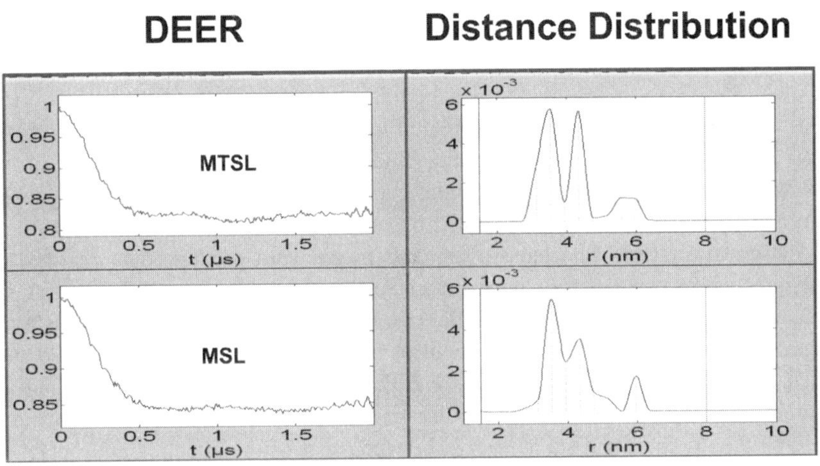

Figure 14. DEER spectra of myosin labeled with MTSSL (a) and MSL (b). Note the similarity in dipolar evolution function and the distance distribution between the two spin labels despite their different size.

3. APPLICATIONS

Structural studies of membrane and muscle proteins by x-ray crystallography or NMR spectroscopy have been limited by their large size, low solubility, and the lack of crystalline samples. Dipolar ESR, cw, DQC, and DEER are now proving to be viable complementary techniques, albeit at low resolution. A few recent studies have demonstrated the potential of DEER for studying the topology and associated structural changes in both motor protein systems and membrane proteins, where high-resolution data are often limited or unavailable. Selected examples illustrating advantages of DEER are discussed in the following section.

3.1. Motor Proteins

Molecular motor proteins are associated with muscle contraction and other cellular events, including intracellular targeting of proteins, transport of vesicles, and cell division. Despite operating by many different mechanisms, they share the common feature of converting the chemical energy of ATP into mechanical motion. Subtle conformational changes in their motor core are then amplified by various mechanical levers to produce movement. This family of proteins is often quite difficult to study because of their large size, the many interacting proteins, the multidomain nature of several of these proteins, and the numerous conformational states possible during the contractile cycle.

3.1.1. Kinesin

One molecular motor, kinesin, is a processive motor protein that uses the energy of ATP hydrolysis to travel along the microtubule network of the cell. It operates as a dimeric construct and is postulated to walk in a "hand-over-hand" manner during ATP hydrolysis by swapping the binding of its two catalytic heads to the microtubules, analogous to "walking" along a narrow path. The two catalytic heads are joined by a "neck-linker," followed by a coiled-coil tail that binds the cargo load for transportation. The hand-over-hand mechanism predicts that changes in the distance between the two neck-linker regions should occur. For kinesin attached with both heads to a microtubule, one would expect a well-defined distance induced by the stereo-specific attachment of the two heads to the microtubule surface. Kinesin that is detached should show the flexibility necessary for finding attachment sites, and kinesin that is walking (one head attached and the other unattached) should show a distance range between that of the attached and detached states.

The application of ESR for measuring the distance between the two heads as the kinesin motor walks along its microtubule track is an excellent example of the sensitivity of the method in measuring distance changes (Sugata et al. 2005). The experimental data for a nitroxide probe attached to residue Cys322 in the neck-linker region of kinesin were consistent with a mechanism where each step of the ATPase cycle is associated with a change in the distance between the two kinesin neck-linkers (Figure 15). When kinesin was free in solution and detached from micotubules (absence of nucleotides), the distance between the two linkers was larger than the 25 Å sensitivity limit of cw-dipolar ESR, but a broad population was detected with DEER, centered at 36 Å, with a width of 10 Å (Figure 15c).

When attached to microtubules, in addition to the above population, kinesin displayed two narrow distance distributions in the range 15–25 Å. Both distance distributions and their fractions were found to be sensitive to the nucleotide state. In the presence of nucleotide analogs AMPPNP or ATPγS, the shorter of the two distances between the neck-linkers closed by as much as 3.9 to 19.5 Å, and the population of interacting spins contributing to the 36 Å distribution significantly decreased. These changes in the distance distributions between the kinesin linkers

a)

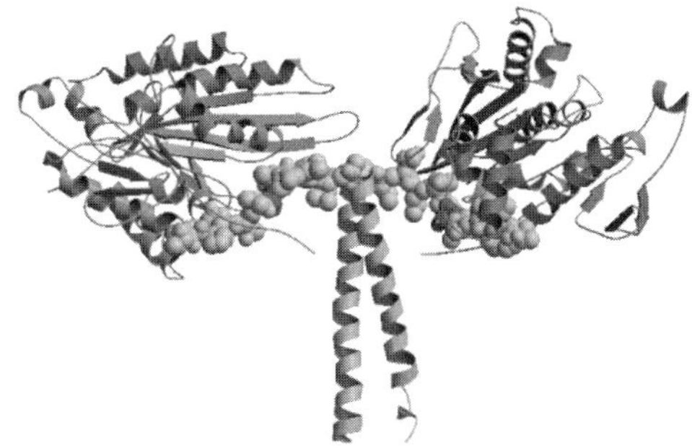

b)

state	Distance (nm)	+/-width (nm)
No nucleotide+MT	2.34	0.22
ADP+Li_2SO_4	2.40	0.23
ADP+MT	2.34	0.20
AMPPNP+MT	1.95	0.25
ATPγS+MT	2.05	0.30

c)

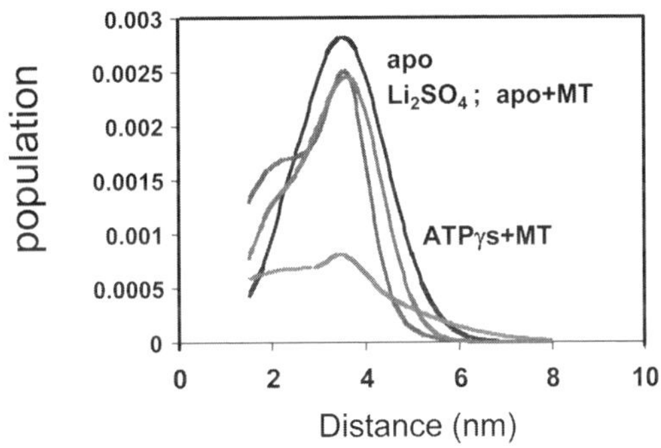

Figure 15. (**a**) The neck linker (sphere representation) joins the two heads of the kinesin dimeric construct. The coiled-coil tail is responsible for cargo transportation (PDB: 3KIN). (**b**) cw-dipolar ESR mean distances and width of distributions for measurements between C332 of the kinesin dimer in the absence (NN) and presence of nucleotide and microtubules (MT). (**c**) DEER distance distribution profile for corresponding states shown in (**b**). The major distance distribution was centered at 36 Å for each state, interpreted as a population of motor domains that are either splayed out in solution or attached only by one domain or loosely to microtubules. Populations for the shorter distances, as measured by cw-dipolar, can be observed as a shoulder peak on the distance distribution profile (see Sugata et al. 2005).

were consistent with the "hand-over-hand" walking mechanism. The open-to-closed transition, of an average of approximately 15 Å between the two adjacent neck-linkers, is coupled to the force-generating changes in kinesin as it walks along the microtubules in a processive manner.

a)

b)

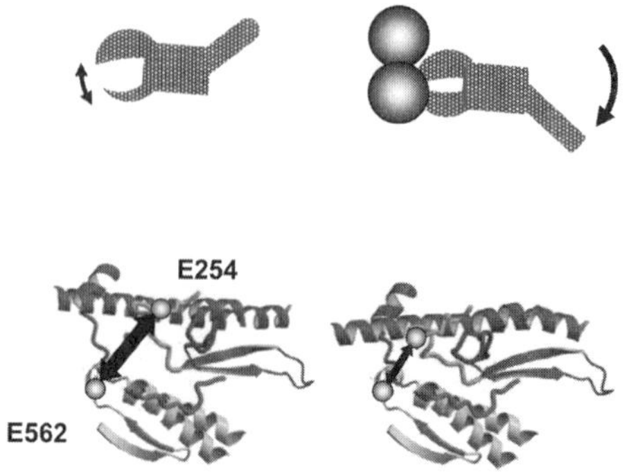

c)

	OPEN	CLOSED
model	22	14
EPR	20	10-12
FRET	21	16

Figure 16. The putative closure of the myosin actin-binding cleft (**left**) allows the binding to actin and rotation of the regulatory domain (**right**), which is the force-generating event in muscle. Distance between residues 254 and 562 across the actin-binding cleft confirmed the closure of the cleft by dipolar ESR and FRET (see Song et al. 2005b). The distances were found to be similar to those found modeled from the non-muscle myosin V crystal structure.

3.1.2. Myosin

The basis of the cross-bridge cycle of muscle contraction is also closely linked to the cycle of binding and hydrolysis of nucleotides by myosin and the binding of myosin to actin filaments. Nucleotide hydrolysis by myosin and binding to actin is hypothesized to be associated with conformational changes in the myosin actin-binding cleft. This cleft is located between the upper and lower 50-kD domains of the catalytic domain of the myosin head and extends into the base of the nucleotide-binding pocket. Closure of the actin-binding cleft allows for tight binding to actin and triggers movement of the elongated regulatory domain, thereby generating force (Rayment et al. 1993) (Figure 16). To directly examine conformational changes across the actin-binding cleft, cw-dipolar and DEER measurements were used to monitor distances between the upper (residue 282 of smooth muscle myosin) and lower (residue 595) domains of the cleft during various nucleotide states of the ATP cycle and in the presence of F-actin. For both cw-dipolar and DEER, only one population centered at 20 Å was detected in the absence of actin (Song et al. 2005b). This was also not significantly altered by the presence of nucleotide analogs. However, upon addition of F-actin, a second shorter distance population with a mean distance of 10–12 Å was detected by cw-dipolar. Thus, binding of actin is required for the closure of the actin-binding cleft of myosin and not the identity of the bound nucleotide.

A similar insight into this mechanism of cleft closure upon actin binding was also observed for the *Dictyostelium* isoform of myosin. For this isoform, two short distance populations of 14 and 23 Å were detected for the interspin distance for residues 416 and 537 in the absence of actin (Song et al. 2005a). The longer of the two distances was also observed by DEER. The presence of actin caused an increase in the 14 Å population by a factor of two, again implying closure of the myosin actin-binding cleft. Similar to the smooth muscle isoform, the average distance distributions for both the long and short distances did not significantly change in the presence of the various nucleotide analogs. Thus, taken together, these results for the two isoforms of myosin demonstrate that actin induces closure of the actin-binding cleft of myosin. This experiment is an example of the need to approach such measurements with the dual methods of both cw-dipolar and DEER in order to extract all conformational states, especially those in the overlap range between the two methods between 15 and 25 Å.

In smooth muscle, force generation is regulated by phosphorylation of the regulatory light chain (RLC) of the myosin head, one on each head. When unphosphorylated, the contractile cycle is blocked by interaction of the two smooth muscle myosin heads, defining the switched OFF state of myosin. Phosphorylation of both RLCs is required for full Mg-ATPase activity and contraction. In the absence of high-resolution crystal structures of the regulated double-headed smooth muscle myosin (SMM) constructs, two very different models for the interaction of the two heads in the OFF state have been proposed. The first model, based on cryo-EM (electron micrography), has an asymmetric interaction between the two motor domains (MD model) of the myosin heads. This MD head–head interaction is abolished upon phosphorylation and the SMM can then interact with actin, thus stimulating ATPase activity (Wendt et al. 2001). A second model, from crosslinking studies, shows a more "symmetric" interaction between the elongated regulatory domains (RD model) (Wahlstrom et al. 2003) (Figure 17).

The mapping of distances by ESR between the RLCs in two adjacent heads of smooth muscle myosin in both the unphosporylated and phosphorylated states can help resolve this discrepancy. Four engineered sites in RLC (15, 38, 59, 96) and one native site (109) were labeled and the distances measured by cw and pulsed methods (Liang et al., unpublished). In the unphosphorylated SMM monomers, the distances were 11.5 Å for residue 38, 32 Å for residue 59, 34/45 Å for residue 96, and >40 Å for residues 109 and 15. The distances between each residue increased beyond the sensitivity limit of DEER upon phosphorylation, indicating that the RLCs are largely disordered upon phosphorylation. This is consistent with the abolishment of the close inter-head interaction between the myosin heads, as proposed by both models. In the unphosphorylated state, the four ESR distances agreed with those predicted for the asymmetric MD model. For example, at residue 59, the RLC–RLC distance was measured by DEER-ESR as 32 Å, consistent with the asymmetric-model predicted distance of 35 Å rather than the symmetric head–head model predicted distance of 75 Å (Figure 17).

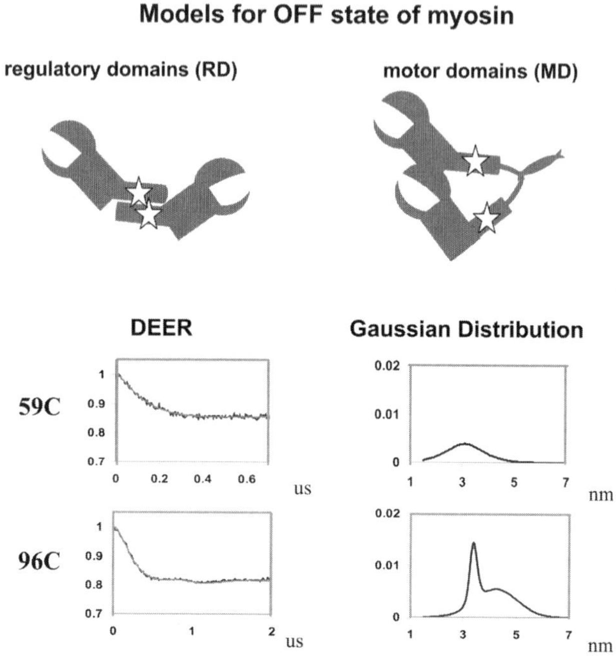

residue	distance in a model		EPR distance
	RD	MD	
59	75	35	32
96	12	40	34/45

Figure 17. Two alternative models for the unphosphorylated OFF state of smooth muscle myosin: **(left)** interaction between the two regulatory domains (RD model); **(right)** interaction between two motor domains (MD model). The measured DEER distances between two pairs of residues (59 and 96) are consistent with the MD model and exclude the RD model.

3.1.3. Troponin

Regulation of contraction of striated muscle is associated with calcium-dependent structural transitions in the muscle thin filament, which is composed of actin, tropomyosin, and the troponin complex. The troponin (Tn) complex is composed of three subunits: troponin C (TnC), which binds calcium (Ca^{2+}); troponin I (TnI), which inhibits acto-myosin activity; and troponin T (TnT), which anchors TnI and TnC to the thin filament. The cw-dipolar and DEER methods were used to observe conformational changes induced by binding of calcium to one subunit, TnC, and propagating to the other two subunits. A popular model for transmission of the signal is movement of a segment of TnI, known as the *switch* peptide. Upon activation with calcium, the TnI *switch* peptide binds to the N lobe hydrophobic cavity of TnC pulling away the TnI *inhibitory* peptide segment from actin. Upon removal of the *inhibitory* peptide from actin, the tropomyosin molecule re-positions on actin to allow the myosin heads to bind, thereby initiating contraction. The crystal structures of the cardiac Tn isoform confirm the interaction of the *switch* peptide with the regulatory N lobe of TnC in the calcium activated state (Fajer 2005). However, there is no structure for the cardiac troponin isoform in the absence of Ca^{2+}, although the closely related skeletal isoform was determined in both Ca^{2+} states (Fajer 2005). The structural differences between the cardiac and skeletal forms are significant and well documented by a number of structural methods. Furthermore, the changes induced in troponin by Ca^{2+} binding are different for both isoforms. Thus, we have tested whether this large proposed movement of the *switch* peptide occurs in the cardiac isoform and in solution. Two spin–spin distances were measured between the *switch* peptide of cardiac TnI (residue 160) to TnI residue 129, located in the TnI/TnT *coiled coil* near the C lobe of TnC, and to a residue (47) in the N lobe of TnC (Figure 18).

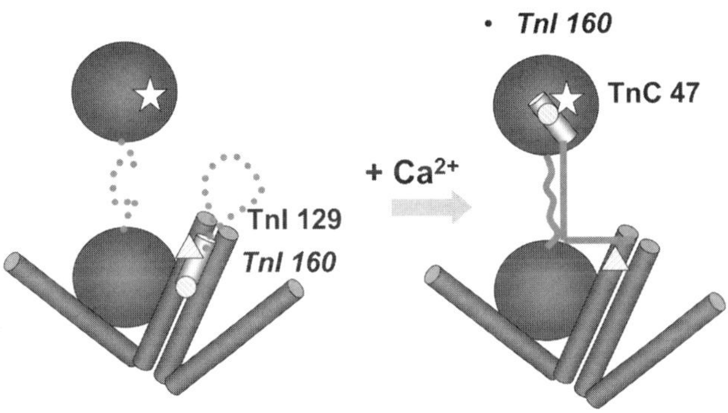

Figure 18. Shift of the TnI switch peptide induced by binding of calcium. In the absence of calcium, the switch peptide of TnI, labeled at residue TnI 160 (*circle*), is in close proximity to residue TnI 129 (*triangle*) located in the TnI/TnT coiled coil region. Upon calcium binding, there is a large-scale conformational transition of the switch peptide to move toward the N lobe of TnC, labeled at residue TnC 47 (*star*).

As expected from the crystal structure of cardiac troponin in the presence of Ca^{2+}, the distance between the *switch* peptide and the N lobe of TnC was short (17 Å), while the distance to the coiled-coil region was significantly longer (47 and 65 Å). Upon removal of calcium, the *switch* peptide moved to within 17 Å of the *coiled-coil*, resulting in a large, greater than 30 Å, movement of the peptide (see Fajer 2005).

Positioning of the N-terminal peptide of TnI (residues 1–47) has also remained somewhat contentious in the literature. Competitive binding of the N-terminal peptide of TnI and TnI peptides encompassing the inhibitory segment, the structural motif that communicates the calcium signal to actin, to the TnC C lobe, has suggested displacement of the N-terminal region of TnI by the inhibitory segment in the absence of Ca^{2+} (Tripet et al. 1997). However, crosslinking data do not support displacement of the N-terminal peptide by the inhibitory segment (Luo et al. 2000). Rather, a close stable interaction of the N terminus of TnI with the C lobe of TnC is required for this region of Tn to behave as a structural anchor. This inconsistency was resolved by measuring distances between the C lobe of cardiac TnC (residue 100) and the N-terminal region of cardiac TnI (residue 51). In the presence of Ca^{2+}, the spin–spin distance was 19 Å (both cw-dipolar and DEER), in very close agreement with the modeled distance from the x-ray structure of cardiac troponin (17 Å). Removal of Ca^{2+}, which results in conformational rearrangement of the *inhibitory* peptide, did not change the distance between these two residues (Fajer 2005). Therefore, the N terminus of TnI does not shift in response to calcium binding but remains in close proximity to the TnC C lobe, as expected for a structural anchor to position Tn on the thin filament.

The recent crystal structure data on Tn complexes in the presence of Ca^{2+} have suggested a "melted" conformation for the TnC central helix for the cardiac isoform and an "extended" rigid helix for the skeletal isoform. TnC belongs to the EF-hand protein family, in which either an extended or melted central helix conformation is feasible and functionally important, especially when complexed with other protein components. For example, calmodulin has an extended conformation that "melts" when bound to myosin light-chain kinase. In the case of cardiac TnC, a large-scale movement of TnC due to melting of the central helix may be required to relay structural changes upon calcium binding (Takeda et al. 2003). However, the reverse was observed in skeletal Tn, where the central helix was better defined as compared to a calcium-free state (Vinogradova et al. 2005).

Three TnC interdomain distances were measured to investigate whether Ca^{2+} regulates the conformation of the TnC central helix of cardiac troponin in solution. The three double TnC mutants (residues 15/136, 15/94, 12/136) were designed to measure the distance between both the N and C lobes of TnC and to distinguish between the extended and melted central helix structures. In isolated TnC, the DEER distances measured between the two lobes of TnC were found to be in perfect agreement with the crystal structures of skeletal TnC in which the central helix is extended. When TnC was complexed with other two troponin subunits, the three distances decreased as compared to those measured in the isolated TnC (Fajer 2005). Thus, the central helix of cardiac TnC is extended in the isolated subunit but is more compact in the troponin complex. The Ca^{2+}-induced changes in the complex did not require further melting of the central helix.

A recent study by Arata and coworkers (Nakamura et al. 2005) measured cardiac TnC interspin distances by DEER between the two native cysteine residues — Cys35 located in the N lobe and Cys84 close to the midpoint of the central helix of

TnC — in isolation and also in reconstituted fibers. They did not detect a large difference in interspin distances for the isolated TnC sample (26 Å) when compared to the reconstituted fiber sample (27 Å). Upon addition of Ca^{2+}, a decrease in the interspin distance of 4 Å was detected in the reconstituted fiber sample.

3.2. Membrane Proteins

ESR has been long applied to the study of the secondary and tertiary structure of membrane proteins due to asymmetric solvation of proteins and the differential solubility of aqueous and gaseous relaxing agents. The following examples highlight recent applications of DEER for studying the topology of proteins within a membrane bilayer and the geometry of the oligomeric states. For a more complete review, see Steinhoff (2004).

3.2.1. Sodium/Proton Antiporter

A good example is the formation of the homooligomer in the integral membrane protein Na^+/H^+ (sodium/proton) antiporter of *E. coli* (NhaA) (Hilger et al. 2005). Electron cryomicroscopy revealed that NhaA forms a dimer in the membrane. The conformation of the dimer state is believed to be associated with the functional pH response of NhaA. The flux of Na^+ increases by three orders of magnitude between pH 7.0 and 8.5. A varying pH may alter the interface between the NhaA monomers in an oligomer, may cause a conformational change within a monomer, or may result in modification in the degree of oligomerization. The measurement of distances between the monomers and of modulation depth as a function of pH helped to distinguish between the above possibilities. The distances between residues 225 were found to be well defined, with a narrow distribution centered at 44 Å. The distance or distribution did not change with pH, suggesting that there was no significant conformational rearrangement between the monomers. The modulation depth was found to vary with pH. For a calibrated cavity, the modulation depth is related to the concentration of interacting spins in an oligomer. Therefore, knowing the concentration of the spin-labeled NhaA protein, the number of NhaA monomers within a complex could be obtained. The limiting number of interacting spins per complex was 2, consistent with a dimer as also found by EM (Figure 19). Thus the relative disposition of the monomers did not change with varying pH, but the degree of dimerization increased with increasing pH.

3.2.2. Light-Harvesting Complex

A second example for monitoring conformational changes, including oligomer formation, was for the plant light-harvesting complex LHCIIb of photosynthesis (Jeschke et al. 2005). LHCIIb is the pigment-associated protein responsible for channeling the energy of captured photons to the photosynthetic reaction centers of photosystems PSI and PSII. The partitioning between the two photosystems is regulated by phosphorylation of Thr-6 of LHCIIb. In the unphosphorylated state,

LHCIIb exists as a trimer, which associates with photosystem PSII. Phosphorylation dissociates the trimers into monomers, which then interact preferentially with photosystem PSI.

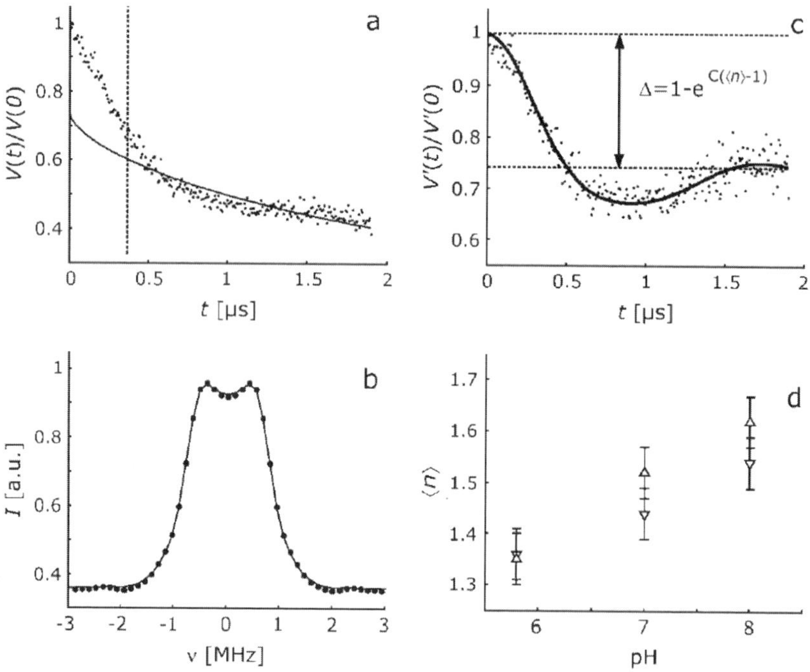

Figure 19. The pH dependence of the average oligomerization for a spin label attached to residue 225 of NhaA. (**a**) Experimental DEER showing planar background fit (solid line) to obtain interacting spins within the same oligomer of NhA. (**b**) Dipolar spectrum obtained by Fourier transformation of (**a**). (**c**) Extraction of modulation depths to obtain an estimate of the number of spins per oligomer. (**d**) The pH dependence of the average number of spin labels per oligomer for increasing (upward pointing triangles) and decreasing (downward pointing triangles) series of pH series. An <*n*> value approaching 2 is consistent with formation of a dimer. Reprinted with permission from Hilger et al. (2005). Copyright © 2005, The Biophysical Society

The site of the Thr-6 phosphorylation-dependent switch in the N-terminal region of LHCIIb was not resolved in the crystal structure of LHCIIb (Liu et al. 2004). Therefore, six intramolecular distances were measured within the LHCIIb monomer, with three of these distances in good agreement with those modeled from the LHCIIb crystal structure (Figure 20). Two distances from the N-terminal domain missing in the crystal structure to two other sites exhibited a bimodal distance distribution, suggesting that the N-terminal domain can exist in two different states in the monomeric form. Taking into account both the inter- and intramolecular distances, the positioning of the flexible N-terminal domain of LHCIIb was modeled by triangulation of distances in both the monomer and trimer states. The two modeled conformations demonstrated the high flexibility of the N terminus of LHCIIb, explaining the absence of electron density in the crystal structure. For the trimeric form, the N-terminal domain was modeled in a location above the core of the protein, extending into the aqueous region on the stromal side. The flexibility of the N terminus is in line with its postulated behavior as a conformational switch.

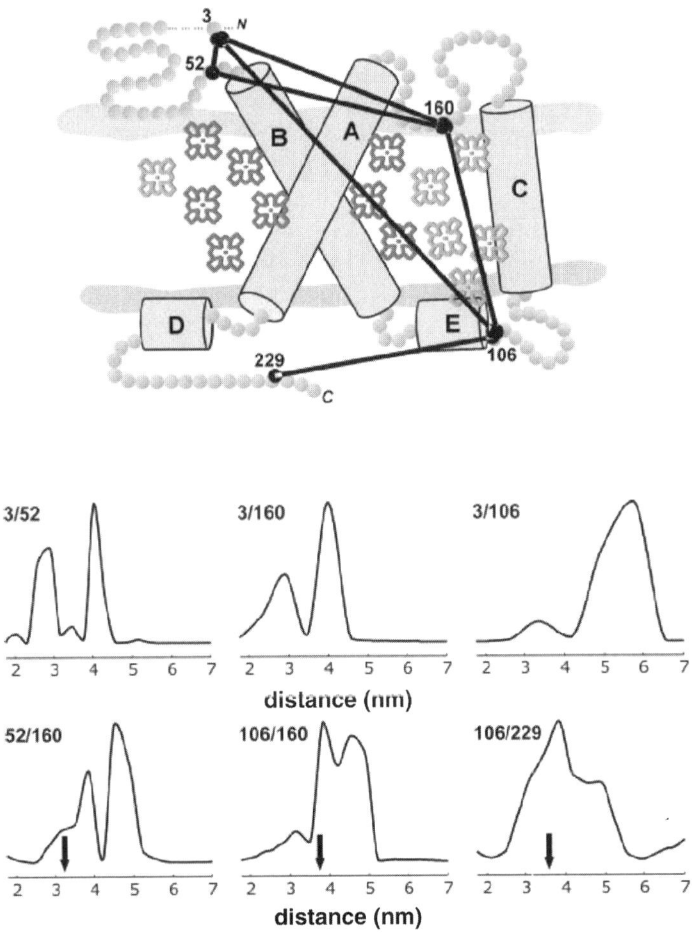

Figure 20. Schematic representation of the 6 intramolecular DEER distances measured in the plant light harvesting complex LHCIIb which has five α–helical domains (A–E). DEER distance distributions for these six interspin measurements are shown below for the monomeric form of LHCIIb. Bold arrows represent distances between amino acids modeled from the LHCIIb crystal structure. Reprinted with permission from Jeschke et al. (2005). Copyright © 2005, American Society for Biochemistry and Molecular Biology.

3.2.3. F_0F_1–ATP Synthase

Dipolar ESR was used to investigate the arrangement of the b-subunits in the F_0F_1-ATP synthase integral membrane protein complex (Steigmiller et al. 2005). F_0F_1-ATP synthase is a large motor protein enzyme that couples ATP synthesis from ADP and phosphate with transmembrane proton (H^+) translocation in bacteria, chloroplasts, and mitochondria. The enzyme consists of two separate motor parts: F_0, the membrane integrated ion pump, and F_1, the hydrophilic ATP synthesizer. Electron microscopy has shown both of these motors to be comprised of eight different subunits, which are connected by a central stalk (γ and ε subunits) and a second thinner peripheral (b-subunit) structural stalk unit. The peripheral b-subunit stalk is also part of the "stator" complex, the structure proposed to behave as a static rigid domain in order for the subsequent movement of the "rotor" part of the enzyme during ATP synthesis. Large conformational changes have been observed to occur in the F_1-ATP synthesizer part of the enzyme upon binding of the non-hydrolysable ATP analog AMPPNP, but little change was observed in F_0, where the b-subunits are located (Bottcher et al. 2000). To date, there are no high-resolution structural data for the b-subunits in intact enzyme.

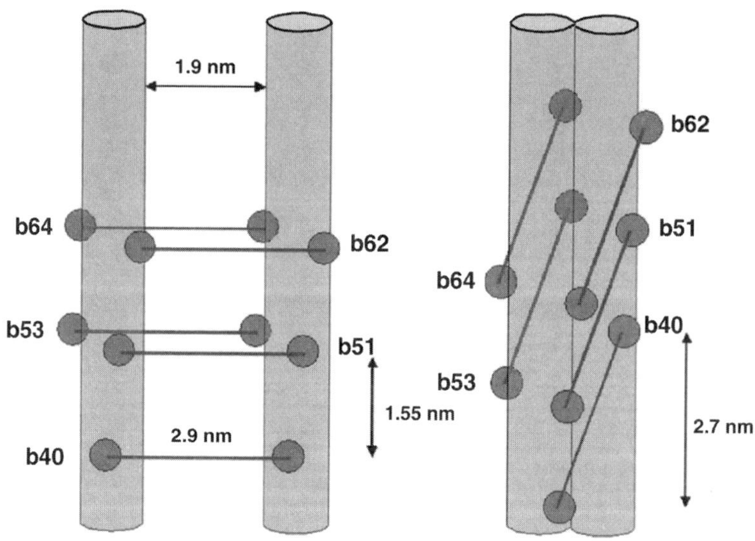

Figure 21. Possible structural models for the arrangement of the two b-subunits of the FoF1–ATP synthase integral membrane complex. DEER distances obtained between five positions on the b-subunits showed the same average distance between spin labels of 29 Å. The first model (**left**) shows the b-subunits in register with no direct interaction between the two helices. An alternate model (**right**) has both helices in direct parallel contact with a helical shift of 27 Å to be in agreement with the DEER obtained interspin distances. Reprinted with permission from Steigmiller et al. (2005). Copyright © 2005, American Society for Biochemistry and Molecular Biology.

The DEER distance measurements in this report centered on the arrangement of two stator b-subunits and their proposed structural role in the F_0F_1-ATP enzyme (Steigmiller et al. 2005). The b-subunits, when not complexed with other subunits of the ATP-synthase, were shown by crosslinking to form a coiled-coil dimer. However, in this study, crosslinking was not observed in the intact enzyme, implying a different arrangement of the b-subunits in the complex. DEER experiments between five different positions on the b-subunits found the same average distance (29 Å), giving a modeled parallel arrangement for the two helices rather than the coiled-coil arrangement as believed for the isolated subunits (Figure 21).

3.2.4. Na$^+$ Symporter

A noteworthy example of the application of DEER for membrane proteins of unknown structure is topology mapping and the ligand-induced conformational changes of PutP (Jeschke et al. 2004). PutP is a member of the large Na^+/substrate symporter family. It is an Na^+/proline transporter that utilizes the Na^+ electrochemical gradient in *E. coli* to drive the coupled uphill transport of a variety of substrates, including sugars, amino acid, and vitamins. This large integral membrane protein is postulated to have 13 transmembrane helical regions, with the N and C terminus on opposite sides of the membrane bilayer. PutP is believed to undergo a series of large conformational transitions exposing the ligand-binding site to one side of the membrane or the other. To date, there is no high-resolution structure of PutP or structural details of the anticipated conformational changes that accompany ligand binding.

Three DEER intraspin distances were reported for PutP in reconstituted proteoliposomes (Figure 22). The first was a long-range distance measured as 48 Å between nitroxide groups labeled in loops 4 and 7, strongly supporting the positioning of these two loop regions on opposite sides of the membrane bilayer. On the other hand, two other shorter distances (22 and 18 Å) were obtained between spin labels attached to loops 2 and 4 as well as loops 2 and 6, implying that these loops are on the same side of the membrane. Furthermore, binding of Na^+ resulted in the dramatic distance increase between loops 2 and 6, suggesting a ligand-induced large conformational change.

4. REFERENCES

Borbat PP, McHaourab HS, Freed JH. 2002. Protein structure determination using long-distance constraints from double-quantum coherence ESR: study of T4 lysozyme. *J Am Chem Soc* **124**:5304–5314.

Borbat PP, Davis JH, Butcher SE, Freed JH. 2004. Measurement of large distances in biomolecules using double-quantum filtered refocused electron spin-echoes. *J Am Chem Soc* **126**:7746–7747.

Bottcher B, Bertsche I, Reuter R, Graber P. 2000. Direct visualisation of conformational changes in EF(0)F(1) by electron microscopy. *J Mol Biol* **296**:449–457.

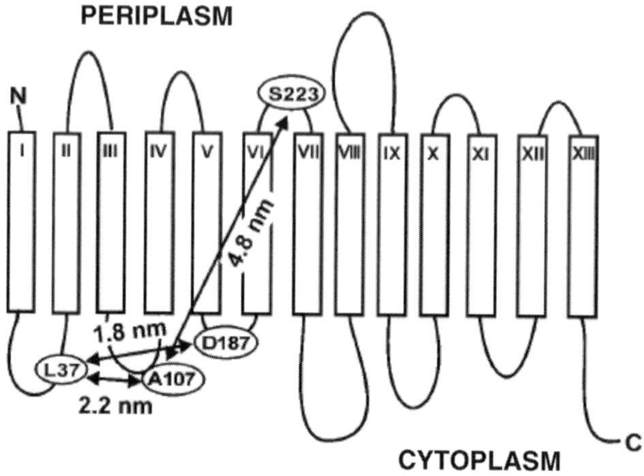

Figure 22. Schematic representation of the PutP Na^+/substrate symporter of *E. coli*, which shows the three mean spin–spin distances, obtained by DEER for the four labeled sites. Reprinted with permission from Jeschke et al. (2004). Copyright © 2004, The Biophysical Society.

Budker V, Du JL, Seiter M, Eaton GR, Eaton SS. 1995. Electron–electron spin–spin interaction in spin-labeled low-spin methemoglobin. *Biophys J* **68**:2531–2542.

Eaton S, Eaton G. 2000. Distance measurements by CW and pulsed EPR. In *Biological magnetic resonance*, pp. 2–21. Ed LJ Berliner, S Eaton, and G Eaton. New York: Kluwer Academic.

Fajer P. 2005. Site-directed spin labeling and pulsed dipolar EPR (DEER) of force activation in muscle. *J Phys Cond Mater* **17**:S1459–S1469.

Hilger D, Jung H, Padan E, Wegener C, Vogel KP, Steinhoff HJ, Jeschke G. 2005. Assessing oligomerization of membrane proteins by four-pulse DEER: pH-dependent dimerization of NhaA Na+/H+ antiporter of *E. coli*. *Biophys J* **89**:1328–1338.

Hubbell WL, Cafiso DS, Altenbach C. 2000. Identifying conformational changes with site-directed spin labeling. *Nat Struct Biol* **7**:735–739.

Jeschke G. 2002. Distance measurements in the nanometer range by pulse EPR. *Chem Phys Chem* **3**:927–932.

Jeschke G. 2004. *DeerAnalysis 2004 user's manual*. Manual is on his website: http://www.mpip-mainz.mpg.de/ ~jeschke/distance.html

Jeschke G, Pannier M, Spiess H. 2000. Double electron–electron resonance. In *Distance measurements in biological systems by EPR*, pp. 493–512. Ed LJ Berliner, GR Eaton. New York: Plenum.

Jeschke G, Koch A, Jonas U, Godt A. 2002. Direct conversion of EPR dipolar time evolution data to distance distributions. *J Magn Reson* **155**:72–82.

Jeschke G, Wegener C, Nietschke M, Jung H, Steinhoff HJ. 2004. Interresidual distance determination by four-pulse double electron-electron resonance in an integral membrane protein: the Na+/proline transporter PutP of *Escherichia coli*. *Biophys J* **86**:2551–2557.

Jeschke G, Bender A, Schweikardt T, Panek G, Decker H, Paulsen H. 2005. Localization of the N-terminal domain in light-harvesting chlorophyll a/b protein by EPR measurements. *J Biol Chem* **280**:18623–18630.

Liang H, Song L, Cremo CR, Fajer P. 2005. Unpublished results.

Liu Z, Yan H, Wang K, Kuang T, Zhang J, Gui L, An X, Chang and W. 2004. Crystal structure of spinach major light-harvesting complex at 2.72 Å resolution. *Nature* **428**:287–292.

Luo Y, Wu JL, Li B, Langsetmo K, Gergely J, Tao T. 2000. Photocrosslinking of benzophenone-labeled single cysteine troponin I mutants to other thin filament proteins. *J Mol Biol* **296**:899–910.

McHaourab HS, Oh KJ, Fang CJ, Hubbell WL. 1997. Conformation of T4 lysozyme in solution: hinge-bending motion and the substrate-induced conformational transition studied by site-directed spin labeling. *Biochemistry* **36**:307–316.

Miick SM, Martinez GV, Fiori WR, Todd AP, Millhauser GL. 1992. Short alanine-based peptides may form 3(10)-helices and not alpha-helices in aqueous solution. *Nature* **359**:653–655.

Milov A, Salikhov K, Shirov M. 1981. Application of ELDOR in electron-spin echo for paramagnetic center space distribution in solids. *Fiz Tverd Tela* (Leningrad) **23**:957–982.

Milov A, Ponomarev A, Tsvetkov and Y. 1984. Electron–electron double resonance in electron spin echo model biradical systemsand the sensitised photolysis of decalin. *Chem Phys Letters* **110**:67.

Nakamura M, Ueki S, Hara H, Arata T. 2005. Calcium structural transition of human cardiac troponin C in reconstituted muscle fibres as studied by site-directed spin labelling. *J Mol Biol* **348**:127–137.

Pannier M, Veit S, Godt A, Jeschke G, Spiess HW. 2000. Dead-time free measurement of dipole–dipole interactions between electron spins. *J Magn Reson* **142**:331–340.

Rabenstein MD, Shin YK. 1995. Determination of the distance between two spin labels attached to a macromolecule. *Proc Natl Acad Sci USA* **92**:8239–8243.

Rayment I, Holden HM, Whittaker M, Yohn CB, Lorenz M, Holmes KC, Milligan RA. 1993. Structure of the actin–myosin complex and its implications for muscle contraction. *Science* **261**:58–65.

Sale K, Song L, Liu YS, Perozo E, Fajer P. 2005. Explicit treatment of spin labels in modeling of distance constraints from dipolar EPR and DEER. *J Am Chem Soc* **127**:9334–9335.

Schiemann O, Piton N, Mu Y, Stock G, Engels JW, Prisner TF. 2004. A PELDOR-based nanometer distance ruler for oligonucleotides. *J Am Chem Soc* **126**:5722–5729.

Song L, Gyimesi M, Bagshaw CR, Malnasi-Csizmadia A, Fajer P.G. 2005a. Closure of actin binding cleft in *Dictostelium d.* muscle myosin: DEER and dipolar CW-EPR. Unpublished results.

Song L, Sweeney HL, Rosenfeld S, Fajer P. 2005b. Closure of actin binding cleft in smooth muscle myosin: DEER and dipolar CW-EPR. Unpublished results.

Steigmiller S, Borsch M, Graber P, Huber M. 2005. Distances between the β-subunits in the tether domain of F(0)F(1)–ATP synthase from *E. coli*. *Biochim Biophys Acta* **1708**:143–153.

Steinhoff HJ. 2004. Inter- and intra-molecular distances determined by EPR spectroscopy and site-directed spin labeling reveal protein–protein and protein–oligonucleotide interaction. *Biol Chem* **385**:913–920.

Sugata K, Song L, Nakamura M, Ueki S, Fajer PG, Arata T. 2005. Distance change between the two neck linkers of kinesin as detected by electron spin resonance spectroscopy. In press.

Takeda S, Yamashita A, Maeda K, Maeda Y. 2003. Structure of the core domain of human cardiac troponin in the Ca(2+)-saturated form. *Nature* **424**:35–41.

Tripet B, Van Eyk JE, Hodges RS. 1997. Mapping of a second actin-tropomyosin and a second troponin C binding site within the C terminus of troponin I, and their importance in the Ca2+-dependent regulation of muscle contraction. *J Mol Biol* **271**:728–750.

Vinogradova MV, Stone DB, Malanina GG, Karatzaferi C, Cooke R, Mendelson RA, Fletterick RJ. 2005. Ca(2+)-regulated structural changes in troponin. *Proc Natl Acad Sci USA* **102**:5038–5043.

Voss J, Salwinski L, Kaback HR, Hubbell WL. 1995. A method for distance determination in proteins using a designed metal ion binding site and site-directed spin labeling: evaluation with T4 lysozyme. *Proc Natl Acad Sci USA* **92**:12295–12299.

Wahlstrom JL, Randall Jr MA, Lawson JD, Lyons DE, Siems WF, Crouch GJ, Barr R, Facemyer KC, Cremo CR. 2003. Structural model of the regulatory domain of smooth muscle heavy meromyosin. *J Biol Chem* **278**:5123–5131.

Wendt T, Taylor D, Trybus KM, Taylor K. 2001. Three-dimensional image reconstruction of dephosphorylated smooth muscle heavy meromyosin reveals asymmetry in the interaction between myosin heads and placement of subfragment 2. *Proc Natl Acad Sci USA* **98**:4361–4366.

CHAPTER 5

Membrane Protein Structure and Dynamics Studied by Site-Directed Spin-Labeling ESR

Enrica Bordignon and Heinz-Jürgen Steinhoff
Universität Osnabrück, Osnabrück, Germany

1. INTRODUCTION

ESR spectroscopy of site-directed spin-labeled biomolecules (Site-Directed Spin Labeling, SDSL) has emerged as a powerful method for studying the structure and conformational dynamics of proteins and nucleic acids under conditions relevant to function (for reviews see, e.g., Feix and Klug 1998; Hubbell et al. 1996; Hubbell et al. 1998, 2002). In this technique a spin-label side chain is introduced at a selected site via cysteine substitution mutagenesis followed by modification of the unique sulfhydryl group with a specific paramagnetic nitroxide reagent. The continuous wave (cw) ESR spectrum yields information about the nitroxide side chain mobility, solvent accessibility, the polarity of its immediate environment, and the distance between the nitroxide and another paramagnetic center in the protein. Hence, ESR data analysis of a series of spin-labeled variants of a given protein allows defining elements of secondary structure, including their solvent exposure, to characterize protein topography and to determine orientations of individual segments of the protein. A complete analysis allows modeling of protein structures with a spatial resolution at the level of the backbone fold (Hubbell et al. 1998, 2000; Koteiche and Mchaourab 1999; Mchaourab and Perozo 2000; Perozo et al. 1998; Wegener et al. 2001a). This method is applicable to any protein with a cloned gene that can be expressed. In particular, it has been shown to be very useful in studying large membrane proteins or protein complexes that are not amenable to NMR methods or do not crystallize. One of the most powerful properties of the method is its sensitivity to molecular dynamics: protein equilibrium fluctuations and conformational changes of functional relevance can be followed on a wide time scale ranging from picoseconds to seconds.

The present chapter summarizes recent progress in the SDSL methodological approach, with special emphasis on application to the study of molecular dynamics, membrane protein structure, and protein–protein interaction.

2. SPIN LABELING

Elucidation of the structure and function of proteins by ESR spectroscopy has become increasingly important in recent years as technological advances have been made in the development of spectrometers and applicable methods. Such advances are also accompanied by novel chemistry approaches to design probes fulfilling desirable requirements in terms of incorporation into proteins, labeling efficiency, similarity with natural amino acid side chains, negligible perturbation of protein structure and function, and dynamical properties chosen ad hoc to monitor the backbone dynamics or to sample the microenvironment of the attached probe.

There are two main approaches to modifying peptides or proteins with paramagnetic spin labels. A very well-established method consists in utilizing the sulfhydryl group of cysteine residues naturally present in the protein under investigation or engineered via site-directed mutagenesis to attach a variety of nitroxides to yield a spin-label side chain. This approach usually requires that the target protein possesses only cysteine residues at the desired sites and that additional cysteines eventually present in the protein must be eliminated (cysteines are usually replaced by serines or alanines). Among the various spin labels utilized in literature works, the spin label (1-oxyl-2,2,5,5-tetramethylpyrroline-3-methyl)methanethiosulfonate (MTS) (Berliner et al. 1982) is most often used in SDSL studies due to its sulfhydryl specificity and its small molecular volume, similar to a tryptophane side chain (Figure 1A). The link between the piperidine-oxyl moiety and the protein backbone renders the label flexible, allowing native folding of the proteins when the spin label is attached (Figure 1D). Moreover, the unique dynamic properties of this spin-label side chain facilitate obtaining detailed structural information from the shape of its ESR spectrum (see §3.1). On the other hand, the structural variability characterizing the MTS spin label (the distance between the C_β and the NO group varies between 4 and 8 Å depending on the binding site) does not unequivocally allow relating of the data obtained directly to the property of the native side chain. Additionally, in the case of distance determination between two labeled positions, only a distance distribution rather than a defined distance can be obtained, entailing the use of a higher number of spin-labeled sites and eventually of conformational searching approaches (Sale et al. 2005) for structural modeling of proteins based on SDSL ESR.

The other approach includes incorporation of a paramagnetic amino acid in a peptide/protein in a step-by-step synthesis by employing nonsense suppressor methodology (e.g.. amber suppressor tRNA chemically aminoacylated with the desired amino acid) (Cornish et al. 1994; Mendel et al. 1995; Shafer et al. 2004) or solid-phase peptide synthesis (SPPS) (Merrifield 1963) based either on a Boc or Fmoc protection strategy. Although the first method might prove generally applicable in the future using unique transfer RNA(tRNA)/aminoacyl–tRNA-synthetase pairs (Chin et al. 2003), currently only few laboratories are equipped to apply the scheme successfully. On the other hand, modern improvement in peptide chemistry (Dawson and Kent 2000) allows SPPS of proteins containing as many as 166

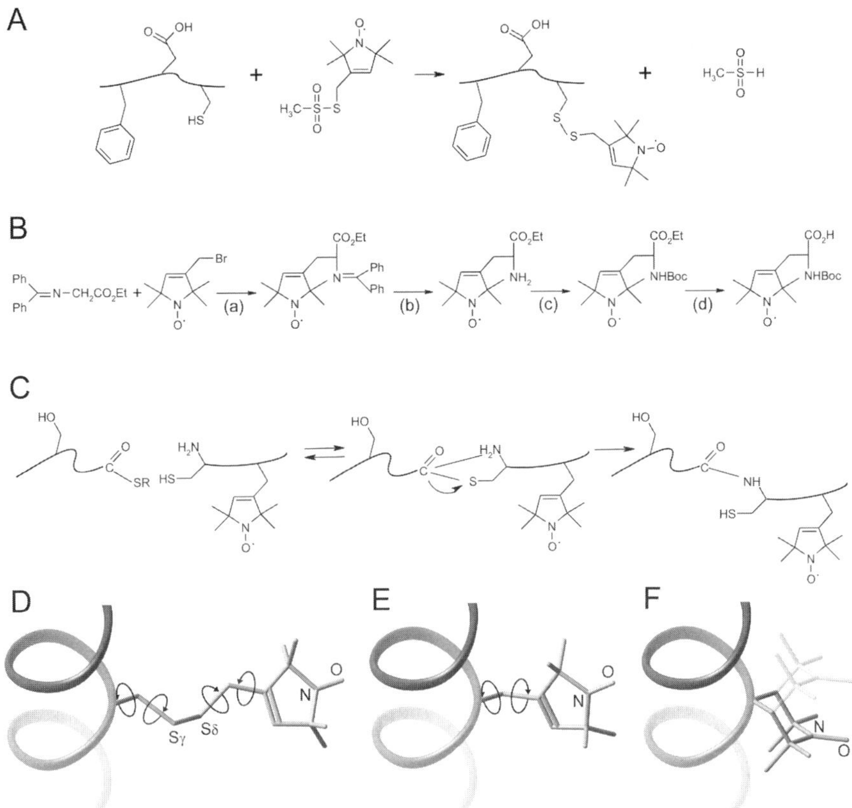

Figure 1. (**A**) Reaction of the methanethiosulfonate (MTS) spin label with a sulfhydryl group, generating the spin label side chain R1. (**B**) Reaction scheme for the synthesis of a novel paramagnetic Boc-protected amino acid (Balog et al. 2004). Reaction conditions are: (a) 10% aq. NaOH, CH_2Cl_2, $BuNHSO_4$; (b) 5% aq. H_2SO_4, EtOH; (c) Boc_2O, THF; (d) 10% aq. NaOH, EtOH. (**C**) Mechanism of native and expressed protein ligation (IPL, EPL). The chemical ligation occurs between a C-terminal thioester and a sulfydryl group of an N-terminal Cys. After rearrangement through an N→S acyl shift a native peptide bond is formed. The reaction can be performed also in the presence of other unprotected cysteine residues because of a first reversible reaction and a second irreversible step. As an example, the paramagnetic amino acid presented in reaction scheme (B) has been incorporated in the fragment containing the N-terminal Cys. (**D**) Conformational space of MTS spin label attached to an α-helix. All flexible bonds within the R1 side chain are highlighted. Rotation around the last bond (–C_ε–C_1–) shows the highest flexibility (Beier and Steinhoff 2006, submitted for publication). (**E**) Conformational space of the novel spin label presented in reaction scheme (B) characterized by a short link between the protein backbone and the piperidine-oxyl moiety. (**F**) Conformational space of the rigid spin label TOAC attached to an α-helix. The flip of the six-membered ring as unique possible degree of freedom is shown in light gray representation.

amino acids (Becker et al. 2003; Kochendoerfer et al. 2003). Furthermore, combining solid-state synthesis with recombinant techniques can provide the tool for introducing unnatural amino acids at sites of choice in large proteins (Hahn and Muir 2005).

2.1. Spin Labeling of Cysteines

A general scheme of the spin labeling procedure proven to yield good labeling efficiency is here briefly described and schematically depicted in Figure 1A. After cysteine substitution mutagenesis, the purified protein is usually stored in the presence of dithiothreitol (DTT) in order to prevent oxidation of the cysteine. Before spin labeling, the protein solution has to be dialyzed against the proper buffer to dilute the concentration of DTT. The protein (concentration adjusted to, e.g., 20 μM) is then usually incubated with 10-fold molar excess of spin label at 4°C overnight. The excess unbound spin label is afterward removed by dialysis, DEAE chromatography, Ni-NTA column (in the case the protein possesses a His-tag) or gel filtration. The spin-labeled protein is concentrated to 10–100 μM and filled into ESR quartz capillaries. At the X band (microwave frequency 9.5 GHz, magnetic field 0.34 T), the use of loop-gap or dielectric resonators provide the necessary sensitivity to yield continuous-wave ESR spectra with a good signal-to-noise ratio using 5 μL of sample within scan times of between 5 and 30 minutes. The spin-labeled cysteine/protein ratio is determined by double integration of the ESR spectra followed by comparison with standard solutions of MTS spin label and determination of the protein concentration. For interspin distance measurements (see §3.4) this ratio has to be close to one. The integrity and functionality of the spin-labeled protein variants have to be checked using independent methods prior to ESR analysis.

2.2. Incorporation of Spin-Labeled Amino Acids into Proteins

For ESR applications a variety of paramagnetic α-, β-, and γ-amino acids have been synthesized. The paramagnetic protein building blocks can be incorporated into peptides or proteins via a step-by-step synthesis by employing nonsense suppressor methodology or Boc/Fmoc-based synthesis either on the solid phase (SPPS) or in solution.

Among the paramagnetic α-amino acids, TOAC (4-amino-1-oxyl-2,2,6,6,-tetramethyl-piperidine-4-carboxylic acid) (Rassat and Rey 1967) is by far the most popular, and it is characterized by only one degree of freedom (the conformation of the six-membered ring, see Figure 1F). It has been repetitively used to study the secondary structure of small peptides in liquid solution (Anderson et al. 1999; Bolin et al. 1999; Hanson et al. 1996a,b), and it has also been successfully incorporated into the α-melanocyte stimulating hormone without loss of function (Barbosa et al. 1999). However, a recent solid-state ESR analysis (Elsaber et al. 2005) of a series of peptides de novo designed to form an α-helix doubly labeled at different sites with TOAC demonstrates that the peptides assume a conformation not corresponding to any of the common helical structures in proteins, highlighting one of the major problems arising by incorporating a rigid unnatural spin-labeled amino acid into peptides or proteins.

Several new N-Boc-protected paramagnetic amino acids with various side chains, having different polarity, orientation, and conformational restrictions, have been recently obtained using O′Donnell synthesis (Balog et al. 2003, 2004; Odonnell et al. 1989). The reaction scheme for the synthesis of a novel spin-label amino acid (Balog et al. 2004) (represented in its conformational space attached to an α-helix in Figure 1E) is presented as an example in Figure 1B.

The chemical synthesis and semisynthesis of proteins with unnatural spin-labeled amino acids relies on an ability to produce the constituent peptides, typically by SPPS. The synthesis of peptides consisting of up to 60 amino acids incorporating a variety of biophysical probes is now routinely performed (Ayers et al. 1999; Kochendoerfer et al. 2004); however, incorporation of spin-labeled amino acids is still challenging in terms of identification of optimal synthesis conditions chemically compatible with the delicate nature of the nitroxide moiety that can be readily protonated under acidic conditions. This problem could be overcome in the case of TOAC by employing Fmoc-based synthesis strategies either on the solid phase in combination with HF cleavage or in solution (Marchetto et al. 1993; Monaco et al. 1999). In the case of pyrroline-based nitroxide amino acids, new strategies for their incorporation into peptides during SPPS have recently been successfully conducted. Additionally, coupling of a pyrroline nitroxide spin label to the ε-amino group of a lysine after complete assembly of the peptide has also been performed yielding a paramagnetic stable compound (Becker et al. 2005).

Combining SPPS with recombinant techniques provides the tool for introducing paramagnetic amino acids potentially at every position in large proteins, and even membrane proteins. The EPL (expressed protein ligation, also named intein-mediated protein ligation IPL) technique can be used to semisynthesize proteins from recombinant and synthetic fragments, thus extending the size and complexity of the protein targets available to chemical engineering. The chemical ligation of two polypeptides requires that one of the fragments possess an N-terminal Cys and the other contain a C-terminal thioester moiety. After rearrangement through an S→N acyl shift, a native peptide bond is formed (Figure 1C). The reaction can be performed even in the presence of other unprotected cysteines. Using this strategy, a spin-labeled RBD protein has been recently synthesized, showing a stable paramagnetic center detected by ESR (Becker et al. 2005).

3. STRUCTURAL INFORMATION DERIVED FROM ESR SPECTRA ANALYSIS

3.1. Nitroxide Dynamics and Motional Freedom

The sensitivity of the ESR spectra to the reorientational motion of the nitroxide side chain attached at specific sites in the protein has been extensively reviewed

(Berliner 1976, 1979; Berliner and Reuben 1989), and the relationship between side chain mobility and protein structure has been explored in detail for T4 lysozyme (Mchaourab et al. 1996). The motion of a nitroxide side chain is characterized by three correlation times: the rotational correlation time for the entire protein, the effective correlation time due to the rotational isomerization about the bonds linking the nitroxide to the backbone, and the effective correlation time for the segmental motion of the backbone with respect to the average protein structure. The first term can be often neglected when considering membrane-bound proteins or soluble proteins with molecular weight above 200 kDa. In these cases the rotational correlation time exceeds 60 ns, which is beyond the sensitivity of the ESR time scale. Estimation of the rotational correlation time for the overall tumbling can be obtained using the Stokes-Einstein equation:

$$\tau_R = \frac{\eta V}{k_B T} = \frac{\eta}{k_B T} \cdot \frac{\bar{V} M}{N_A}, \tag{1}$$

where η is the viscosity of the solution, k_B is the Boltzmann constant, T is the absolute temperature, and V is the volume of the protein, which can be written as the product of $\bar{V}$, the partial specific volume, and M, the molecular mass, divided by Avogadro's number, N_A. For most proteins dissolved in pure water at 293 K, $\bar{V} \cong$ 0.73 $g^{-1}cm^3$ (Voet and Voet 2004). In the case of small-size proteins, the addition of sucrose (30% w/v) to the aqueous solution can be used to increase the viscosity by about a factor of three, thus minimizing the spectral effects due to overall tumbling of the molecule (Mchaourab et al. 1996; Qin et al. 2003).

The effective correlation time due to the reorientation of the spin label is expected to be a complex function of the spin label molecular structure and the primary, secondary, tertiary, and eventually quaternary structure of the protein under investigation, while the effective correlation time due to backbone motion is related to backbone flexibility, and thus to the secondary structure of the protein.

Generally the term "mobility" is used to define the effects on ESR spectral features due to the motional rate, amplitude, and anisotropy of the overall reorientational motion of the nitroxide label. Weak interaction between the nitroxide and the rest of the protein as found for helix surface sites or loop regions results in a high degree of mobility (Figure 2A, position 164). In this case the apparent hyperfine splitting and the line width of the ESR spectrum are as small as illustrated in Figure 2B. In turn, if the residual motion is restricted due to strong interaction of the nitroxide group with neighboring side chains or backbone atoms as found for tertiary contact or buried sites (Figure 2A, position 171), the apparent hyperfine splitting and the line width are increased (Figure 2B). In general, the resulting spectra cannot be simulated with a simple isotropic model of motion. Due to the interaction of the nitroxide with neighboring protein atoms, the motion has to be anisotropic, as was shown by molecular dynamics simulations (Steinhoff and Hubbell 1996; Steinhoff et al. 2000a). Additionally, a distribution of motional states can be concluded from those spectra that exhibit more than one component. In spite of the complex relation between nitroxide dynamics and spectral line width, the line width of the central line ($m_I = 0$), ΔH_0, has been found to be correlated with the structure of the binding site environment and has been used as a simple semi-empirical mobility parameter (Hubbell et al. 1996; Mchaourab et al. 1996).

A second semi-empirical parameter that has been found to correlate with the spectral line width is the spectral breadth, represented by the second moment, $\langle H^2 \rangle$:

$$\langle H^2 \rangle = \frac{\int (B - \langle H \rangle)^2 S(B) dB}{\int S(B) dB}, \tag{2}$$

which is defined based on the first moment $\langle H \rangle$ (the mean or geometrical center of the spectrum):

$$\langle H \rangle = \frac{\int B\, S(B) dB}{\int S(B) dB}, \tag{3}$$

where B is the magnetic field, and $S(B)$ is the absorption spectrum of the spin-labeled protein.

The numerical values of ΔH_0 and $\langle H^2 \rangle$ increase for any particular motional geometry as the frequency of nitroxide reorientational motion is reduced. In order that these experimental parameters are increasing functions of mobility, the inverse of both quantities will be used as "mobility" parameters. At X-band frequencies both parameters depend on the degree of averaging of the anisotropic **g** and hyperfine tensors. In the presence of multicomponent spectra mobility parameter ΔH_0^{-1} will be dominated by the most mobile component unless it is present in extremely small amounts; on the other hand, the second moment will be biased toward the less mobile component.

The plot of these mobility parameters versus residue number reveals secondary structure elements through the periodic variation of mobility as the nitroxide sequentially samples surface, tertiary, or buried sites. The assignment of α-helices, β-strands, or random structures from the data is straightforward.

A more quantitative interpretation of the experimental data in terms of dynamic mechanisms and local tertiary interaction requires simulation of the ESR spectra. Simulations can be performed on the basis of dynamic models developed by Freed and coworkers (Barnes et al. 1999; Borbat et al. 2001; Freed 1976) that show excellent agreement with the corresponding experimental spectra of solvated

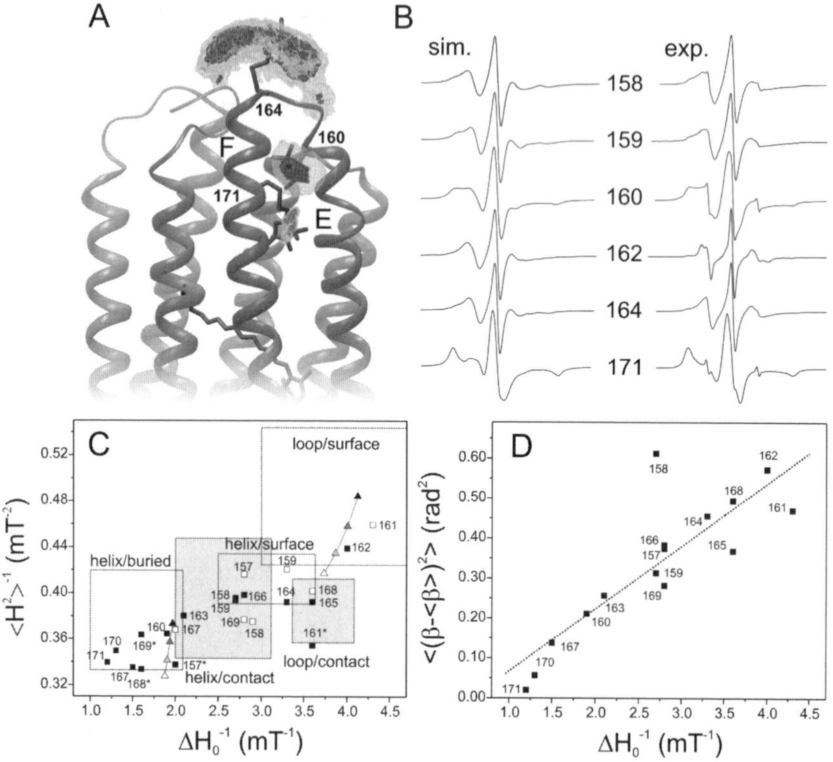

Figure 2. (**A**) Ribbon representation of bacteriorhodopsin x-ray structure from Essen et al. (1998). The retinal chromophore is depicted in stick representation. Three positions subjected to site-directed spin labeling are shown (A160R1, R164R1, F171R1). Side chain R1 (indicating the methanethiosulfonate spin label) is depicted in stick representation. The gray clouds define more than 90% of the complete accessible space of the nitroxide nitrogen as obtained from an MD simulation of 6 ns at 600 K, whereas the dark gray contours delimit areas with distinctly higher residence probability. Position 164 shows the largest cloud, indicating a higher degree of reorientation of the nitroxide ring during the MD simulation. The motion of the spin label attached at position 160 is slightly restricted due to interaction with the neighboring residues. On the other side the strongly restricted motion of the spin label attached at position 171, which is buried in the interior of the protein, is visualized by the small size of the cloud. (**B**) Comparison between simulated and experimental X-band ESR spectra for a set of positions in the BR (bacteriorhodopsin) loop region under investigation. The structural-based MD simulation approach used to simulate the spectra has been described in detail elsewhere (Beier and Steinhoff 2006, submitted for publication). The ESR spectra are depicted in the usual 1st derivative. The anisotropy of the hyperfine interaction is partially averaged due to the spin label motion for positions 158, 159, 162, and 164. The appearance of a prominent component characterized by higher line width and hyperfine extremes for position 160 is indicative for interaction of the spin label with neighboring residues restricting its reorientational motion. On the other side a powder-like spectrum results in the case of highly restricted motion (position 171), characterized by prominent positive and negative peaks at the extremes of the spectrum. (*C*) Plot of $\langle H^2\rangle^{-1}$ (inverse of the second moment) versus ΔH_0^{-1} (inverse of the central line width) extracted from the experimental spectra of the set of positions investigated in the BR loop region. The black squares indicate the values extracted from the ESR spectra of trimeric BR in purple membrane lipids. The values obtained for several positions for the monomeric solubilized BR

(in Triton X-100) are represented as open squares. The asterisks denote positions where quaternary interactions are present in the purple membrane samples and show distinct different spectra in the solubilized samples. The topological regions in the plot have been drowned according to (Isas et al. 2002; Mchaourab et al. 1996). The triangles connected by a dotted line represent mobility data for two positions characterized by high and low mobility. The effect of the variation of A_{zz} from 3.4 (apolar microenvironment) to 3.7 (polar microenvironment) mT is shown (black: 3.4, dark gray: 3.5, light gray: 3.6, white: 3.7 mT). The mobility parameters are extracted from two spectra simulated using an isotropic Brownian diffusion model (Freed 1976). In the simulations all the parameters except A_{zz} were kept constant. This allows to visualize independently the effect of a polarity change on the position of a point in the plot $\langle H^2 \rangle^{-1}$ vs ΔH_0^{-1}. **(D)** Plot of the mean square fluctuation amplitude (variance) of Euler angle β calculated from the trajectories of an MD simulation of 6 ns at 600 K (Beier, unpublished) versus ΔH_0^{-1} (inverse of the central line width) for the set of positions investigated in the BR loop region. The line width values are obtained from spectra from trimeric BR in purple membrane lipids and from solubilized monomeric BR (positions 158, 159, 162, and 164). The linear regression of the data is shown as a dotted line to highlight the correlation observed.

spin labels or spin-labeled lipids and proteins and give very valuable insight into the dynamics of these systems. On the other hand, simulations of ESR spectra can also be performed on the basis of molecular dynamics (MD) simulations. This approach facilitates the study of the influence of the structure in the vicinity of the spin label binding site on the ESR spectral line shape and allows one to verify, refine, or suggest structural models on the grounds of experimental ESR spectra of singly spin-labeled proteins (Steinhoff et al. 1996, 2000a).

The strategy for detecting and interpreting the experimental data is illustrated with an example of structural details obtained for the E–F loop region of bacteriorhodopsin (BR). BR is a light-driven pump that translocates protons across the cell membrane of archaeal and halobacteria. Figure 2A shows the structure determined by x-ray diffraction. It consists of seven transmembrane helices that bury the chromophore retinal (Essen et al. 1998). The structure of the E–F loop was found to be only poorly resolved in several of the reported x-ray diffraction experiments, most probably due to structural disorder or its high flexibility (Luecke et al. 1998; Luecke et al. 1999; Pebay-Peyroula et al. 1997). A set of positions from the cytoplasmic end of helix-E, through the E–F loop to the cytoplasmic edge of helix-F (positions 157–171), have been subjected to site-directed spin labeling and investigated in terms of mobility of the attached spin label to define the structural properties of the E–F loop (Pfeiffer et al. 1999).

Figure 2A shows three MTS spin labels bound at different positions in BR (160, 164, 171) and visualizes the motional behavior of the nitroxide group as determined from MD simulation (6 ns, 600 K). The gray clouds define more than 90% of the accessible space of the nitroxide nitrogen, whereas the black contours delimit areas with distinctly higher residence probability.

A selection of experimental spectra obtained for trimeric BR complexes in purple membranes is shown in Figure 2B together with the simulated spectra obtained from the molecular dynamics simulation approach based on the BR monomeric structure determined by Essen et al. (1998) (PDB code entry 1BRR). The agreement between experimental and simulated spectra shows how the MD simulation performed on protein sites of known x-ray structure can be a useful tool to study in detail the influence of the protein microenvironment in the vicinity of the spin label binding site on the ESR spectral line shape. A detailed analysis of the quality of the simulation with respect to the experimental ESR data can facilitate refinement of dynamic domains that are usually not properly defined in x-ray structures (as in the case of the E–F loop in BR) or the modeling of unknown structural domains when x-ray data are not available.

The correlation between the inverse of the second moment, $\langle H^2 \rangle^{-1}$, and the inverse of the central line width, ΔH_0^{-1}, extracted from the experimental data is presented in Figure 2C. Side chains from different topographical regions of the protein can be classified on the basis of the x-ray structures of T4 lysozyme and annexin 12 (Hubbell et al. 1996; Isas et al. 2002; Mchaourab et al. 1996). This general classification of the regions of the plot accommodating buried, surface-exposed, or loop residues is presented in Figure 2C. Since this coarse classification is mainly derived from α-helical structures, subdivisions might be necessary — for example, when investigating β-sheet structures, random coil, or coiled-coil structures.

It is interesting to observe that the ability to distinguish among different topological classes is better expressed in the line width rather than in the second moment. For example, a value for $\langle H^2 \rangle^{-1}$ of 0.38 mT^{-2} spans three groups in the plot of Figure 2C, namely the buried, helix/contact, and loop/contact, which are on the contrary well separated when considering the line width parameter.

In addition to the mobility correlation due to the dependence on the degree of averaging of the anisotropic hyperfine and **g** tensors, the line width and second moment are also sensitive to the value of the A_{zz} hyperfine tensor component, which is known to vary for spin labels sampling different water/protein/lipid interfaces (see §3.3). The effect of the variation of A_{zz} from 3.4 (apolar microenvironment) to 3.7 (polar microenvironment) mT is shown in Figure 2C for two sites characterized by high and low mobility. The mobility parameters are extracted from two spectra simulated using an isotropic Brownian diffusion model (Freed 1976). In these simulations all the parameters except A_{zz} were kept constant. This allows to visualize independently the effect of polarity changes on the position of a point in the plot of $\langle H^2 \rangle^{-1}$ versus ΔH_0^{-1}. It is evident that the higher the mobility, the more relevant are the variations induced by polarity changes on both mobility parameters. In fact, increasing the A_{zz} value from 3.4 to 3.7 mT decreases $\langle H^2 \rangle^{-1}$ by 14 or 12% and decreases ΔH_0^{-1} by 10 or 5% for the mobile or restricted sites, respectively. The second moment is found to be more sensitive than the line width to the polarity variations due to the fact that A_{zz} changes mainly influence the outer extremes of the ESR spectra.

The investigated positions of BR are found in different regions of the plot presented in Figure 2C, ranging from buried to loop/surface topology. Evidence from solubilization experiments suggests that the spectra obtained in the purple membrane samples for positions 157, 161, 168, and 169 are biased by the contribution of quaternary contacts between BR molecules in the trimeric complex (Beier and Steinhoff 2006, submitted for publication). Despite the valuable information obtained from the ESR spectra on the quaternary structure, the correlation between the mobility parameters and the structural properties of the monomeric BR loses its relevance. To overcome this problem, the mobility parameters obtained from detergent-solubilized monomeric BR are analyzed for the affected positions.

The highest mobility is observed for positions 161 and 162, which are located in the loop/surface region of the plot. The following positions, 164 and 165, characterized by slightly restricted motion, are located in the loop/contact region of the plot. The preceding (158, 159) and following (166, 167) positions are characterized by mobility values, indicating tertiary contacts with neighboring residues (helix/contact region of the plot). Interestingly, restricted motion dominates the spectra of positions 160 and 163, which are in proximity to the most mobile sites. This is strong evidence for a loop conformation with these two nitroxide side chains oriented toward the protein. The mobility data suggest an arrangement of residues 160 to 165 within a single turned loop (Pfeiffer et al. 1999).

Molecular dynamics studies performed on site-directed spin-labeled proteins provide the means to extract detailed information on the reorientation dynamics of the nitroxide ring and its interaction with the microenvironment (LaConte et al. 2002; Steinhoff and Karim 1993). From an analysis of the transformation matrix, defined in terms of three Eulerian angles α, β, γ, rotating the nitroxide coordinate system x, y, z (shown in Figure 4) into the laboratory coordinate system (X, Y, Z), the reorientation of the nitroxide ring with respect to the direction of the external B field (Z axis) can be inferred. The variance of angle β during a 6-ns MD simulation at 600 K for a set of positions in BR is plotted versus the experimentally determined ΔH_0^{-1} values in Figure 2D. Although the variance of β does not reflect the complexity of the anisotropic reorientational motion of the nitroxide, the clear correlation revealed in Figure 2D proves once more that the information on the local protein structure is coded in the ESR line width.

The set of ESR spectra simulated using the MD simulation approach based on the structural model obtained from x-ray diffraction data by Essen and coworkers (1998) nicely agrees with the experimental data. This result validates this x-ray structural model and provides evidence that the loop structure found under physiological conditions is retained in the crystal state.

3.2. Solvent Accessibility of the Attached Nitroxides

The motional analysis of elements of secondary and tertiary structure can be supplemented by measuring the accessibility of the nitroxide toward a paramagnetic probe selectively partitioning in different environments to define the topology of the natural side chain with respect to the protein/water/membrane boundaries. By definition, the "accessibility" of R1 in a macromolecule is proportional to the Heisenberg exchange frequency (W_{ex}) of the nitroxide with a paramagnetic exchange reagent present in solution. The paramagnetic exchange reagents employed in the measurements of accessibility may be metal ion complexes, molecular oxygen, organic radicals, and even a second nitroxide. Polar metal complexes (Ni(II) complexes or chromium oxalate (CrOx)) preferentially partition into the aqueous phase, whereas the product of concentration and diffusion coefficient of apolar oxygen has been shown to increase with distance from the membrane/water interface and exhibits a maximum value in the center of the membrane bilayer (Altenbach et al. 1994; Marsh et al. 2006). Due to the fact that the Heisenberg exchange mechanism leads to variations in the relaxation properties of the spin label attached at the investigated sites, each method capable to determine the variation of T_{1e} (spin–lattice relaxation time) or T_{2e} (spin–spin relaxation time) of a spin label bound to a protein upon addition of the paramagnetic exchange reagent is in principle suitable to unravel the topology of the nitroxide with respect to different interfaces. In general, if the longitudinal relaxation time of the reagent is smaller than the encounter complex lifetime ($T_{1R} < \tau_c$), the Heisenberg exchange leads to equal changes in T_{1e} and T_{2e}:

$$W_{ex} = \Delta\left(\frac{1}{T_{1e}}\right) = \Delta\left(\frac{1}{T_{2e}}\right). \quad (4)$$

For a bimolecular encounter between a small nitroxide moiety in solution (N) and an exchange reagent (R) the Heisenberg exchange frequency experienced by the nitroxide is

$$W_{ex} = k_{ex} C_R \ , \quad (5)$$

where k_{ex} is the exchange constant and C_R is the concentration of the reagent (R). In the strong exchange limit, when Heisenberg exchange is diffusion controlled, k_{ex} (expressed in L mol^{-1} s^{-1} unit) can be written as (Altenbach et al. 2005)

$$k_{ex} = P_{max} f k_D = P_{max} f\left(4\pi\ N_A (D_N + D_R) r_C \cdot 10^3\right), \quad (6)$$

where P_{max} is the maximum exchange efficiency ($P_{max} = 1$ if the conditions of strong exchange limit and $T_{1R} < \tau_c$ are fulfilled), f is the dimensionless steric factor, k_D is the diffusion-controlled rate constant (expressed in L mol^{-1} s^{-1}), N_A is Avogadro's number, D is the diffusion constant for the nitroxide (N) and reagent (R) (expressed in m^2 s^{-1}), and r_C is the collision radius (expressed in m, the sum of the effective radii of N and R).

In the case of a spin label bound to a protein, D_N becomes the diffusion constant of the protein, which is small compared to D_R and can be neglected in the equation. Effects induced by the local protein environment and the interactions with neighboring residues can reduce the effective collision frequency of the nitroxide toward the reagent. In this case W_{ex} can be expressed as

$$W_{ex} = \rho f 4\pi \frac{N_A}{1000} D_N r_C C_R \ , \quad (7)$$

where ρ is a proportionality factor, which takes into consideration the above mentioned effects due to the presence of the protein (Altenbach et al. 2005).

Among the possible methods to determine W_{ex}, those based on T_{1e} determination are the most common. Multiquantum ESR is a recently developed method that could be exploited to measure relative accessibilities in proteins (Mchaourab and

Hyde 1993). ELDOR (electron electron double resonance) can also be employed to measure absolute values of W_{ex} when the exchange reagent is a second nitroxide (Feix et al. 1984; Shin and Hubbell 1992). Saturation recovery ESR (SR-ESR) provides the means to measure T_{1e} directly as well as W_{ex} (Altenbach et al. 1989b; Nielsen et al. 2004; Percival and Hyde 1975; Yin et al. 1987), and it has been recently applied to the study of T4 lysozyme (Pyka et al. 2005). Its applicability might be of relevance to identify multiple states of the side chain that are usually visible in the cw ESR spectra and to characterize them in terms of relaxation properties. The cw power saturation technique has been traditionally the technique of choice to estimate relaxation rates through the calculation of accessibility parameter Π in the presence of paramagnetic reagents soluble in different environments. Determination of Π in combination with nitroxide scanning has been shown to be a valuable tool in obtaining structural information for soluble and membrane proteins (Cuello et al. 2004; Hubbell et al. 1998). In this case the cw ESR signal height is monitored as a function of incident microwave power. The observed rise to a maximum value and the following decrease of the signal vs. the square root of the incident power gives the so-called saturation curve, from which a parameter ($P_{1/2}$) related to the relaxation rates of the nitroxide can be determined (Figure 3B). In the following a brief description of the theoretical background of cw saturation method and its application is presented.

For a single homogeneous Lorentzian line the peak-to-peak amplitude of the first derivative absorption spectrum, Y', has the form (Poole 1983)

$$Y' \propto \frac{H_1}{\left(1+H_1^2\gamma^2 T_{1e}T_{2e}\right)^{3/2}} = \frac{\Lambda\sqrt{P}}{\left(1+\Lambda^2 P\gamma^2 T_{1e}T_{2e}\right)^{3/2}}, \tag{8}$$

where H_1 is the microwave magnetic field component in Gauss (which is proportional to the square root of the microwave power, P, with Λ depending on the properties of the resonator), γ is the gyromagnetic ratio of the electron, and T_{1e} and T_{2e} are the spin–lattice and spin–spin relaxation times, respectively. From the above equation it is clear that Y' is proportional to $\sqrt{P}$ for low microwave (MW) power and to $1/P$ for higher MW power. Saturation occurs when the amplitude of the signal deviates from the square root dependency on power, and it is due to the inability of the spin system to relax to equilibrium at a rate competitive with that at which power is absorbed. The relaxation time is measured by T_{1e}. An easily extractable parameter related to the relaxation quantity $T_{1e}T_{2e}$ is $P_{1/2}$, defined as the MW power at which the signal amplitude reaches half of the theoretical value reachable in the absence of saturation:

$$Y'(P_{1/2}) = \frac{\Lambda\sqrt{P_{1/2}}}{2}. \tag{9}$$

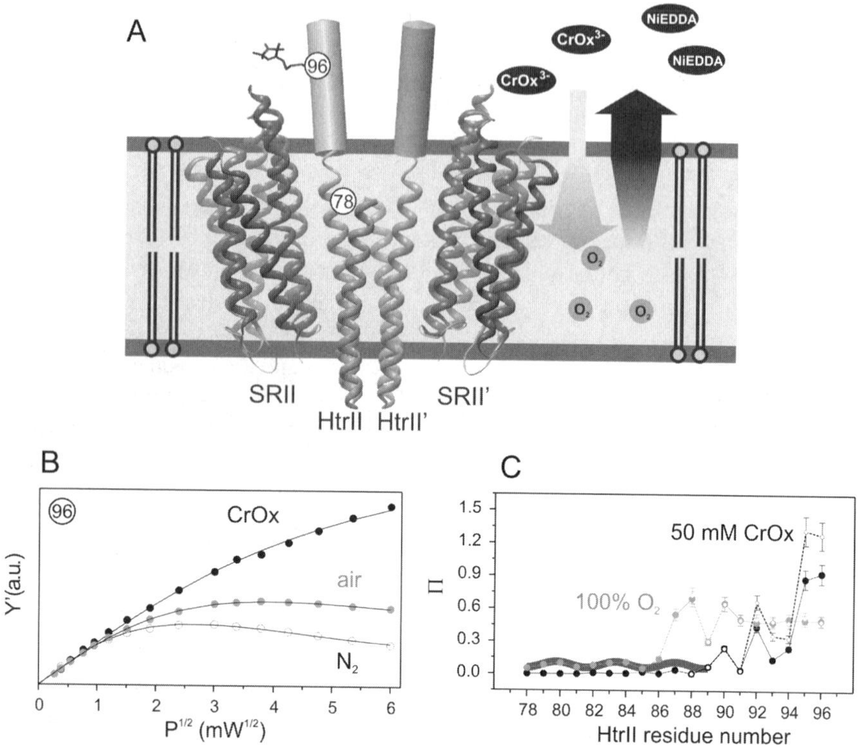

Figure 3. (**A**) Cartoon representation of NpSRII/NpHtrII 2:2 complex embedded in the lipid bilayer. The ribbon represents the x-ray structure (PDB code entry 1H2S) of the transmembrane domain of the complex; a schematic representation of the predicted α-helical AS-1 sequence protruding perpendicularly to the membrane plane into the cytoplasmic phase is also depicted (positions 83–101). For the sake of clarity, the AS-1 cylinder is proportionally stretched according to the length of the sequence. Two putative helix-breaking residues (G83 and G84) are represented as the gray lines connecting the helices TM2 to the AS-1 cylinders. The black and gray arrows indicate the gradient of concentration of the water-soluble reagents (CrOx and NiEDDA) and the lipid-soluble reagent (O_2), respectively. (**B**) Plot of the saturation curves (peak-to-peak intensity of the central ESR line (Y') vs. the square root of microwave power) obtained for position 96 in the transducer in NpSRII/NpHtrII complexes reconstituted in purple membrane lipids. The curves are obtained in the presence of N_2, air, and CrOx 50 mM. The $P_{1/2}$ values obtained by fitting of the data are 7.5, 16, and 95 mW, respectively (the estimated error is 10%). The accuracy in determination of the $P_{1/2}$ values decreases for very accessible sites. In these cases it is recommended to repeat the saturation experiment, decreasing the concentration of the reagent and subsequently scaling the Π values obtained to the usual concentration of the reagent. (**C**) Accessibility parameters Π_{Crox} (black circles) and Π_{O_2} (gray circles) calculated according to Eq. (15). The Π_{air} values obtained experimentally were multiplied by 5 to scale them to 100% O_2. The open circles in the same color code represent the values calculated according to Eq. (14) for positions 88–96. Deviations in the Π values due to changes in the line width are negligible in the presence of air through all the positions observed but become evident in the presence of CrOx 50 mM for the most mobile positions (92–96). The dark gray line highlights the 3.6 periodicity in Π_{O2} values.

The expression of $P_{1/2}$ for a homogeneous line can then be written as

$$P_{1/2} = \frac{\left(2^{2/3} - 1\right)}{\Lambda^2 \gamma^2 T_{1e} T_{2e}}. \quad (10)$$

Rewriting Eq. (8) accordingly, one obtains

$$Y' \propto \frac{\Lambda \sqrt{P}}{\left(1 + (2^{\varepsilon} - 1)P / P_{1/2}\right)^{\varepsilon}}, \quad (11)$$

with $\varepsilon = 3/2$. An analogous expression can be obtained in the inhomogeneous saturation limits with $\varepsilon = 1/2$. To fit the experimental data of Y' vs. $\sqrt{P}$, the following equation is used:

$$Y' = \frac{I \sqrt{P}}{\left(1 + (2^{\varepsilon} - 1)P / P_{1/2}\right)^{\varepsilon}}, \quad (12)$$

with I, ε, and $P_{1/2}$ as adjustable parameters (Figure 3B). In the presence of exchange paramagnetic reagents colliding with the nitroxide, variations in the relaxation times of the nitroxide are reflected in changes of $P_{1/2}$ values. Thus, calculating the $P_{1/2}$ values in the presence and absence of reagents leads to an estimation of the Heisenberg exchange frequency, which in turn is proportional to the topological properties of the nitroxide bound to a protein:

$$\Delta P_{1/2} = P_{1/2}^{R} - P_{1/2}^{0} \propto \frac{1}{T_{1e}^{R} T_{2e}^{R}} - \frac{1}{T_{1e}^{0} T_{2e}^{0}} \propto \frac{W_{ex}}{T_{2e}^{0}}. \quad (13)$$

The equation has been approximated due to the fact that being $T_{2e} << \tau_C$, the spin–spin relaxation time is not affected by the presence of the exchange reagent and $T_{2e}^{R} \simeq T_{2e}^{0}$. For different spin labels with comparable T_{2e}, $\Delta P_{1/2}$ reflects the relative collision frequency, or their relative "accessibility" toward the exchange reagent (Altenbach et al. 1989, 1994). To eliminate the dependency on T_{2e}, the $\Delta P_{1/2}$ values are usually divided by the ESR spectral line width ΔH_0, thus obtaining (Farahbakhsh et al. 1992)

$$\Delta P'_{1/2} = \frac{\Delta P_{1/2}}{\Delta H_0} \propto W_{ex}. \quad (14)$$

In the case a change in the ESR line width is observed upon addition of exchanging reagent ($T_{2e}^{R} \neq T_{2e}^{0}$), the approximation used in Eq. (13) is no longer valid. To overcome this problem, which usually occurs in systems with high nitroxide mobility at concentrations of CrOx > 25 mM or NiEDDA > 10 mM, one can decrease the concentration of the exchange reagent until the effect on the line width is negligible. An alternative approach consists in dividing $P_{1/2}{}^{R}$ and $P_{1/2}{}^{0}$ by $\Delta H_0{}^{R}$ and $\Delta H_0{}^{0}$, respectively, to eliminate the T_2 dependency of $\Delta P_{1/2}$:

$$\Delta P'_{1/2} = \frac{\Delta P^{R}_{1/2}}{\Delta H^{R}_{0}} - \frac{\Delta P^{0}_{1/2}}{\Delta H^{0}_{0}}. \tag{15}$$

To account for instrumental variability, normalization of this quantity with a standard reference (std) such as a dilute 2,2-diphenyl-1-picrylhydrazyl (DPPH) powder in KCl leads to dimensionless parameter Π:

$$\Pi = \Delta P'_{1/2} \frac{\Delta H(\mathrm{std})}{P_{1/2}(\mathrm{std})} \propto W_{\mathrm{ex}}. \tag{16}$$

The deviations in Π values calculated utilizing the two formulas for $\Delta P'_{1/2}$ (Eqs. (14) and (15)) are negligible for nitroxides characterized by restricted motional behavior but become significant for very mobile nitroxides (Figure 3C).

In general, the use of air (21% O_2) and NiEDDA (10 mM) as exchange reagents to monitor the accessibility of the bound nitroxide toward lipid and water phase, respectively, fulfills the $T_{2e}{}^{R} \cong T_{2e}{}^{0}$ criterion. The neutral NiEDDA is a more suitable exchange reagent with respect to CrOx (negatively charged) to monitor accessibility of lipid/water interfacial region. Contribution of Coulomb interactions might in fact lead to variations of the accessibility parameter in the presence of charged membrane surfaces or protein segments and even high salt concentrations (Nielsen et al. 2004).

Experimentally, to obtain a cw power saturation curve and extract the Π values, the sample is loaded into gas-permeable TPX capillaries. Loop-gap resonators are preferentially used that allow sample volumes less than 10 µL and provide a homogeneous B_1. The samples are first deoxygenated for 20 minutes by fluxing nitrogen around the sample capillary. The saturation curve is then detected in the presence of N_2. Subsequently, for oxygen accessibility experiments, nitrogen is replaced by air (21% oxygen) or 100% O_2. For CrOx or NiEDDA accessibility experiments a negligible volume of a concentrated stock solution is added to the sample to obtain the desired final concentration. The peak-to-peak amplitude of the central ESR line is measured at different incident microwave power levels and plotted vs. $\sqrt{P}$ (Figure 3B). To extract the $P_{1/2}$ values, homemade fitting programs can be used.

As an example, the accessibility analysis performed on a segment of the transducer protein (HtrII) associated with the photophobic receptor sensory rhodopsin II (SRII) from *Natronomonas pharaonis* (Np) reconstituted in purple membrane lipids is presented (Bordignon et al. 2005) (Figure 3). It has been shown that the NpSRII/NpHtrII (*Natronomonas pharaonis* sensory rhodopsin II/*Natronomonas pharaonis* halobacterial transducer II) complex is present in a 2:2 stoichiometry in membrane, and x-ray data are available for NpSRII in complex with the transmembrane region of NpHtrII up to residue 82. Following the transmembrane region, constituted by two helices (TM1 and TM2), the transducer contains a linker membrane adjacent region (starting with a conserved signal transduction domain called HAMP (Appleman et al. 2003)) and a long (~2 nm) cytoplasmic signaling domain. Light activation of NpSRII leads to a displacement of helix-F, which in turn triggers a rotation/screw-like motion of TM2 in NpHtrII. This conformational change is thought to be transmitted through the HAMP domain to the cytoplasmic signaling domain of the transducer. The architecture and function of the HAMP domain are still unknown. Based on primary sequence homologies, it is predicted to consist of two helical amphipathic sequences (AS-1 and AS-2) connected by a sequence of undefined secondary structure.

The accessibility data obtained for 20 spin-labeled samples with the spin labels located at the end of the transmembrane part of TM2 (positions 78–82, for which x-ray data are available) and in the first part of the AS-1 sequence of the HAMP domain (83–96, for which no structural information is available) are here described and discussed in terms of structural and topological information obtainable. A set of saturation curves obtained in the presence of N_2, air, and 50 mM CrOx for position 96 is presented as an example in Figure 3B. From a comparison of the saturation curves, it is apparent that the highest accessibility is toward CrOx (the maximum of the curve is not reached at the maximum power measured). The calculated Π values are depicted as a function of sequence position in Figure 3C. For better visualization, the Π_{air} values have been multiplied by 5 to scale them to 100% O_2, according to their proportionality to the exchange reagent concentration. The low Π values found for both CrOx and O_2 for side chains 78–86 indicate residues in a densely packed protein–protein interface. The low mobility values measured for the same region (data not shown) corroborate this observation. The clear periodicity of 3.6 residues revealed by the oxygen accessibility from positions 78 to 89 strongly supports the idea that the first part of the AS-1 sequence is the α-helical extension of TM2. The statistical relevance of the accessibility properties in a nitroxide scanning experiment is a useful tool for structural analysis (Hubbell et al. 1998). Interestingly, higher oxygen accessibility values are observed for the positions oriented toward NpSRII, indicating a dense packing of residues in the transducer–transducer interface.

For positions 87 to 94, the Π_{Crox} values gradually increase in line with the notion that the side chains become more exposed to the bulk water (a gradual increase in the spin label mobility is also observed). This finding provides strong evidence that region 87–94 of the transducer is protruding out of the protein–protein interface into the cytoplasmic phase. Accordingly, the Π_{O2} values are also increasing. The faster increase of O_2 than CrOx accessibility might be ascribed both to the smaller steric hindrance of the oxygen molecule, which can easily penetrate the interfacial regions, and to the presence of the charged membrane bilayer surface that might interfere with the CrOx diffusion. The periodicity observed in this region cannot be directly ascribed to a helical pattern, although it might be present. The reason for this failure to recognize possible secondary structure elements might be the fact that for interfacial regions the accessibility/mobility data lose their statistical significance (Cuello et al. 2004).

Positions 95 and 96 show Π_{Crox} values approaching those typical for water-exposed residues. The almost constant level of Π_{O2} (~0.6) is also in line with residues fully exposed to water. The high mobility observed suggests a very dynamic region lacking fixed structural features.

The results obtained from the accessibility analysis show the potentiality of this method, capable to validate the x-ray structural data available so far up to position 82 (closely packed α-helix) and defining the secondary and tertiary structure of the following region, where no structural information was previously available (83–89 α-helix; 86–94 first segment protruding in the water phase; 95–96 water-exposed residues).

3.3. Polarity of the Nitroxide Microenvironment

Characterizing certain regions of the proteins according to their polarity and proticity profiles or identifying hydrophobic barriers along ion channels in transmembrane proteins may provide the means to obtain structural and topological details of proteins and to understand specific biological processes on the molecular level. The observed dependencies of hyperfine component A_{zz} and **g** tensor component g_{xx} to the polarity and proticity of the spin label microenvironment make SDSL ESR a useful technique in this respect. Generally, a polar environment shifts tensor component A_{zz} to higher values, whereas tensor component g_{xx} is decreased. While the A_{zz} component can be easily obtained from cw X-band ESR spectra of spin-labeled proteins in frozen samples, the principal **g**-tensor components and their variation due to solute–solvent interactions can be determined with high accuracy using high-field ESR techniques due to the enhanced Zeeman resolution of rigid-limit spectra of disordered samples (Burghaus et al. 1992; Huber and Törring 1995; Prisner et al. 1995). In a sequence of a regular secondary structure with anisotropic solvation, the water density and hence the tensor component values are a periodic function of residue number and can be used similarly to the accessibility values for water soluble paramagnetic reagents (see §3.2) to extract structural and topological information on the protein under investigation. Moreover, the polarity of the protein environment surrounding the label may reveal detailed information on protein folding.

Theoretically, both g_{xx} and A_{zz} are expected to be linearly dependent on $\rho_\pi{}^O$, the π spin density on the oxygen atom of the nitroxide group. The hyperfine component A_{zz}, given by (Plato et al. 2002),

$$A_{zz} = Q_\pi^N \rho_\pi^N \tag{17}$$

is linearly dependent on $\rho_\pi{}^O$ on account of sum condition

$$\rho_\pi^O + \rho_\pi^N \cong 1, \tag{18}$$

which follows from the fact that the spin density is practically fully confined to the nitroxide group. For g_{xx} apart from a direct proportionality to $\rho_\pi{}^O$, there is an additional dependence on the specific properties of the oxygen lone-pair orbitals. This originates from an approximate expression for g_{xx} derived for organic π-radicals (Stone 1963):

$$g_{xx} \cong g_e + \frac{2\varsigma(\mathrm{O})\rho_\pi^{\mathrm{O}} c_{ny}^2}{\Delta E_{n\pi^*}}, \tag{19}$$

where $g_e = 2.0023$ is the free-electron g value; $\zeta(\mathrm{O})$ is the oxygen spin–orbit coupling parameter; ρ_π^{O} is the π spin density on the oxygen $2p_z$ atomic orbital; c_{ny} is the LCAO coefficient of the $2p_y$ atomic orbital contributing to the oxygen lone-pair MO; and $\Delta E_{n\pi^*} = E_{\pi^*} - E_n$ is the $n \rightarrow \pi^*$ excitation energy.

The polarity dependence of ρ_π^{O} follows from the variation of ρ_π^{N} due to the interaction between the permanent electric dipole induced by the charge displacement in the NO π bond and the various intermolecular fields in the NO bond region (which can be described by an average local field E, E_x is the electric field component along the NO bond) (Griffith et al. 1974; Plato et al. 2002):

$$\Delta\rho_\pi^{\mathrm{N}} = C_1 E_x \quad (C_1 > 0)\,. \tag{20}$$

Moreover, the lone-pair orbital energy E_n is known to be sensitive to the electrostatics, i.e., the polarity of the environment, and particularly sensitive to H bonding of the lone-pairs to water or to polar amino acid residues in the surrounding medium. Furthermore, H bonding can also affect the partial electron population c_{ny}^2 of the lone-pair orbitals. Thus any change in $\delta\Delta g_{xx} = g_{xx} - g_e$ can be ascribed to three contributions:

$$\frac{\delta\Delta g_{xx}}{\Delta g_{xx}} \cong -\frac{\rho_\pi^{\mathrm{N}}}{\rho_\pi^{\mathrm{O}}}\frac{\delta A_{zz}}{A_{zz}} - \frac{\delta\Delta E_{n\pi^*}}{\Delta E_{n\pi^*}} + \frac{\delta c_{ny}^2}{c_{ny}^2}\,. \tag{21}$$

Qualitatively, Eq. (21) correctly predicts the observed negative slope of the g_{xx} vs. A_{zz} plot (Möbius et al. 2005; Plato et al. 2002; Steinhoff et al. 2000b; Wegener et al. 2001b). The last two terms account for additional vertical displacement.

As an example, the 95-GHz high-field ESR analysis of the polarity profile along the putative proton channel of bacteriorhodopsin (BR) is presented and discussed (Figure 4). For the purpose of polarity analysis, W-band ESR spectra have to be recorded at low temperatures to avoid motional averaging of the anisotropic tensors. In the temperature regime below 200 K the reorientational correlation time of an otherwise unrestricted spin-label side chain exceeds 100 ns (Steinhoff et al. 1989), i.e., the nitroxide may be considered as immobilized on the ESR time scale. Librational motion of small amplitude still prevails and may lead to small deviations of the measured tensor values from their rigid limit values due to partial motional averaging. However, this effect is small (e.g., less than 2% for A_{zz} at 200 K (Steinhoff et al. 1989)) and is further minimized by decreasing the temperature for data collection to 120 K. Spectra of selected BR mutants spin labeled with MTS in

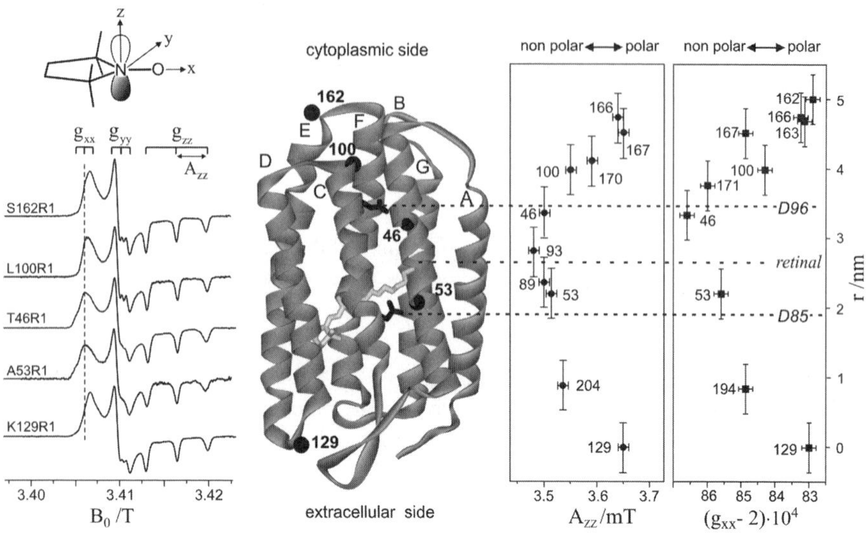

Figure 4. (A) Experimental high-field ESR spectra (1st derivative, $T = 200$ K, 95 GHz) for a set of spin-labeled BR variants (the C_α atoms of the chosen residues are indicated as a ball representation in (B)). The hyperfine splitting in the g_{xx} and g_{yy} regions is not completely resolved. The vertical line marks the g_{xx} position of T46R1. The variation of g_{xx} reflects the change of polarity in the microenvironment of the spin label when moving through the BR proton channel from the cytoplasmic to the extracellular side. The molecular structure of the "probe head" of the spin label MTS is depicted on top, together with the *x, y, z* coordinate system in use. (**B**) Structural model of BR. The chromophore retinal and residues D96 and D85 participating in proton transfer are indicated in stick representation. (**C**) The magnitude of tensor elements A_{zz} and g_{xx} of the spin labels are plotted as a function of the nitroxide location in the protein with respect to position 129 (see (B)). The A_{zz} values are determined by fitting of the X-band spectra detected at 170 K (Steinhoff et al. 1999); the g_{xx} values are extracted from the W-band spectra detected at 200 K (Steinhoff et al. 2000b). The vertical bars indicate the range of the possible location of the NO group (0.7 nm); the horizontal bars indicate 2σ errors for A_{zz} and g_{xx} of about $\pm 10^{-2}$ mT and $\pm 2 \times 10^{-5}$, respectively. The dotted horizontal lines indicate the positions of the proton donor D96, retinal, and D85. The values of g_{xx} and A_{zz} can be treated as a polarity index, and thus the plotted data reflect the hydrophobic barrier the proton has to overcome on its way through the BR proton channel.

the proton channel are shown in Figure 4A. The variation of g_{xx} with the nitroxide binding site is clearly revealed by the shift of the position of the low-field maximum. The variations of g_{xx} and A_{zz} with the nitroxide binding site can be measured with high precision. Plots of g_{xx} and A_{zz} vs. nitroxide position along the proton channel reveal a characteristic variation in the polarity of the nitroxide microenvironment (Figure 4C). Residue S162 is located in the E–F loop at the cytoplasmic surface (for the residue positions see Figure 4B), whereas residue K129 is positioned in the D–E loop on the extracellular surface. The high polarity in the environment of these residues ($A_{zz} \approx 3.7$ mT, $g_{xx} \approx 2.0083$) is clear evidence that the nitroxides are accessible to water, which is in agreement with the structural details.

The environmental polarity of the nitroxide at positions 100, 167, and 171 is significantly less and reaches its minimum at position 46 between proton donor D96 and the retinal. The plot directly reflects the hydrophobic barrier that the proton has to overcome on its way through the proton channel. The question if the structural changes occurring in BR upon illumination are a prerequisite for vectorial proton transport and how they modify this hydrophobic barrier remains to be elucidated. SDSL together with high-field ESR might provide the tools to unravel this issue.

Together with the polarity analysis discussed above, characterization of the microenvironment of the spin-label side chain in terms of protic and aprotic properties (Möbius et al. 2005; Steinhoff et al. 2000b) can be obtained from an analysis of the plot g_{xx} vs. A_{zz}. Furthermore, the polarity variation with bilayer depth across the lipid-exposed surface of a transmembrane α-helix might be used to locate a spin-labeled site with respect to the lipid–water interface, thus providing valuable information on the orientation of membrane-associated proteins or membrane anchors.

3.4. Inter-Nitroxide Distance Measurements

The simultaneous exchange of two native amino acids by cysteines followed by modification with MTS spin labels allows determination of inter-residual distances and thus provides a strategy for deducing proximity of selected secondary structural elements (for reviews see Berliner et al. 2000; Hustedt and Beth 1999). Together with mobility, accessibility, and polarity analysis, interspin distance determination provides a unique means for deducing structural details in proteins of unknown structure.

The spin–spin interaction between two spin labels attached to a protein is composed of static dipolar interaction, modulation of the dipolar interaction by the residual motion of the spin-label side chains, and exchange interaction. The static dipolar interaction leads to considerable broadening of the cw (continuous wave) ESR spectrum if the interspin distance is less than 2 nm. For unique orientations of the nitroxides relative to each other as found for spin labels introduced at buried sites in the rigid limit, a rigorous solution of the spin Hamiltonian of the system can be obtained. Spectra simulations yield the distance between the nitroxides and the Euler angles describing their relative orientation and that of the interspin vector relative to the magnetic field (Hustedt et al. 1997; Hustedt et al. 1999). However, the most frequently encountered case in SDSL consists of nitroxides that adopt a statistical distribution of distances and relative orientations. Values of the interspin distances can be determined from a detailed line shape analysis of spectra measured below 200 K (Rabenstein and Shin 1995; Steinhoff et al. 1991; Steinhoff et al. 1997) or in solutions of high viscosity (Altenbach et al. 2001) using spectra convolution or deconvolution techniques. The ESR spectra simulation program DIPFIT (Steinhoff et al. 1991, 1997), e.g., employs automated routines to determine best-fit parameters and considers a Gaussian distribution of interspin distances and variable contributions of singly spin-labeled protein.

Since the line width of the spectra is a steep function of the interspin distance (see Figure 5A, upper spectra), empirical or semi-empirical parameters as spectral amplitude ratios (Likhtenshtein 1976) or spectral second moment values are often sufficient to answer structural questions (Mchaourab et al. 2000; Radzwill et al. 2001). The lower limit for reliable distance determination using the above methods is given by the increasing influence of exchange interaction with decreasing interspin distances due to partial overlap of the nitrogen π-orbitals of the two interacting nitroxides. This effect makes quantification of interspin distances of less than 0.8 nm difficult (Closs et al. 1992; Fiori and Millhauser 1995).

A simplified quantitative interspin distance determination can be performed using the difference between the spectral second moments (see Eq. (2)) calculated for the doubly and the singly labeled protein:

$$\langle \Delta H^2 \rangle = \frac{\int (B-\langle H \rangle)^2 S_D(B)dB}{\int S_D(B)dB} - \frac{\int (B-\langle H \rangle)^2 S_S(B)dB}{\int S_S(B)dB}, \qquad (22)$$

where S_I is the absorption spectrum of the doubly spin-labeled protein sample ($I = D$) or the spectrum without spin–spin interaction determined from the superposition of the two corresponding singly spin-labeled protein samples ($I = S$). This difference does not depend on exchange interaction and is less sensitive to incomplete spin labeling in comparison to the analysis of spectral amplitude ratios. In the case of a radical pair, the relation between the difference of second moments and the interspin distance, r, yields:

$$r = 2.32\left\langle \Delta H^2 \cdot 10^8 \right\rangle^{-1/6} \text{nm}, \qquad (23)$$

with $\langle \Delta H^2 \rangle$ given in T^2. The upper limit for distance determination using the method of second moments is between 1.5 and 1.7 nm, depending strongly on the quality of the baseline.

Convolution and deconvolution techniques were shown to be applicable for inter-nitroxide distances up to 2 nm and provide valuable information on distance distributions. Recent advances in pulsed ESR techniques lead to protocols that expand this range from 2 to 8 nm (Borbat and Freed 1999; Pannier et al. 2000). Hence, cw ESR and pulsed ESR methods perfectly complement each other and provide interspin distances in the range between 0.8 and 8 nm.

The method of interspin distance determination has been successfully applied to a number of proteins, including rhodopsin (Altenbach et al. 1996; Cai et al. 1997), lactose permease (Voss et al. 1998), the KcsA potassium channel (Gross et al. 1999; Perozo et al. 1998), and alpha-crystallin (Koteiche et al. 1998, 1999). These applications demonstrate the capability and reliability of the SDSL method in resolving protein structure at the level of the backbone fold.

As an example for the elucidation of protein–protein interaction using dipole-dipole interaction, the structural model of the transmembrane region of the complex consisting of the photophobic receptor sensory rhodopsin II from *N. pharaonis* (NpSRII) and its transducer (NpHtrII) is presented (already introduced in §3.2). The model derived from ESR data (Wegener et al. 2001a) is shown in Figure 5B and compared to the x-ray structure subsequently released for the transmembrane part of the complex (Gordeliy et al. 2002) (depicted in light grayscale in Figure 5C). The x-ray data for the transducer, however, are still incomplete, being that only the transmembrane region was successfully crystallized. The crystal structure ends in fact at position 82 in transducer helix TM2, closer to the membrane–water interface, leaving the rest of the protein (which consists of more than 500 amino acids) completely undetermined. Primary sequence analogies suggest that the long cytoplasmic tail of the transducer shows similarities with the coiled-coil helical bundle of the chemoreceptors from *E. coli* (Kim et al. 1999); however, the structure of the membrane adjacent region of the protein, which is thought to be the signal transduction domain, is completely undetermined. To obtain information on this region of the transducer, a nitroxide scanning starting from the last position in TM2 for which x-ray data are available (L82) has been performed. The recent structural model derived from the ESR data for the region of the transducer beyond position 82 is also briefly presented (Figures 5C–D, dark grayscale) (Bordignon et al. 2005).

The structural analysis of the transmembrane part of the complex was based on interspin distances for 26 pairs of spin labels introduced into transmembrane helices TM1 or TM2 of the transducer and helices F or G of the receptor. Considerable spin–spin interactions observed for the singly labeled transducer with the nitroxide side chain located at position 78 (Figure 5A) or 82 suggested a dimeric arrangement of TM2 and TM2′. The strongest interaction corresponding to the closest distances (less than 1.2 nm) between NpSRII and NpHtrII are observed between positions 25^{TM1}–210^{G}, 81^{TM2}–210^{G}, as well as between 81^{TM2}–211^{G}, indicating that both TM1 and TM2 are close to helix G. Distances in the range from 1.2 to 1.6 are determined for the following five TM2-NpSRII pairs: 80^{TM2}–158^{F}, 80^{TM2}–157^{F}, 80^{TM2}–210^{G}, 80^{TM2}–211^{G}, and 81^{TM2}–158^{F}. According to these interspin distance values, TM2 has to be located between C-terminal helices F and G. The dimer formation of the NpSRII/NpHtrII complex in membranes is suggested to orient the respective positions 22 of TM1 and TM1′ toward each other. These findings and the distance values determined between the spin-labeled positions of TM1, 21, 23, 24, and 25 with 210^{G} or 158^{F} are in agreement with this arrangement. These results revealed a quaternary complex between two copies of NpHtrII and NpSRII each (2:2 complex), with apparent twofold symmetry. The model proposed based on the ESR data is presented in Figure 5B. The structure of NpSRII is based on a crystal structure of bacteriorhodopsin (Essen et al. 1998) validated for the investigated positions in NpSRII. Transducer TM1 and TM2 helices are modeled as canonical α-helices. The arrangement of NpSRII and NpHtrII proposed based on ESR distance determination (Wegener et al. 2001a) is in agreement with the one found in the crystal (Figure 5C, light grayscale) (Gordeliy et al. 2002; Klare et al. 2004a), proving the reliability of the SDSL ESR methodology to unravel structural details of proteins. Moreover, the agreement between the model and the x-ray structure proves that arrangement of the transmembrane part of the complex observed in

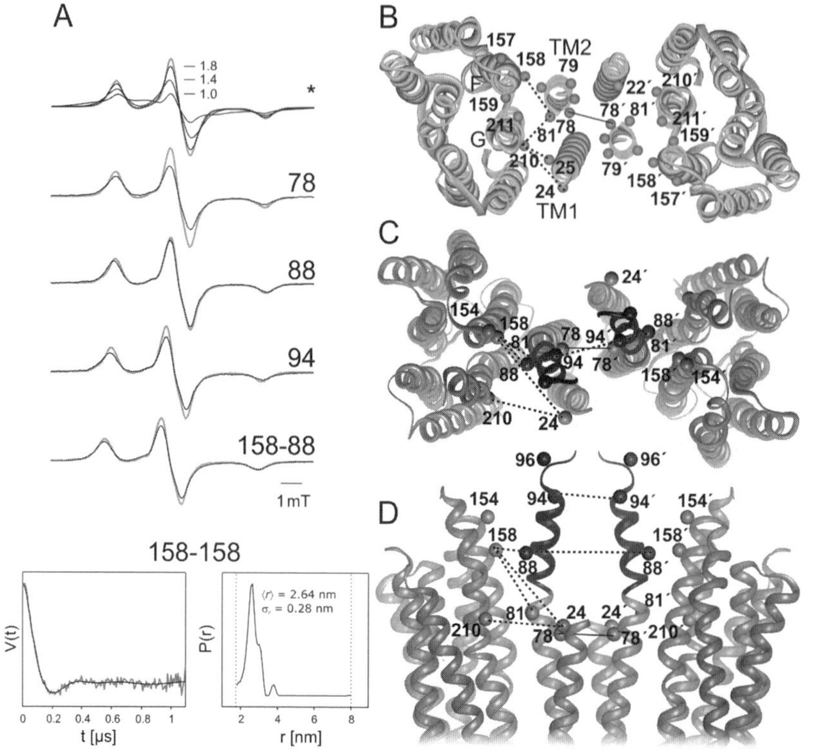

Figure 5. (**A**) The upper spectra marked with an asterisk are reference powder spectra simulated with the DIPFIT program. The parameters used are g_{xx} = 2.0086, g_{yy} = 2.0066, g_{zz} = 2.0026, A_{xx} = 0.52 mT, A_{yy} = 0.45 mT, and A_{zz}= 3.5 mT. The spectra were convoluted with a field-independent line shape function composed of a superposition of 44% Lorentzian and 56% Gaussian of 0.3 and 0.39 mT width, respectively. The fraction of the singly spin-labeled component has been chosen to be 30%. Clearly, the lower the amount of singly labeled protein present in the sample, the more pronounced the spectral changes due to dipolar broadening appear. To visualize the decrease of the central line amplitude in the spin-normalized spectra due to dipolar interaction, interspin distances of >2 (gray line), 1.8, 1.4 and 1.0 (black lines) nm are shown. The following spectra denoted with the number of the residue at which the spin label is attached are spin-normalized ESR spectra (T = 160 K) of singly and doubly spin-labeled mutants of the NpHtrII truncated at position 157 in complex with the NpSRII reconstituted in purple membrane lipids. The ESR spectra (black line) are compared with reference spectra without dipolar broadening (gray line). The strength of dipolar interaction and the resulting line broadening is visible by the decrease in amplitude of the central line with respect to the reference spectra. At the bottom, the DEER time domain data obtained after subtraction of the decay curve due to a 3D homogeneous distribution of spins for spin-labeled position 158 in the F-helix of NpSRII is depicted. To dilute the spin system under analysis, and focus mainly on the intra-dimer interactions, a high lipid/protein ratio has been chosen for this sample. Under this condition it is possible to suppress peaks otherwise present at distances >3.5 nm in the usual lipid/protein ratio used for cw experiments (40-fold excess). The distance distribution is obtained by Tikhonov regularization in the range from 2 to 8 nm with a regularization parameter of 100 (Jeschke 2002). The mean distance of 2.6 nm obtained between the two nitroxides attached at

residue S158 in the 2:2 complex agrees with the distance calculated from the crystal structure between the oxygen atoms of the respective serine residues (3 nm). These results once more confirm the presence of a 2:2 complex in membrane. Moreover, they show that DEER experiments can be performed to screen protein/lipid ratio conditions able to induce further aggregation of the complex. (**B**) Schematic model of the transmembrane region of the proposed 2:2 complex of NpSRII with its transducer NpHtrII (view from the cytoplasm) (Wegener et al. 2001a). The structural model of NpSRII is based on a BR-crystal structure (Essen et al. 1998), validated for side chains 157–160 (F-helix) and 210–213 (G-helix). Transmembrane helices TM1 and TM2 of NpHtrII are modeled as canonical α-helices. These components are oriented with respect to each other based on dipolar spin–spin interactions and derived distance constraints. C_β atoms from selected residues are represented as balls. The dotted lines connecting them highlight selected distances derived from ESR data. The continuous line denotes the strong interaction occurring between positions 78 in TM2 and TM2′. (**C**) Top view of the HAMP domain model up to position 96 (Bordignon et al. 2005). The NpSRII/NpHtrII transmembrane complex obtained from the x-ray data is depicted in light gray ribbon representation. The AS-1 sequence proposed by ESR data is colored in dark grayscale. C_β atoms from selected residues are represented as balls. The AS-1 segment from G83 to A94 is proposed to be the helical extension of TM2. To account for the high dynamical properties observed, the residues starting from A95 are represented in an extended conformation, although from the ESR data a highly dynamic helical segment cannot be excluded. A structural kink is proposed at residue G83 that leads to the loose coiled-coil motif suggested. (**D**) Side view of the model.

membrane reconstituted samples suspended in aqueous buffer is retained in the crystal.

Figure 5C shows the structural model suggested from analysis of the membrane adjacent region of the transducer in the 2:2 complex in membrane (the side view of the complex is presented in Figure 5D). The study was carried out determining interspin distances between selected residues in the cytoplasmic edge of helix-F (positions 154, 157, 158) and 8 selected residues in the transducer AS-1 sequence beyond position 82 (positions 88–95). Moreover, being the transducer present as a dimer in the complex under investigation, the distances determined between 17 singly spin-labeled transducer mutants gave information on the relative topology of the two transducers in the complex (Bordignon et al. 2005) (examples of spin normalized low-temperature spectra from selected mutants are presented in Figure 5A).

In the 88–94 region of the AS-1 sequence, interspin interactions are detectable starting from position 91 (1.9 nm) and become more evident for positions 93 and 94, where the fitting of the simulated spectra to the experimental ones yields distance values of 1.7–1.8 nm. It is worth noting that the clear periodicity of 3.6 residues revealed by the oxygen accessibility from positions 78 to 89 strongly supports the idea that the first part of the AS-1 sequence is the α-helical extension of TM2 (see §3.2, Figure 3). In the case of a straight continuation of the TM2 helix perpendicular to the membrane, positions 93 and 94 are expected to be located within the interface between the neighboring transducers, but should not show stronger interactions than those observed for positions 89 and 90. The fact that closer contacts are observed for positions 93 and 94 indicates a possible inclination of the membrane adjacent part of the transducers toward each other.

Despite the highly dynamic nature of region 95–101 (characterized by very high values of the mobility parameter in the 4–5 mT^{-1} range, not shown) the two transducers interact with each other in the complex showing distances in the 1.8 nm range. The lack of periodicity in the distances is to be expected due to the dynamic properties of this region.

To investigate the relative orientation of the AS-1 sequence of NpHtrII with respect to the cognate NpSRII, distance measurements between residues at the cytoplasmic end of helix-F (157^F, 158^F) and positions 88–94 of the transducer were performed. Additionally, proximity relations between 154^F, located at the cytoplasmic edge of helix-F in the receptor, and positions 91–95 in the transducer have been determined. It has to be pointed out that the system under investigation is a complex of four proteins and a double mutation involving the receptor and the transducer leads to several spin–spin interactions. In particular, dipolar interactions between both receptor–transducer and transducer–transducer positions can lead to the appearance of dipolar broadening in the low-temperature cw ESR spectra.

In order to focus only on the dipolar broadening due to the receptor–transducer interaction, the spectra of the doubly labeled variants must be compared to the spectra of the respective singly labeled transducer. The additional decrease in intensity in the spin-normalized spectra indicates the presence of considerable dipolar interaction between the spin-labeled transducer–receptor positions.

The closest distances to 157^F are observed for positions 88, 89, and 90. In agreement with these findings, the closest distances to 158^F are also observed for positions 88 (Figure 5A) and 89, whereas the double mutant NpSRII-S158R1/NpHtrII-S91R1 shows weak interactions. These observations suggest an α-helical arrangement of the transducer residues where the helical turn 88, 89, and 90 is located in the vicinity of residues K157 and S158 in helix-F. In the case of AS-1 being a straight prolongation of helix TM2, close proximity relations are to be expected between the following positions in the transducer (91–95) and 154^F, located one helix turn above 157^F and 158^F. On the contrary, the spectra obtained for these doubly labeled variants do not show any additional broadening with respect to the singly labeled transducers, thus indicating a more complex topology of the transducer dimer. Additionally, the very similar distances observed between position 158^F and positions 88–90 in the AS-1 sequence (1.4–1.6 nm) and 80–81 located two helical turns down in TM2 (1.4–1.5 nm) (Wegener et al. 2001a) are also a clear indication that the AS-1 sequence is not the straight prolongation of TM2; otherwise, one would have expected much closer interactions between S158R1 and the 88–90 region than those actually detected. Hence, the AS-1 region following residue S91 is proposed to be in closer contact with the neighboring transducer rather than to the receptor, corroborating the conclusions drawn from the distance values determined from the singly labeled transducer molecules.

The AS-1 sequence investigation in terms of interspin interaction allows the proposal of a topological model for this region in the 2:2 complex. Overall, these findings strongly suggest the inclination of the transducers in the membrane adjacent region to a direction opposite to the receptor, which lead to formation of a loose coiled–coil motif (Figure 5C–D).

4. DETECTION OF CONFORMATIONAL CHANGES

One attractive feature of site-directed spin labeling is the ability to time resolve changes in any of the parameters discussed above. Hence, changes in the protein secondary structure, protein tertiary fold, or domain movements can be followed with up to 100 μs resolution with conventional ESR instrumentation and detection scheme (field modulation). Important examples found in the literature include the detection of rigid-body helix motion in both rhodopsin and bacteriorhodopsin (Farahbakhsh et al. 1993; Farrens et al. 1996; Steinhoff et al. 1994; Thorgeirsson et al. 1997), domain movements in T4 lysozyme (Hubbell et al. 1996), structural reorganization in colicin E1 upon membrane binding (Shin et al. 1993), and conformational changes during signal transfer from sensory rhodopsin NpSRII to the transducer NpHtrII (Wegener et al. 2001a). The strategy for triggering, detecting, and interpreting conformational changes is illustrated with two examples. The first shows detection of a conformational change occurring in rhodopsin upon photoisomerization of the retinal. The second presents the time-resolved detection of changes in mobility and interspin distances in NpSRII and NpHtrII upon light excitation and subsequent isomerization of the retinal, revealing the magnitude and entity of helix motion and signal transfer process.

Rhodopsin, shown in the resting state (Figure 6A, dark gray) is the photoreceptor protein of vertebrate retina. Photoisomerization of the bound retinal chromophore triggers a rigid-body helix tilt of a single helix, as demonstrated by a 0.6-nm increase in the distance between nitroxide spin labels attached to sites 63 and 241 (Figure 6A, light gray). The interspin distance was measured in the resting and stabilized activated states at cryogenic temperature using DEER (Jeschke 2002). The dipolar evolution functions and the corresponding distance distributions are given for the resting and light-activated states (Figure 6A, right panel). The interspin distance calculated from the crystal structure of the resting state is shown to be in agreement with the ESR determined distance (Altenbach and Hubbel, unpublished).

If the protein conformations of interest are of transient nature, the determination of the properties of the respective trapped intermediate states can be accompanied by direct detection of the kinetic difference spectra under physiological conditions. This is illustrated in the case of the NpSRII/NpHtrII complex reconstituted in membrane (Figure 6B). Difference spectra between the photoactivated state and the dark-adapted state can be determined during a single *B*-field scan by subtraction of

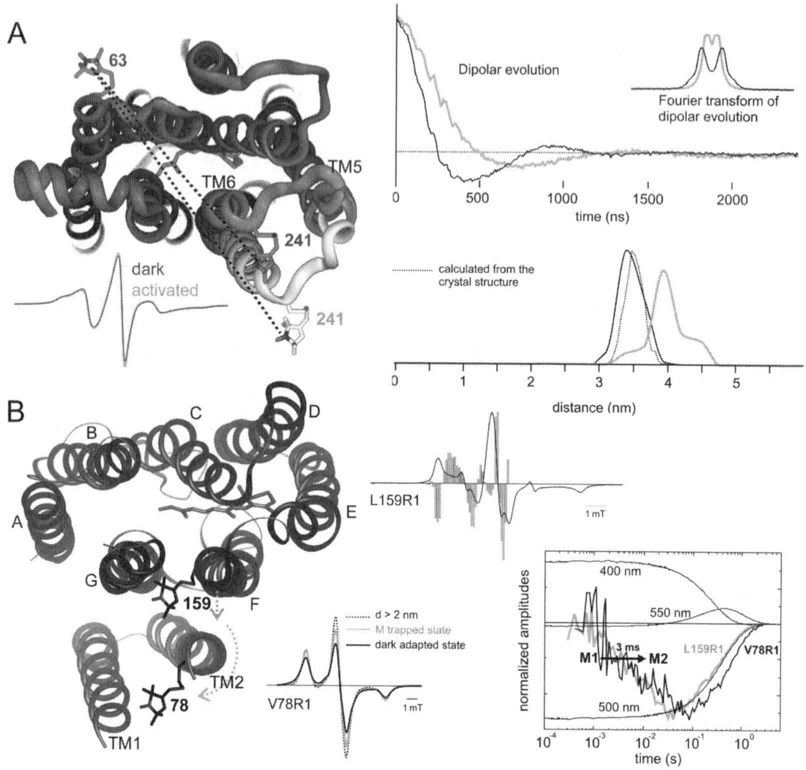

Figure 6. (**A**) Resting (dark gray) and activated (light gray) states of rhodopsin. The two spin labels attached to positions 63 and 241 are shown in stick representation. In the right panel the dipolar evolution functions and the corresponding distances are presented (dark gray: resting state; light gray: activated state). The distance distribution calculated from the crystal structure is also presented as a dotted line, and it is shown to be in agreement with the DEER data. The increase in the distance between positions 63 and 241 by 0.6 nm is clear evidence for a dramatic rearrangement of helix TM6 after excitation. Based on ESR data, a rigid-body helix tilt of TM6 is proposed in the activated state of rhodopsin (Altenbach and Hubbell, unpublished). (**B**) View onto the cytoplasmic side of the sensory rhodopsin/transducer complex. For sake of clarity, only half of the 2:2 complex is presented, consisting of one NpSRII and one NpHtrII moieties. The two MTS spin labels attached to positions 159^{F} and 78^{TM2} in helix-F of the receptor and in the TM2 helix of the transducer, respectively, are shown in stick representation. The upper right panel shows the cw ESR spectrum of the spin label attached at position 159^{F} in the complex reconstituted in purple membrane lipids. The spin label at position 159^{F} serves as reporter group for transient structural changes. The light-minus-dark difference spectrum (stick representation) is calculated from the transient ESR traces obtained for each *B* field position investigated (to allow comparison with the cw spectrum, the difference spectrum is multiplied by 100). Light excitation is achieved via an excimer-pumped dye laser tuned at 500 nm and the ESR transients are recorded for 4 seconds after light excitation, with a pre-trigger of 100 ms. To obtain a good signal-to-noise ratio a lock in time constant of 5 ms is used and an average of 500–1000 ESR transients is performed for each *B* field value with *B* field intervals of 0.1 mT. The pre-trigger signal intensity is used to obtain the cw spectrum presented. The height of each stick of the difference spectrum represents the amplitude of the ESR transient, obtained by fitting with a single exponential decay. The difference spectrum thus obtained is found in

agreement with previous published difference spectra utilizing the methods of subtraction of signals averaged within two succeeding sampling intervals starting with the light flash of a lamp (Wegener et al. 2000, 2001a). The right bottom panel depicts ESR transients (noisy lines) and the corresponding optical traces (continuous lines) recorded at 500 (ground state), 550 (O state), and 400 nm (M state) of NpSRII/NpHtrII reconstituted in purple membrane (Klare et al. 2004b). The ESR transients are recorded at fixed *B* field values where a maximum of intensity is observed for NpSRII/NpHtrII-V78R1 (black trace), and for NpSRII-L159R1/NpHtrII (gray trace) after activation with light. The time constant of the lock in is 100 μs, to allow detection of the rise time of the structural changes monitored by ESR. The structural rearrangement clearly occurs synchronously in the receptor and in the transducer within the M1→M2 transition (spectroscopically silent) and it is characterized by a 3 ms time constant. Return to the dark-adapted structure follows recovery of the ground state for position 159, but it is delayed for the transducer. The transient increase of mobility (inferred by the difference spectrum) of the spin label side chain at position 159 suggests an outward displacement of helix-F of the receptor. The ESR signal changes observed for position 78 in the complex uncover a transient distance increase between positions 78 of the two transmembrane helices of the transducer, TM2 and TM2′, as inferred from analysis of the low-temperature spectra of the M-trapped state shown on the left (Wegener et al. 2001a).

the signals averaged within two succeeding sampling intervals starting with the light flash. The lengths of the sampling intervals are set to values longer than the recovery time of the initial state. Following this scheme, the signal recorded during the first sampling interval is an average of the signals of all photo-intermediates including the initial state. The signal determined during the second sampling interval represents the recovered pure initial state. A second more quantitative approach considers the photocycle time constants obtained via optical transient spectroscopy and fit the ESR data utilizing the same set of time constants to allow direct comparison between the optical cycle and the structural changes occurring. The difference spectra reveal the entity and sign of the mobility changes occurring, as for example, in the case of 159^F, where a transient increase in the mobility is observed upon activation of NpSRII by a light flash (Figure 6B, upper right panel). The conformational changes occurring in NpSRII upon light excitation observed in the cytoplasmic edge of helix-F are transferred also to the transmembrane part of helix TM2 in the transducer, as exemplified by the ESR transient traces obtained for positions 159^F and 78^{TM2} (Figure 6B, lower right panel). Comparison of the ESR with the optical traces recorded at different wavelength identifies M as the signaling state where the observed conformational changes occur. With the reformation of the ground state, the reactions characteristic for the receptor seems to be decoupled from those of the transducer. The resetting movement of helix-F into the original position seems to precede the recovery of the TM2 position.

To further refine the structural changes inferred by the kinetic data, low-temperature ESR spectra were recorded for several receptor/transducer and transducer/transducer positions in the dark-adapted and M-trapped states revealing distance changes occurring upon light activation (Figure 6B). The data obtained suggest a model of signal transduction where the light-induced F-helix displacement in NpSRII leads to a rotation/screw-like motion of helix TM2 in the transducer. The recently published x-ray structure of the late M state in NpSRII/NpHtrII indeed shows a rotation of TM2 in the signaling state, according to the ESR interpretation (Moukhametzianov et al. 2006).

5. ACKNOWLEDGMENTS

We gratefully acknowledge the support of the Deutsche Forschungsgemeinschaft in the frame of the Schwerpunktprogramm SPP1051 and the Sonderforschungs-bereich SFB 431. We thank W. L. Hubbell for providing us the data shown in Figure 6A.

6. REFERENCES

Altenbach C, Flitsch SL, Khorana HG, Hubbell WL. 1989a. Structural studies on transmembrane proteins, 2: spin labeling of bacteriorhodopsin mutants at unique cysteines. *Biochemistry* **28**:7806–7812.

Altenbach C, Froncisz W, Hyde JS, Hubbell WL. 1989b. Conformation of spin-labeled melittin at membrane surfaces investigated by pulse saturation recovery and continuous wave power saturation electron-paramagnetic resonance. *Biophys J* **56**:1183–1191.

Altenbach C, Greenhalgh DA, Khorana HG, Hubbell WL. 1994. A collision gradient method to determine the immersion depth of nitroxides in lipid bilayers: application to spin-labeled mutants of bacteriorhodopsin. *Proc Natl Acad Sci USA* **91**:1667–1671.

Altenbach C, Yang K, Farrens DL, Farahbakhsh ZT, Khorana HG, Hubbell WL. 1996. Structural features and light-dependent changes in the cytoplasmic interhelical E-F loop region of rhodopsin: a site-directed spin-labeling study. *Biochemistry* **35**:12470–12478.

Altenbach C, Oh KJ, Trabanino RJ, Hideg K, Hubbell WL. 2001. Estimation of inter-residue distances in spin labeled proteins at physiological temperatures: experimental strategies and practical limitations. *Biochemistry* **40**:15471–15482.

Altenbach C, Froncisz W, Hemker R, Mchaourab H, Hubbell WL. 2005. Accessibility of nitroxide side chains: absolute Heisenberg exchange rates from power saturation EPR. *Biophys J* **89**:2103–2112.

Anderson DJ, Hanson P, McNulty J, Millhauser G, Monaco V, Formaggio F, Crisma M, Toniolo C. 1999. Solution structures of TOAC-labeled trichogin GA IV peptides from allowed (g approximate to 2) and half-field electron spin resonance. *J Am Chem Soc* **121**:6919–6927.

Appleman JA, Chen LL, Stewart V. 2003. Probing conservation of HAMP linker structure and signal transduction mechanism through analysis of hybrid sensor kinases. *J Bacteriol* **185**:4872–4882.

Ayers B, Blaschke UK, Camarero JA, Cotton GJ, Holford M, Muir TW. 1999. Introduction of unnatural amino acids into proteins using expressed protein ligation. *Biopolymers* **51**:343–354.

Balog M, Kalai T, Jeko J, Berente Z, Steinhoff HJ, Engelhard M, Hideg K. 2003. Synthesis of new conformationally rigid paramagnetic alpha-amino acids. *Tetrahedron Lett* **44**:9213–9217.

Balog MR, Kalai TK, Jeko J, Steinhoff HJ, Engelhard M, Hideg K. 2004. Synthesis of new 2,2,5,5-tetramethyl-2,5-dihydro-1H-pyrrol-1-yloxyl radicals and 2-substituted-2,5,5-trimethylpyrrolidin-1-yloxyl radicals based alpha-amino acids. *Synlett* **14**:2591–2593.

Barbosa SR, Cilli EM, Lamy-Freund MT, Castrucci AML, Nakaie CR. 1999. First synthesis of a fully active spin-labeled peptide hormone. *FEBS Lett* **446**:45–48.

Barnes JP, Liang ZC, Mchaourab HS, Freed JH, Hubbell WL. 1999. A multifrequency electron spin resonance study of T4 lysozyme dynamics. *Biophys J* **76**:3298–3306.

Becker CFW, Hunter CL, Seidel R, Kent SBH, Goody RS, Engelhard M. 2003. Total chemical synthesis of a functional interacting protein pair: the protooncogene H-Ras and the Ras-binding domain of its effector c-Raf1. *Proc Natl Acad Sci USA* **100**:5075–5080.

Becker CFW, Lausecker K, Balog M, Kalai T, Hideg K, Steinhoff HJ, Engelhard M. 2005. Incorporation of spin-labelled amino acids into proteins. *Magn Reson Chem* **43**:S34–S39.

Berliner LJ, ed. 1976. *Spin labeling: theory and applications*. New York: Academic Press.

Berliner LJ, ed. 1979. *Spin labeling II: theory and applications*. New York: Academic Press.

Berliner LJ, Reuben J, eds. 1989. *Spin labeling theory and applications*, Vol. 8: *Biological magnetic resonance*. New York: Plenum Press.

Berliner LJ, Grunwald J, Hankovszky HO, Hideg K. 1982. A novel reversible thiol-specific spin label: papain active-site labeling and inhibition. *Anal Biochem* **119**:450–455.

Berliner LJ, Eaton SS, Eaton GR, eds. 2000. Distance measurements in biological systems by EPR: biological magnetic resonance. New York: Kluwer Academic/Plenum.

Bolin KA, Anderson DJ, Trulson JA, Thompson DA, Wilken J, Kent SBH, Gantz I, Millhauser GL. 1999. NMR structure of a minimized human agouti related protein prepared by total chemical synthesis. *FEBS Lett* **451**:125–131.

Borbat PP, Freed JH. 1999. Multiple-quantum ESR and distance measurements. *Chem Phys Lett* **313**:145–154.

Borbat PP, Costa-Filho AJ, Earle KA, Moscicki JK, Freed JH. 2001. Electron spin resonance in studies of membranes and proteins. *Science* **291**:266–269.

Bordignon E, Klare JP, Doebber M, Wegener AA, Martell S, Engelhard M, Steinhoff HJ. 2005. Structural analysis of a HAMP domain: the linker region of the phototransducer in complex with sensory rhodopsin II. *J Biol Chem* **280**:38767–38775.

Burghaus O, Rohrer M, Gotzinger T, Plato M, Möbius K. 1992. A novel high-field high-frequency EPR and ENDOR spectrometer operating at 3 mm wavelength. *Meas Sci Technol* **3**:765–774.

Cai KW, Langen R, Hubbell WL, Khorana HG. 1997. Structure and function in rhodopsin: topology of the C-terminal polypeptide chain in relation to the cytoplasmic loops. *Proc Natl Acad Sci USA* **94**:14267–14272.

Chin JW, Cropp TA, Anderson JC, Mukherji M, Zhang ZW, Schultz PG. 2003. An expanded eukaryotic genetic code. *Science* **301**:964–967.

Closs GL, Forbes MDE, Piotrowiak P. 1992. Spin and reaction dynamics in flexible polymethylene biradicals as studied by EPR, NMR, and optical spectroscopy and magnetic-field effects: measurements and mechanisms of scalar electron–spin spin coupling. *J Am Chem Soc* **114**:3285–3294.

Cornish VW, Benson DR, Altenbach CA, Hideg K, Hubbell WL, Schultz PG. 1994. Site-specific incorporation of biophysical probes into proteins. *Proc Natl Acad Sci USA* **91**:2910–2914.

Cuello LG, Cortes DM, Perozo E. 2004. Molecular architecture of the KvAP voltage-dependent K+ channel in a lipid bilayer. *Science* **306**:491–495.

Dawson PE, Kent SBH. 2000. Synthesis of native proteins by chemical ligation. *Annu Rev Biochem* **69**:923–960.

Elsaber C, Monien B, Haehnel W, Bittl R. 2005. Orientation of spin labels in de novo peptides. *Magn Reson Chem* **43**:S26–S33.

Essen LO, Siegert R, Lehmann WD, Oesterhelt D. 1998. Lipid patches in membrane protein oligomers: crystal structure of the bacteriorhodopsin-lipid complex. *Proc Natl Acad Sci USA* **95**:11673–11678.

Farahbakhsh ZT, Altenbach C, Hubbell WL. 1992. Spin labeled cysteines as sensors for protein lipid interaction and conformation in rhodopsin. *Photochem Photobiol* **56**:1019–1033.

Farahbakhsh ZT, Hideg K, Hubbell WL. 1993. Photoactivated conformational changes in rhodopsin: a time-resolved spin-label study. *Science* **262**:1416–1419.

Farrens DL, Altenbach C, Yang K, Hubbell WL, Khorana HG. 1996. Requirement of rigid-body motion of transmembrane helices for light activation of rhodopsin. *Science* **274**:768–770.

Feix JB, Klug CS. 1998. Site directed spin labeling of membrane proteins and peptide-membrane interactions. In *Spin labeling: the next millennium*, Vol. 14, pp. 251–281. Ed LJ Berliner. New York: Plenum.

Feix JB, Popp CA, Venkataramu SD, Beth AH, Park JH, Hyde JS. 1984. An electron–electron double-resonance study of interactions between [N-14]stearic and [N-15]stearic acid spin-label pairs: lateral diffusion and vertical fluctuations in dimyristoylphosphatidylcholine. *Biochemistry* **23**:2293–2299.

Fiori WR, Millhauser GL. 1995. Exploring the peptide 3(10)-helix-reversible-arrow-alpha-helix equilibrium with double-label electron-spin-resonance. *Biopolymers* **37**:243–250.

Freed JH. 1976. Theory of slow tumbing ESR spectra for nitroxides. In *Spin labeling: theory and applications*, pp. 53–132. Ed LJ Berliner. New York: Academic Press.

Gordeliy VI, Labahn J, Moukhametzianov R, Efremov R, Granzin J, Schlesinger R, Büldt G, Savopol T, Scheidig AJ, Klare JP, Engelhard M. 2002. Molecular basis of transmembrane signalling by sensory rhodopsin II-transducer complex. *Nature* **419**:484–487.

Griffith OH, Dehlinger PJ, Van SP. 1974. Shape of hydrophobic barrier of phospholipid bilayers (evidence for water penetration in biological-membranes). *J Membr Biol* **15**:159–192.

Gross A, Columbus L, Hideg K, Altenbach C, Hubbell WL. 1999. Structure of the KcsA potassium channel from Streptomyces lividans: a site-directed spin labeling study of the second transmembrane segment. *Biochemistry* **38**:10324–10335.

Hahn ME, Muir TW. 2005. Manipulating proteins with chemistry: a cross-section of chemical biology. *Trends Biochem Sci* **30**:26–34.

Hanson P, Martinez G, Millhauser G, Formaggio F, Crisma M, Toniolo C, Vita C. 1996a. Distinguishing helix conformations in alanine-rich peptides using the unnatural amino acid TOAC and electron spin resonance. *J Am Chem Soc* **118**:271–272.

Hanson P, Millhauser G, Formaggio F, Crisma M, Toniolo C. 1996b. ESR characterization of hexameric, helical peptides using double TOAC spin labeling. *J Am Chem Soc* **118**:7618–7625.

Hubbell WL, Mchaourab HS, Altenbach C, Lietzow MA. 1996. Watching proteins move using site-directed spin labeling. *Structure* **4**:779–783.

Hubbell WL, Gross A, Langen R, Lietzow MA. 1998. Recent advances in site-directed spin labeling of proteins. *Curr Opin Struct Biol* **8**:649–656.

Hubbell WL, Cafiso DS, Altenbach C. 2000. Identifying conformational changes with site-directed spin labeling. *Nat Struct Biol* **7**:735–739.

Huber M, Törring JT. 1995. High-field EPR on the primary electron-donor cation-radical in single-crystals of heterodimer mutant reaction centers of photosynthetic bacteria: first characterization of the g-tensor. *Chem Phys* **194**:379–385.

Hustedt EJ, Beth AH. 1999. Nitroxide spin–spin interactions: applications to protein structure and dynamics. *Ann Rev Biophys Biomol Struct* **28**:129–153.

Hustedt EJ, Smirnov AI, Laub CF, Cobb CE, Beth AH. 1997. Molecular distances from dipolar coupled spin-labels: the global analysis of multifrequency continuous wave electron paramagnetic resonance data. *Biophys J* **72**:1861–1877.

Isas JM, Langen R, Haigler HT, Hubbell WL. 2002. Structure and dynamics of a helical hairpin and loop region in annexin 12: a site-directed spin labeling study. *Biochemistry* **41**:1464–1473.

Jeschke G. 2002. Distance measurements in the nanometer range by pulse EPR. *Chem Phys Chem* **3**:927–932.

Kim KK, Yokota H, Kim SH. 1999. Four-helical-bundle structure of the cytoplasmic domain of a serine chemotaxis receptor. *Nature* **400**:787–792.

Klare JP, Bordignon E, Engelhard M, Steinhoff HJ. 2004a. Sensory rhodopsin II and bacteriorhodopsin: light activated helix F movement. *Photochem Photobiol Sci* **3**:543–547.

Klare JP, Gordeliy VI, Labahn J, Büldt G, Steinhoff HJ, Engelhard M. 2004b. The archaeal sensory rhodopsin II/transducer complex: a model for transmembrane signal transfer. *FEBS Lett* **564**:219–224.

Kochendoerfer GG, Chen SY, Mao F, Cressman S, Traviglia S, Shao HY, Hunter CL, Low DW, Cagle EN, Carnevali M, Gueriguian V, Keogh PJ, Porter H, Stratton SM, Wiedeke MC, Wilken J, Tang J, Levy JJ, Miranda LP, Crnogorac MM, Kalbag S, Botti P, Schindler-Horvat J, Savatski L, Adamson JW, Kung A, Kent SBH, Bradburne JA. 2003. Design and chemical synthesis of a homogeneous polymer-modified erythropoiesis protein. *Science* **299**:884–887.

Kochendoerfer GG, Jones DH, Lee S, Oblatt-Montal M, Opella SJ, Montal M. 2004. Functional characterization and NMR Spectroscopy on full-length Vpu from HIV-1 prepared by total chemical synthesis. *J Am Chem Soc* **126**:2439–2446.

Koteiche HA, Mchaourab HS. 1999. Folding pattern of the alpha-crystallin domain in alpha A-crystallin determined by site-directed spin labeling. *J Mol Biol* **294**:561–577.

Koteiche HA, Berengian AR, Mchaourab HS. 1998. Identification of protein folding patterns using site-directed spin labeling: structural characterization of a beta-sheet and putative substrate binding regions in the conserved domain of alpha A-crystallin. *Biochemistry* **37**:12681–12688.

LaConte LEW, Voelz V, Nelson W, Enz M, Thomas DD. 2002. Molecular dynamics simulation of site-directed spin labeling: experimental validation in muscle fibers. *Biophys J* **83**:1854–1866.

Likhtenshtein GI. 1976. *Spin labeling methods in molecular biology*. New York: John Wiley & Sons.

Luecke H, Richter HT, Lanyi JK. 1998. Proton transfer pathways in bacteriorhodopsin at 2.3 Å resolution. *Science* **280**:1934–1937.

Luecke H, Schobert B, Richter HT, Cartailler JP, Lanyi JK. 1999. Structure of bacteriorhodopsin at 1.55 Å resolution. *J Mol Biol* **291**:899–911.

Marchetto R, Schreier S, Nakaie CR. 1993. A novel spin-labeled amino-acid derivative for use in peptide-synthesis: (9-Fluorenylmethyloxycarbonyl)-2,2,6,6-tetramethylpiperidine-N-oxyl-4-amino-4-carboxylic acid. *J Am Chem Soc* **115**:11042–11043.

Marsh D, Dzikovski BG, Livshits VA. 2006. Oxygen profile in membranes. *Biophys J* **90**:L49–L51.

Mchaourab HS, Hyde JS. 1993. Dependence of the multiple-quantum EPR signal on the spin-lattice relaxation-time: effect of oxygen in spin-labeled membranes. *J Magn Reson Ser B* **101**:178–184.

Mchaourab HS, Perozo E. 2000. Determination of protein folds and conformational dynamics using spin labeling EPR spectroscopy. In *Distance measurements in biological systems by EPR*, pp. 155–218. Ed LJ Berliner, SS Eaton, GR Eaton. New York: Kluwer.

Mchaourab HS, Lietzow MA, Hideg K, Hubbell WL. 1996. Motion of spin-labeled side chains in T4 lysozyme, correlation with protein structure and dynamics. *Biochemistry* **35**:7692–7704.

Mendel D, Cornish VW, Schultz PG. 1995. Site-directed mutagenesis with an expanded genetic-code. *Ann Rev Biophys Biomol Struct* **24**:435–462.

Merrifield RB. 1963. Solid phase peptide synthesis, 1: synthesis of a tetrapeptide. *J Am Chem Soc* **85**:2149–2154.

Möbius K, Savitsky A, Wegener C, Plato M, Fuchs M, Schnegg A, Dubinskii AA, Grishin YA, Grigor'ev IA, Kühn M, Duché D, Zimmermann H, Steinhoff HJ. 2005. Combining high-field EPR with site-directed spin labeling reveals unique information on proteins in action. *Magn Reson Chem* **43**:S4–S19.

Monaco V, Formaggio F, Crisma M, Toniolo C, Hanson P, Millhauser G, George C, Deschamps JR, Flippen-Anderson JL. 1999. Determining the occurrence of a 3(10)-helix and an alpha-helix in two different segments of a lipopeptaibol antibiotic using TOAC, a nitroxide spin-labeled C-alpha-tetrasubstituted alpha-amino acid. *Bioorg Med Chem* **7**:119–131.

Moukhametzianov R, Klare JP, Efremov R, Baeken C, Göppner A, Labahn J, Engelhard M, Büldt G, Gordeliy VI. 2006. Development of the signal in sensory rhodopsin and its transfer to the cognate transducer. *Nature* **440**:115–119.

Nielsen RD, Canaan S, Gladden JA, Gelb MH, Mailer C, Robinson BH. 2004. Comparing continuous wave progressive saturation EPR and time domain saturation recovery EPR over the entire motional range of nitroxide spin labels. *J Magn Reson* **169**:129–163.

O'Donnell MJ, Bennett WD, Wu SD. 1989. The stereoselective synthesis of alpha-amino-acids by phase-transfer catalysis. *J Am Chem Soc* **111**:2353–2355.

Pannier M, Veit S, Godt A, Jeschke G, Spiess HW. 2000. Dead-time free measurement of dipole–dipole interactions between electron spins. *J Magn Reson* **142**:331–340.

Pebay-Peyroula E, Rummel G, Rosenbusch JP, Landau EM. 1997. X-ray structure of bacteriorhodopsin at 2.5 angstroms from microcrystals grown in lipidic cubic phases. *Science* **277**:1676–1681.

Percival PW, Hyde JS. 1975. Pulsed EPR spectrometer, 2. *Rev Sci Instrum* **46**:1522–1529.

Perozo E, Cortes DM, Cuello LG. 1998. Three-dimensional architecture and gating mechanism of a K+ channel studied by EPR spectroscopy. *Nat Struct Biol* **5**:459–469.

Pfeiffer M, Rink T, Gerwert K, Oesterhelt D, Steinhoff HJ. 1999. Site-directed spin-labeling reveals the orientation of the amino acid side-chains in the E-F loop of bacteriorhodopsin. *J Mol Biol* **287**:163–171.

Plato M, Steinhoff HJ, Wegener C, Törring JT, Savitsky A, Möbius K. 2002. Molecular orbital study of polarity and hydrogen bonding effects on the g and hyperfine tensors of site directed NO spin labelled bacteriorhodopsin. *Mol Phys* **100**:3711–3721.

Poole CP. 1983. *Electron spin resonance*. New York: Wiley.

Prisner TF, Vanderest A, Bittl R, Lubitz W, Stehlik D, Möbius K. 1995. Time-resolved W-band (95 GHz) EPR spectroscopy of Zn-substituted reaction centers of rhodobacter-sphaeroides R-26. *Chem Phys* **194**:361–370.

Pyka J, Ilnicki J, Altenbach C, Hubbell WL, Froncisz W. 2005. Accessibility and dynamics of nitroxide side chains in T4 lysozyme measured by saturation recovery EPR. *Biophys J* **89**:2059–2068.

Qin PZ, Hideg K, Feigon J, Hubbell WL. 2003. Monitoring RNA base structure and dynamics using site-directed spin labeling. *Biochemistry* **42**:6772–6783.

Rabenstein MD, Shin YK. 1995. Determination of the distance between 2 spin labels attached to a macromolecule. *Proc Natl Acad Sci USA* **92**:8239–8243.

Radzwill N, Gerwert K, Steinhoff HJ. 2001. Time-resolved detection of transient movement of helices F and G in doubly spin-labeled bacteriorhodopsin. *Biophys J* **80**:2856–2866.

Rassat A, Rey P. 1967. Nitroxides, 23: preparation of amino-acid free radicals and their complex salts. *Bull Soc Chim Fr* **3**:815–818.

Sale K, Song LK, Liu YS, Perozo E, Fajer P. 2005. Explicit treatment of spin labels in modeling of distance constraints from dipolar EPR and DEER. *J Am Chem Soc* **127**:9334–9335.

Shafer AM, Kalai T, Liu SQB, Hideg K, Voss JC. 2004. Site-specific insertion of spin-labeled L-amino acids in Xenopus oocytes. *Biochemistry* **43**:8470–8482.

Shin YK, Hubbell WL. 1992. Determination of electrostatic potentials at biological interfaces using electron electron double-resonance. *Biophys J* **61**:1443–1453.

Shin YK, Levinthal C, Levinthal F, Hubbell WL. 1993. Colicin-E1 binding to membranes: time-resolved studies of spin-labeled mutants. *Science* **259**:960–963.

Steinhoff HJ. 2002. Methods for study of protein dynamics and protein–protein interaction in protein ubiquitination by electron paramagnetic resonance spectroscopy. *Front Biosci* **7**:C97–C110.

Steinhoff HJ, Hubbell WL. 1996. Calculation of electron paramagnetic resonance spectra from Brownian dynamics trajectories: application to nitroxide side chains in proteins. *Biophys J* **71**:2201–2212.

Steinhoff HJ, Karim C. 1993. Protein dynamics and EPR-spectroscopy: comparison of molecular dynamic simulations with experiments. *Ber Bunsenges Phys Chem* **97**:163–171.

Steinhoff HJ, Lieutenant K, Schlitter J. 1989. Residual motion of hemoglobin-bound spin labels as a probe for protein dynamics. *Z Naturforschung C: J Biosci* **44**:280–288.

Steinhoff HJ, Dombrowsky O, Karim C, Schneiderhahn C. 1991. Two-dimensional diffusion of small molecules on protein surfaces: an EPR study of the restricted translational diffusion of protein-bound spin labels. *Eur Biophys J Biophy Lett* **20**:293–303.

Steinhoff HJ, Mollaaghababa R, Altenbach C, Hideg K, Krebs M, Khorana HG, Hubbell WL. 1994. Time-resolved detection of structural changes during the photocycle of spin-labeled bacteriorhodopsin. *Science* **266**:105–107.

Steinhoff HJ, Radzwill N, Thevis W, Lenz V, Brandenburg D, Antson A, Dodson G, Wollmer A. 1997. Determination of interspin distances between spin labels attached to insulin: Comparison of electron paramagnetic resonance data with the x-ray structure. *Biophys J* **73**:3287–3298.

Steinhoff HJ, Pfeiffer M, Rink T, Burlon O, Kurz M, Riesle J, Heuberger E, Gerwert K, Oesterhelt D. 1999. Azide reduces the hydrophobic barrier of the bacteriorhodopsin proton channel. *Biophys J* **76**:2702–2710.

Steinhoff HJ, Müller M, Beier C, Pfeiffer M. 2000a. Molecular dynamics simulation and EPR spectroscopy of nitroxide side chains in bacteriorhodopsin. *J Mol Liq* **84**:17–27.

Steinhoff HJ, Savitsky A, Wegener C, Pfeiffer M, Plato M, Möbius K. 2000b. High-field EPR studies of the structure and conformational changes of site-directed spin labeled bacteriorhodopsin. *Biochim Biophys Acta Bioenerg* **1457**:253–262.

Stone AJ. 1963. Gauge invariance of g tensor. *Proc Roy Soc London A* **271**:424–434.

Thorgeirsson TE, Xiao WZ, Brown LS, Needleman R, Lanyi JK, Shin YK. 1997. Transient channel-opening in bacteriorhodopsin: an EPR study. *J Mol Biol* **273**:951–957.

Voet D, Voet JG. 2004. *Biochemistry*, 3rd ed. New York: John Wiley & Sons.

Voss J, Hubbell WL, Kaback HR. 1998. Helix packing in the lactose permease determined by metal–nitroxide interaction. *Biochemistry* **37**:211–216.

Wegener AA, Chizhov I, Engelhard M, Steinhoff HJ. 2000. Time-resolved detection of transient movement of helix F in spin-labelled pharaonis sensory rhodopsin II. *J Mol Biol* **301**:881–891.

Wegener AA, Klare JP, Engelhard M, Steinhoff HJ. 2001a. Structural insights into the early steps of receptor-transducer signal transfer in archaeal phototaxis. *EMBO J* **20**:5312–5319.

Wegener C, Savitsky A, Pfeiffer M, Möbius K, Steinhoff HJ. 2001b. High-field EPR-detected shifts of magnetic tensor components of spin label side chains reveal protein conformational changes: the proton entrance channel of bacteriorhodopsin. *Appl Magn Reson* **21**:441–452.

Yin JJ, Pasenkiewiczgierula M, Hyde JS. 1987. Lateral diffusion of lipids in membranes by pulse saturation recovery electron-spin-resonance. *Proc Natl Acad Sci USA* **84**:964–968.

CHAPTER 6

HIGH-FIELD ESR SPECTROSCOPY IN MEMBRANE AND PROTEIN BIOPHYSICS

Tatyana I. Smirnova and Alex I. Smirnov
North Carolina State University,
Raleigh, North Carolina, USA

1. INTRODUCTION

The main emphasis of this chapter, as well as of the whole volume, is on site-directed spin-labeling (SDSL) and spin-labeling ESR as an alternative and a complementary structural technique for NMR and x-ray diffraction of proteins and other complex biomolecular systems. The spin labeling method is particularly powerful for elucidating local protein folds and the transmembrane structure of membrane proteins as well as for studying conformational dynamics of proteins and protein–lipid interactions.

While applications of ESR to membrane and protein biophysics continue to grow at a rather rapid pace, the majority of these experiments is still carried out at the X-band (9–10 GHz) frequency. A search of recent literature showed that only relatively few spin labeling studies utilize Q-band (35 GHz) and higher frequencies. However, high-field/high-frequency spin-labeling ESR utilizing microwaves above ca. 90 GHz is uniquely positioned to offer new quantitative and qualitative information on biomolecules. With an increase in magnetic field, the ESR spectra from a nitroxide spin label undergo significant changes. The main transformations in the shapes of the spectra occur at the resonant magnetic fields of ca. 1.1 T. Above these fields the magnitude of the Zeeman anisotropy in the spin Hamiltonian prevails over the anisotropy of the nitrogen hyperfine interaction. With a further increase in the magnetic field/frequency the *g*-factor resolution continues to increase. Thus, while at the magnetic fields of conventional X-band ESR (0.3 T, 9 GHz) the nitroxide spectrum is determined by an axial hyperfine term and an averaging from molecular motion, at magnetic fields above 3.4 T (95 GHz, W-band), the rhombic Zeeman term provides new information on protein structure and dynamics that is inaccessible by traditional ESR spectroscopy.

The technical problems of developing high-field ESR were first addressed successfully by Professor Yakov S. Lebedev and coworkers from the Institute of Chemical Physics (Moscow). Using a 140-GHz (5 T) ESR spectrometer, they were clearly the first who explored the enhanced sensitivity of spin-label high-field ESR spectra to molecular motion and local polarity, and pioneered many other useful applications (Grinberg et al. 1976, 1983; Lebedev 1990). However, in some sense, their pioneering studies of spin labels with high-field ESR were a bit ahead of the supporting biotechnology: at that time mutation of proteins was expensive and a very time-consuming task. Some researchers even questioned the utility of labeling proteins with nitroxides, because only a few proteins have native unpaired cysteines and thus can be labeled with nitroxides.

Although the Moscow 140-GHz high-field ESR spectrometer was clearly state of the art at that time, the sensitivity for lossy liquid aqueous samples was often not sufficient for routine spin-labeling experiments with fully hydrated proteins and membranes at ambient temperatures. Nevertheless, Lebedev and coworkers carried out several benchmark spin-labeling high-field ESR studies and were the true pioneers of this method.

Nowadays, high-field ESR is undergoing rapid development powered by the efforts of many research groups to advance the field. In our view, high-field ESR in general and its application to proteins and other complex biomolecules and assemblies in particular have tremendous potential, and we confidently expect to see ever-increasing activity in the future. The intention of this chapter is to (i) review the basic features of ESR experiments at high magnetic fields with an emphasis on nitroxide spin labeling in applications to membrane and membrane protein biophysics, (ii) review what kind of new data on protein structure and dynamics can be obtained from such experiments, (iii) examine how these high field experiments are carried out, and (iv) outline areas of research for future exploration.

2. ANALYSIS OF HIGH-FIELD ESR SPECTRA OF SPIN LABELS

We shall begin by outlining the principal features of ESR at high magnetic fields — namely, enhancements in spectral resolution and sensitivity. We will then detail the physical aspects of high-field ESR experiments with spin labels by considering the spin Hamiltonian and the magnitudes of contributing anisotropies. With these considerations in hand, we will proceed to qualitative analysis of spin-label spectra at slow and rigid limits, discuss intermediate and fast molecular motion, and the use of high-field ESR to analyze the spectra of macroscopically aligned samples. It should be noted here that it is not the purpose of this chapter to provide a comprehensive treatment of ESR line shape theory for nitroxide spin labels. Some of the theoretical aspects of spin-labeling ESR have been discussed in other chapters of this volume. There are also a number of excellent texts that deal with slow-motion ESR line shape theory, including several that detail the stochastic Liouville equation approach (Schneider and Freed 1989; Gamliel and Levanon 1995). Rather than that, our goals here are to discuss the basic features of high-field ESR of spin labels in application to membrane protein biophysics and to outline the avenues of further developments in the field.

2.1. Basic Features of High Field ESR

The main reasons for developing ESR at a high magnetic field could be appreciated by considering the very basics of magnetic resonance. Electron spin resonance is a physical phenomenon based on resonance absorption of electromagnetic radiation by a system possessing unpaired electronic spins. In order for the electromagnetic energy to be absorbed, its frequency, ν, should satisfy the Bohr condition:

$$\Delta E = h\nu \,, \tag{1}$$

where ΔE is the energy difference between the spin energy levels for which transition is allowed according to quantum mechanical rules. The condition given by Eq. (1) is fundamental for all spectroscopic methods that allow one to learn about energy levels and their populations for a system of interest. For example, infrared spectra of biological molecules allow one to study vibrational modes of specific molecular bonds that are determined by the molecular structure. What makes the magnetic resonance unique compared with, for example, optical spectroscopic methods, is that for a given molecular system splittings ΔE between the energy levels are not constant but can be varied to a rather large degree by applying an external magnetic field. It should also be noted that in the absence of a magnetic field the quantum states of an $S = 1/2$ spin system are degenerate and no magnetic resonance is observed.

The degeneracy of electronic spin quantum states is lifted in the presence of magnetic field B according to

$$E_{m_s} = -\gamma_e \hbar B m_s = g_e \mu_B B m_s \,, \tag{2}$$

where γ_e is the magnetogyric ratio of the free electron, m_s is projection of the angular momentum quantum number, g_e is the free electron g-factor, and μ_B is the Bohr magneton. Thus, in order to observe magnetic resonance for electronic spin $S = 1/2$ the frequency of electromagnetic radiation should be "in tune" with the applied magnetic field:

$$\omega = 2\pi\nu = \gamma_e B \,. \tag{3}$$

This frequency is also called the Larmor frequency, after Sir Josef Larmor, a British mathematician who derived a frequency of precession of a particle having a magnetic moment in an external magnetic field.

Equation (3) means that, in principle, for the same $S = 1/2$ spin system ESR could be observed at any electromagnetic frequency if a proper magnetic field is applied. In practice, several physical and technical considerations, such as availability of high-field magnets with sufficient field stability and homogeneity and/or electromagnetic field devices as well as sensitivity considerations, affect the choice of resonant frequency for ESR experiments. One of the notable examples is the first magnetic resonance experiment carried out by Evgeny Zavoisky in 1944 using a radiofrequency generator operating from 0.5 to 17 MHz (corresponding wavelength from 200 to 18 m) and a solenoid electromagnet (Kochelaev and Yablokov, 1995). It should be noted that the electron Larmor frequency of 17 MHz corresponds to just 0.6 mT of magnetic field. In his doctoral dissertation Zavoisky documented perhaps the first multifrequency ESR experiment, which was carried out with solid $CuSO_4 \cdot 5H_2O$ (Zavoisky 1944). Specifically, Zavoisky demonstrated that the magnetic field corresponding to the maximum of ESR absorption for this sample increased linearly from 6 to 17 MHz in accord with Eq. (3). The extreme hardship of the Second World War and very limited scientific equipment available to Zavoisky in his laboratory at Kazan State University (Russia) certainly contributed to the choice of resonant conditions for these first ESR experiments.

After his initial studies, Zavoisky quickly recognized that better sensitivity could be achieved by increasing the resonant field/frequency. In his follow-up experiments he employed microwave sources at 133 MHz, 460 MHz, and 2.75 GHz. In order to dissolve some skepticism expressed by Moscow scientists regarding his initial results, he assembled and installed an X-band (9–10 GHz, 10-cm wavelength) ESR apparatus at the Institute of Physical Problems in Moscow. This spectrometer utilized the best microwave equipment available at that time: an American klystron, a German microwave cable, and the rest of the components were the very best that could be found by a scientist in postwar Russia (Kochelaev and Yablokov, 1995).

Commercial ESR spectrometers developed during the fifties and sixties relied on components of the microwave radar technology that emerged during the war. The microwave frequencies were also subdivided onto separate bands that received somewhat confusing names (L-, X-, Q-, W-band, etc.). One of the main reasons for such subdivisions was the limited bandwidth of microwave components and the need for standardization. The microwave frequency of 8–12 GHz (X-band) was found to be particularly useful for tracking moving objects with radars (X-band radars are still used by traffic police throughout the world). The same microwave frequency is also found in the most versatile and widely available ESR spectrometers. Currently, the great majority of all ESR spectrometers still operate within this frequency range. Basically, there are two reasons for this. The first lies in the availability of low-phase noise microwave sources, high-sensitivity detectors, and other high-performance microwave components. The other is the somewhat prohibitive cost of generating a homogeneous magnetic field above that required for X-band ESR. Indeed, the Larmor electronic frequency of 8–12 GHz corresponds to 0.28–0.43 T. Such a field could be easily generated with iron-core electromagnets at a still reasonable cost. However, fields above ca. 2.5 T require superconducting magnets with a homogeneity of at least 10 ppm over a sphere of 1 cm in diameter. These magnets become progressively more expensive especially for magnetic fields above 9 T because of the high cost of the wires (multifilament and, preferably, twisted niobium-tin alloys) that remain superconducting at those fields.

Nowadays, ESR is performed at various combinations of magnetic fields and microwave frequencies. ESR imaging of small animals is typically carried out at 0.25–1.2 GHz because the penetration depth of microwave into a tissue quickly falls above 1 GHz. The majority of spin labeling and other biophysical experiments still relies on X-band ESR. ESR spectrometers of this microwave frequency range provide the best concentration sensitivity. In addition, conventional ESR requires typical sample volumes of 0.2 ml or less (depending on the resonator employed). These quantities of solutions of spin-labeled biomolecules are easy to prepare and handle. In addition to X-band, commercial ESR spectrometers were available at ca. 35 GHz (Q-band) for the last two decades. More recently, commercial CW and time-domain ESR were introduced at ca. 95 GHz (W-band) (Hofer et al. 2004). Currently, a few laboratories around the world are advancing ESR instrumentation beyond 95 GHz, and experiments at resonant frequencies as high as 700 GHz have already been demonstrated (Brunel et al. 2004). It should be mentioned here that ESR spectrometers become progressively more expensive with increasing resonant frequency because of the high cost of both superconducting magnet systems operating at high magnetic fields and the millimeter-wave components employed.

One would like to ask a question: what drives the development of ESR to even higher resonant frequencies and what kind of benefits can one expect in application of high-field and multifrequency ESR to the field of membrane and protein biophysics?

First of all, an increase in the magnetic field should yield gains in both sensitivity and resolution in magnetic resonance experiments. These two principal gains are the major driving forces behind high-field NMR. The most advanced commercial NMR spectrometers now operate at magnetic fields of ca. 21 T (corresponding proton resonance frequency of 900 MHz). One of the most noteworthy uses of these spectrometers is to solve the structure of large protein molecules from solution NMR spectra. At lower field/frequency such spectra have poor resolution, and this complicates spectral assignment and solving for unique structure(s).

It should be noted that the line width of solution NMR spectra is very small compared to the narrowest lines observed in ESR experiments. The anisotropy of magnetic interactions contributing to nuclear spin Hamiltonian is typically small and is readily averaged out by the molecular tumbling that many molecules undergo in solution. This is not generally the case for ESR. Thus, the arguments of gaining resolution with increasing magnetic field may not always hold for ESR.

In a continuous wave experiment, the magnetic field separation, ΔB, between ESR signals from two paramagnetic centers with g-factors g_1 and g_2 is given by

$$\Delta B = \frac{h\nu}{\mu_B}\left(\frac{1}{g_1}-\frac{1}{g_2}\right). \tag{4}$$

Then in order to (partially) resolve these two signals in an experiment, separation ΔB should exceed twice the line width, $\Delta B_{1/2}$, i.e. $\Delta B \geq 2\Delta B_{1/2}$. Therefore in order to predict the frequency dependence of the spectral resolution in an ESR experiment one has to know the frequency dependence of $\Delta B_{1/2}$. One could argue that the line width of a sample is determined by a distribution of individual g-factors characterized by width $\delta(g)$. Such a spread in g-factors could arise, for example, from a distribution in local magnetic fields of a molecular nature, often called a "*g*-strain" effect. For such samples $\Delta B_{1/2} \propto \delta(g)\nu$, and, therefore, both signal separation ΔB and line width $\Delta B_{1/2}$ will grow proportionally to the resonant field/frequency, yielding no gains in resolution with increasing magnetic field.

Fortunately, for most paramagnetic systems *g*-strain is not the major component of the line width. It is now well documented that for organic free radicals, including nitroxide spin probes and related spin traps adducts, an inhomogeneous line width is dominated not by *g*-strain but by unresolved hyperfine interactions with neighboring magnetic nuclei. These interactions give rise to inhomogeneous broadening of about $\Delta B_{\text{hyp}} \approx 0.1\text{–}0.2$ mT and are independent of magnetic field. Then, if one assumes that the line width of these paramagnetic species is essentially field/frequency independent, the spectral lines will become resolved at a resonant frequency ν when

$$\Delta B = \frac{h\nu}{\mu_B}\left(\frac{1}{g_1} - \frac{1}{g_2}\right) \geq 2\Delta B_{\text{hyp}}, \tag{5}$$

where intentionally neglected all the other sources of line width broadening such as incomplete averaging of *g*-factor anisotropy by molecular motion. These issues will be discussed further later in the chapter.

Equation (5) defines conditions of a high-field/high-frequency ESR experiment for a spin system whose spectra are determined by field-independent (hyperfine interactions) and field-dependent features arising from the *g*-factor.

For many organic free radicals including nitroxides the isotropic electronic *g*-factor is very close to that of a free electron, $g_e \approx 2.0023$, and the relative differences rarely exceed $\Delta g/g \approx 0.001$. Then $g_1 \approx g_2 \approx g_e$, and

$$\Delta B \approx \frac{h\nu}{g_e^2 \mu_B}\Delta g = B_0 \frac{\Delta g}{g_e}. \tag{6}$$

where B_0 is the resonant magnetic field. At conventional X-band ($\nu \approx 9.5$ GHz) ESR, these typical relative differences in *g* factors, $\Delta g/g \approx 0.0001\text{–}0.001$, translate into 0.03–0.3 mT shifts in the magnetic field position of resonance lines. These differences are rather difficult to resolve for lines with $\Delta B_{\text{hyp}} \approx 0.1\text{–}0.2$ mT in width. However, at a tenfold higher frequency of 95 GHz (W-band) the same range of $\Delta g/g$ corresponds to 0.3–3.0 mT. Thus, for a large portion of organic free radicals, conditions of a high-field ESR experiment should be already fulfilled at the W-band resonant frequency.

Let us now discuss the second principal advantage of increasing the magnetic field in ESR experiments: an increase in sensitivity. In continuous wave ESR, the signal is detected by observing absorption of microwave power by the sample at resonance. This power, W, is proportional to the product of the electromagnetic field quantum energy, $h\nu$, and the difference in populations of electronic spin energy levels n_α and n_β ($\Delta n = (n_\beta - n_\alpha)$):

$$W = P \cdot (h\nu) \cdot \Delta n, \tag{7}$$

where P is the transition probability and $|\alpha\rangle$ and $|\beta\rangle$ are Dirac notations for the corresponding spin states. When the spin system is in thermal equilibrium at a temperature T, the ratio of energy level populations is given by a Boltzman distribution:

$$\frac{n_\alpha}{n_\beta} = e^{-\Delta E / kT} = e^{-\gamma_e \hbar B_0 / kT} . \tag{8}$$

In the high-temperature limit, $kT >> h\nu$, and

$$\Delta n = n_\beta - n_\alpha \approx N\gamma_e \hbar B_0 / 2kT , \tag{9}$$

where N is the total number of spins in the sample. With the help of time-dependent perturbation theory, it can be shown that transition probability P is given by

$$P \propto \gamma_e^2 B_1^2 g(\nu) , \tag{10}$$

where $g(\nu)$ is the line shape factor and B_1 is the magnitude of the electromagnetic field oscillating at frequency ν and directed perpendicular to B_0. Then the power of electromagnetic radiation W absorbed by the sample can be approximated as:

$$W \propto \gamma_e^4 B_0^2 B_1^2 N g(\nu) / T / \tag{11}$$

It should be noted that this approximation is only valid when population difference Δn does not deviate significantly from that at equilibrium (i.e., when B_1 is low).

The simple estimate given by Eq. (11) demonstrates that the power of electromagnetic radiation absorbed by the sample has a quadratic dependence upon magnetic field B_0; thus, development of high-field ESR instrumentation is highly desirable from the point of view of increased sensitivity. Specifically, this equation predicts that the minimum number of detectable spins should depend upon resonant frequency ν as ν^{-2}. This means that high-field ESR will be advantageous for studying small quantities of biological samples and protein crystals of rather small (submillimeter) sizes. It should be noted that the crystal size requirements for studying with W-band ESR are essentially the same as for x-ray crystallography. This is a very important consideration because in many cases these crystals cannot be grown any larger. Thus, one could study essentially the same single-protein crystals with high-field ESR and x-ray (see, e.g., Groenen and Schmidt (2004) and the references therein).

Now, if we will be applying the same simplified considerations to predict the frequency dependence of minimally detectable spin concentrations by ESR, we can easily conclude that no gains will be achieved with increasing magnetic field. Indeed, if the linear dimensions of the ESR resonators and the samples are all scaled inversely proportional to the frequency, then the sample volume decreases as ν^{-3}. This exceeds the sensitivity gain. However, more careful analysis carried out by accounting for other physical and technical factors, such as, for example, resonator design and the noise characteristics of microwave components, shows that some small gains can be achieved under certain conditions. The reader is directed to the comprehensive analysis of the frequency/field dependence of ESR sensitivity given by Eaton et al. (1998) and Rinard et al. (1999). One of the conclusions of such an analysis was that for samples of constant size but with resonator liner dimensions scalable as ν^{-1} and the same B_1, the minimum detectable number of spins, $N_{\min}$, decreased as

$$N_{\min} \propto \nu^{-11/4}. \quad (12)$$

Typically, $N_{\min}$ is defined as that number of spins that will yield an ESR signal with a signal-to-noise ratio equal to 1. In addition, $N_{\min}$ is also normalized by the line width of the species (differences in line shape are usually ignored) and also by the modulation amplitude and bandwidth of the detection system. The custom is to report $N_{\min}$ in spin/G at 1 Hz bandwidth (1 Hz bandwidth corresponds to a 1-s integration time constant) and a modulation amplitude of 1 G. In this chapter we are using an SI system, so $N_{\min}$ is reported in spin/mT at 1 Hz bandwidth (1 G = 0.1 mT).

The experimentally demonstrated sensitivity was found to be close to these predictions (Rinard et al. 1999). Depending on assumptions with regard to electromagnetic (microwave) field power and B_1 on the sample, the same theory predicts only modest gains or even losses in concentration sensitivity upon increasing the magnetic field. It should be noted that for lossy dielectric samples, such as aqueous solutions of spin-labeled proteins and membranes, the best theoretical sensitivity is difficult to achieve because of a strong dependence of signal intensity on sample geometry. Also, it should be noted that the characteristics of microwave components usually fall off in the millimeter wave region when compared to the conventional X-band. These considerations lead some skeptics to believe that high-field ESR should have generally poor sensitivity compared to conventional ESR. Indeed, the first high-field ESR spectrometer that operated above conventional X- and Q-band ESR frequencies demonstrated a sensitivity of only 5×10^{10} spin/mT at 1 Hz bandwidth when the sample was at 4.2 K (Slade and Ingram 1969). This spectrometer operated at 70 GHz (4 mm wavelength) and utilized a cylindrical cavity. Clearly, this first-of-its-kind apparatus did not demonstrate any sensitivity gains with an increase in the resonance field: after correcting for population level differences one could find that its sensitivity was actually worse than that of the best X-band equipment available today (ca. 10^{11} spin/mT when measured at room temperature).

However, a rapid progress in millimeter microwave devices and advances in designing tiny microwave resonators for high-field ESR made it possible to achieve much better sensitivity. For example, Lebedev and coworkers demonstrated that $N_{\min} \approx 4 \times 10^8$ spin/mT (1 Hz bandwidth) could be detected with a 150-GHz CW ESR spectrometer that they developed (Lebedev 1990). The commercial Bruker ELEXSYS E600/680 W-band (95 GHz) spectrometer has a detection limit of $N_{min} \approx 2 \times 10^8$ spin/mT (Hofer et al. 2004). It should be noted that the numbers quoted above were measured for non-lossy samples and that the best sensitivity for liquid aqueous samples is still somewhat lower. Some of the sensitivity losses are likely to be caused by using non-optimal sample holders for aqueous samples. Typically, these samples are drawn into thin (i.d. = 0.1–0.2 mm) capillary tubes that are then placed in cylindrical high-field ESR resonators. Such sample geometry is known to be non-optimal for aqueous samples when used with cylindrical ESR cavities at X-band. Nevertheless, exceptional concentration sensitivity has been demonstrated even for liquid samples that are almost purely aqueous. For example, researchers from the University of Illinois reported a concentration sensitivity of a home-built W-band ESR spectrometer as low as 70 nM/mT on an aqueous solution of perdeuterated Tempone (2,2',6,6'-tetramethyl-4-piperidone-1-oxyl) at room temperature without extrapolating to the maximum available power (Clarkson et al. 2001). The optimum size of the capillary for an aqueous sample was found to be just 0.15 mm in inside diameter. This translates into about a 70-nL volume inside the sensitivity region of the cavity. Although the width of the spin-label ESR spectra corresponding to the slow and intermediate motion regimes increases greatly with magnetic field, this sensitivity is still sufficient to carry out spin-labeling ESR experiments with samples even at 20–100 μM concentrations and a $<10^{-12}$ M sample quantity. The latter becomes a rather useful feature of high-field ESR in experiments when only small amounts of spin-labeled proteins and/or nucleic acids are available. As the size of single-mode resonators scales down with the wavelength of the mm waves, so too the optimum diameter of sample capillaries. At about 200 GHz and above, both the resonators and sample capillaries become difficult to handle. At these frequencies multi-mode resonators and/or resonators of an open-type design (such as the Fabry-Perot) are more common and convenient. The optimal volume for aqueous samples in such resonators is typically 0.1–1.0 μL, and the concentration sensitivity can be exceptionally high, reaching the μM range, as demonstrated by the Cornell ACERT ESR Center (Earle and Freed 1999), and, more recently, by researchers from NHMFL (van Tol et al. 2005). Some new interesting possibilities for high-field ESR resonators and aqueous sample arrangements were reviewed by Budil and Earle (2004). We anticipate that optimization of the sample and resonator design could further improve the sensitivity of high-field ESR for aqueous samples in the near future.

2.2. The Spin Hamiltonian and the Rigid-Limit ESR Spectra of Nitroxide Spin Labels at High Magnetic Fields

A general spin Hamiltonian for a nitroxide spin label is given by:

$$H = \left(\mu_B \vec{B} \cdot g \cdot \hat{S} + \mu_n \vec{B} \cdot g_n \cdot \hat{I}\right) + h\hat{S} \cdot A \cdot \hat{I} + (\text{other terms}), \tag{13}$$

where the two terms in parentheses describe electronic and nuclear Zeeman interactions, μ_n is the nuclear magneton, g_n is the nuclear g-factor, and the third term represents the hyperfine interaction of the electronic spin with the neighboring nitrogen nucleus (I = 1 for ^{14}N- or I = ½ for ^{15}N-substituted labels). Typically, the nuclear Zeeman term is not considered unless the experiments are carried out at very high frequencies such as 250 GHz and above (Earle et al. 1993). Thus, in the absence of spin–spin interactions, the ESR spectrum of a nitroxide is mainly determined by an electronic **g**-matrix and a hyperfine tensor ***A***. The latter two terms have different symmetry and different magnetic field dependences. In particular, for a typical nitroxide, hyperfine tensor ***A*** is almost axial with $A_z > A_y \approx A_x$, while the **g**-matrix has a clear rhombic character: $g_z < g_y < g_x$ (see also Table 1 for typical magnetic parameters of a nitroxide radical in a non-polar solvent).

Table 1. Magnetic Parameters for Perdeuterated Tempone in Toluene-d$_8$

	Data Set A	Data Set B
g_x	$2.00936 \pm 3\times10^{-5}$	$2.00939 \pm 3\times10^{-5}$
g_y	$2.00633 \pm 3\times10^{-5}$	$2.00626 \pm 3\times10^{-5}$
g_z	$2.00233 \pm 3\times10^{-5}$	$2.00232 \pm 3\times10^{-5}$
$A_x\times(2\pi/\gamma_e)^a$, mT	0.41 ± 0.02	0.46 ± 0.02
$A_y\times(2\pi/\gamma_e)$, mT	0.51 ± 0.02	0.49 ± 0.02
$A_z\times(2\pi/\gamma_e)$, mT	3.36 ± 0.02	3.38 ± 0.02

[a] Note that the units for the components of hyperfine coupling tensor ***A*** in Eq. (13) are Hz. For convenience, characteristic components of this tensor are often reported in units of magnetic field. For converting hyperfine coupling constants A_i from Hz to mT, one should multiply A_i by $2\pi/\gamma_e$. Here γ_e is the magnetogyric ratio of the free electron as defined from Eq. (2). Data set A was reported by Earle and coworkers (1993), while data set B was verified by Smirnov (2002). Note that in the above-mentioned references, the factor of 2π was omitted from this conversion because γ_e was defined as $\gamma_e/2\pi$.

To illustrate the different contributions of these two anisotropic terms to experimental ESR spectra at conventional (X-band) and high frequencies, let us first consider ESR spectra of a typical spin label (such as Tempone, with magnetic parameters given in Table 1) doped to a single crystal. It is a custom in the literature to choose a right-hand Cartesian molecular coordinate system $\{x, y, z\}$ for a nitroxide in a way that the x-axis will coincide with the direction of the N–O bond and the z-axis directed along the 2pπ orbital. In this molecular coordinate system, the axes coincide with the principal axes of the **g**-matrix and the ***A***-tensor because of the symmetry of the 2pπ orbital. Higher-order effects cause a small — a few degrees — tilt between the principal axis of the **g**-matrix and ***A***-tensor. Typically, this tilt angle could be neglected, and so the all off-diagonal elements of the **g**-matrix and the ***A***-tensor. All magnetic parameters listed in Table 1 were derived under the assumption that the **g**-matrix and ***A***-tensor are collinear. We will denote the diagonal elements (or principal values) of the magnetic parameter matrices as g_x, A_x, and so on. Let us now direct magnetic field $\vec{B}$ along the z-axis of the spin label. If we neglect the nuclear Zeeman term and the effects of spin–spin interactions, the spin Hamiltonian given by Eq. (13) is simplified to

$$H = \mu_B g_z B \hat{S}_z + hA_z \hat{I}_z \hat{S}_z \,, \tag{14}$$

where $\hat{S}_z$ and $\hat{I}_z$ are projections of the electronic and the nuclear spin operators on the z-axis. Then, if high field condition $\mu_B g_z B >> A_z$ is fulfilled, the energy levels in the zero approximation are

$$U(m_S, m_I) = \mu_B g_z B m_S + hA_z m_I m_S \,, \tag{15}$$

where the spin quantum numbers take values m_S= +1/2, –1/2, and m_I= 1, 0, –1. For allowed electronic transitions, we have selection rule $\Delta m_s = \pm 1$ and $\Delta m_I = 0$. Then the ESR spectrum should consist of three resonance lines. The resonance conditions for each of the m_I= 1, 0, –1 lines are given by

$$\Delta U(m_I) = h\nu = \mu_B g_z B + hA_z m_I \,. \tag{16}$$

If we assume that the relaxation effects are independent of m_I, then the continuous wave ESR spectrum should appear as three equal lines that are split by A_z (measured in Hz) with the middle line positioned at $B_0 = h\nu/\beta g_z$. The same results are obtained for the x- and y-orientations of magnetic field $\vec{B}$ in the nitroxide molecular coordinate frame. These theoretical spectra are shown in the middle of Figure 1, with the magnetic field orientation labeled next to them as x, y, and z. These spectra were calculated for ν_0 = 9.0 GHz (upper part of Figure 1) and for ν_0 = 94.4 GHz (bottom part of Figure 1).

At ν_0 = 9.0 GHz (X-band) the small difference between g_x, g_y, and g_z causes only a fraction of an mT shift between the middle (m_I= 0) nitrogen hyperfine components of the single-crystal ESR spectra at the three principal axis orientations (Figure 1). This shift is comparable with A_x and A_y when expressed in magnetic field units (by multiplying A_x and A_y in Hz by $2\pi/\gamma_e$), causing the spectra to overlap (partially). To avoid such an overlap and to assist in identifying the spectra, we have drawn them one under another with the magnetic field scale given at the top of Figure 1.

Upon increasing the resonant frequency about tenfold to ν_0 = 94.4 GHz (W-band), the hyperfine splitting for the spectra at each principal axis orientation will not change, but the magnetic field positions will be shifted proportionally to B_0 (Eq. (6)) and thus to ν_0. At ν_0 = 94.4 GHz this will lead to complete separation of all three spectra. This allowed us to position all the spectra horizontally without any overlap (Figure 1, bottom).

The gain in spectral resolution achieved for nitroxides at the W-band is further illustrated on an example of a powder pattern spectrum. Such a spectrum is observed for a spin-labeled sample with no macroscopic order and no averaging of magnetic parameters due to spin-exchange interaction or molecular motion. Typically, such conditions are satisfied for diluted spin-labeled biological samples at

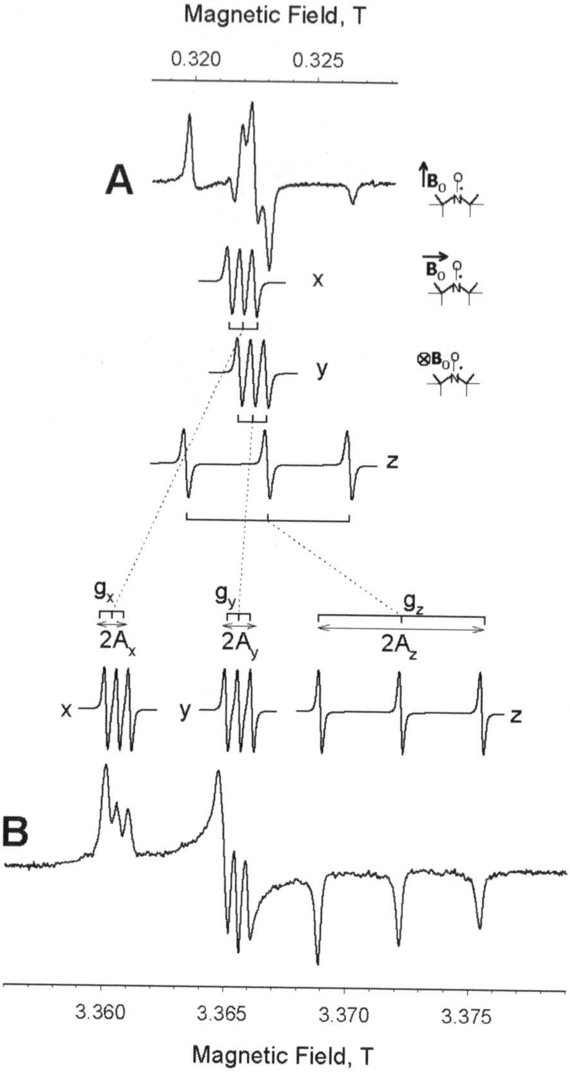

Figure 1. Illustration of transformations of single crystal and powder average nitroxide ESR spectra from (**A**) X-band, 9.0 GHz to (**B**) W-band, 94.4 GHz. The very top and the very bottom show experimental X- (**A**, top) and W-band (**B**, bottom) ESR spectra from a 0.05 mM solution of perdeuterated 2,2′,6,6′-tetramethyl-4-piperidone-1-xyl (Tempone or PDT) in toluene-d_8. At the temperature of both experiments of 130 K the spectra approached the rigid limit. The middle section of the figure shows theoretical ESR spectra of the same nitroxide with magnetic field $\vec{B}$ directed along the nitroxide principal axes x, y, and z, as depicted by cartoons to the right of the X-band spectra. Approximate positions of the principal axis orientations of the $\boldsymbol{g}$-matrix and nitrogen hyperfine $\boldsymbol{A}$-tensor are indicated by bars, and the magnitudes of the A_i components (converted to magnetic field units by multiplying by $2\pi/\gamma_e$) are given by arrows.

low temperatures. The term "powder spectrum" implicates that such a spectrum could be measured when a single crystal is ground into a fine powder so the microcrystallites take on multiple and random orientations with respect to the external magnetic field. Because all local molecular motions in such a sample are essentially frozen, the same spectrum is often called a rigid-limit spectrum.

For a single microcrystal oriented at polar angles $\{\theta, \phi\}$ with respect to $\vec{B}_0$, the resonance condition is written as

$$\Delta U = h\nu = \mu_B g(\theta,\phi)B + hA(\theta,\phi)m_I \,, \tag{17}$$

where

$$g(\theta,\phi) = g_x \sin^2\theta\cos^2\phi + g_y \sin^2\theta\sin^2\phi + g_z \cos^2\theta \,, \tag{18}$$

$$A(\theta,\phi) = \left(A_x \sin^2\theta\cos^2\phi + A_y \sin^2\theta\sin^2\phi + A_z \cos^2\theta\right)^{1/2} . \tag{19}$$

Equations (17)–(19) define the resonance condition for each particular orientation of the microcrystal in the external magnetic field. For sufficiently fine powder in the absence of aligning forces, one expects all the microcrystals to be oriented randomly. Then the resulting ESR spectrum can be found by adding together the individual ESR spectra for all possible orientations of $\vec{B}_0$ with respect to the nitroxide molecular coordinate system. In practice, the powder line shape is calculated digitally using various approximations of the spherical coordinate grid (Bernhard and Fouse 1989; Wang and Hanson, 1995) because one cannot derive an analytical expression for such a spectrum.

Least-squares analysis of powder line shapes for deriving magnetic parameters of nitroxides as well as other paramagnetic centers is widely available with the help of several commercial and free software programs. Many of the programs available work equally well. Because the website addresses for companies and the individual research groups change frequently, we decided not to provide the readers with specifics. Rather, readers are invited to use the programs provided with this volume.

It should be noted here that computer-assisted least-squares simulation analysis is a necessity when magnetic parameters cannot be determined directly from experimental ESR spectra because of the overlapping contributions of principal axes components. This is exemplified by a powder X-band ESR spectrum from a nitroxide Tempone shown in Figure 1A. Specifically, while the outer peaks of this spectrum are clearly identifiable with $m_I = -1$ and $m_I = +1$, the nitrogen hyperfine components of the single-crystal spectrum at the magnetic field orientation along the nitroxide z-axis (see the corresponding spectrum below), contributions from the x- and y-principal axis orientations all overlap in the middle of the spectrum. Thus, from such a powder average the x- and y-orientations of the nitroxide in the magnetic field are essentially indistinguishable.

Increasing the magnetic field, a high-field limit is reached under which the anisotropy of the Zeeman term exceeds that of the hyperfine tensor and all principal axis orientations become resolved. For a typical nitroxide radical, the latter condition is well satisfied at ESR frequencies of 94 GHz and above. This is illustrated in Figure 1B, which shows an experimental ESR spectrum from the same nitroxide (perdeuterated Tempone) at 94 GHz (W-band). From this comparison, the advantages of high magnetic fields for spin labeling ESR are becoming clear: all components of the *g*-matrix are well resolved. The lines corresponding to nitrogen hyperfine components may also be resolved in some cases, such as the one shown in Figure 1B. Thus, one could say that in the high-field limit the spin-label ESR powder spectrum contains easily identifiable elements of the single-crystal spectrum, so it appears to be single crystal-like. Such a high resolution allows one to accurately identify magnetic parameters (both $\boldsymbol{g}$ and $\boldsymbol{A}$) directly from the spectrum and then refine their values by least-squares simulations. It should be noted here that because of low *g*-factor resolution at the X-band, the principal components of the $\boldsymbol{g}$-matrix for many nitroxides could only be determined from least-squares simulations and, in many cases, with less accuracy than from high-field ESR.

Thus, our **first** conclusion drawn from this discussion is that magnetic parameters of spin labels can be determined from high-field ESR spectra with a greater accuracy than from conventional X-band. As will be further discussed in §6 of this chapter, this superior *g*-factor resolution makes it possible to study polarity and hydrogen bonding in biomolecules using spin-labeling high-field ESR methods.

2.3. Basics of Relaxation Theory

One of the principal advantages of spin-labeling ESR over NMR and optical methods is that it allows one to carry out detailed studies of molecular rotations and librations on a model-dependent basis over a very broad range of correlation times. Generally, the range of measurable correlation times spreads from ca. 10^{-11} to 10^{-7} s for conventional CW ESR but expands to ca. 10^{-3} s if nonlinear saturation transfer methods are employed (Thomas et al. 1976).

In order to start our discussion of high-field ESR spectra of spin labels in various motional averaging regimes, we would like to review briefly the physical origin of electronic relaxation and anisotropic broadening in ESR spectra. Let us consider a molecule bearing an electronic spin and undergoing stochastic motion. One of the results of such a motion is a stochastic reorientation of a molecule with respect to the external magnetic field. Thus, for such a molecule polar angles $\{\theta, \phi\}$ in Eqs. (17)–(19) are not constants but time-dependent functions: $\theta = \theta(t)$ and $\phi = \phi(t)$. Analogously, the spin Hamiltonian for such a system becomes time dependent because magnetic parameters are replaced by time-dependent matrices: $\boldsymbol{g} = \boldsymbol{g}(t)$ and $\boldsymbol{A} = \boldsymbol{A}(t)$. It is convenient to break such a time-dependent spin Hamiltonian into isotropic H_0 and anisotropic $H_1(t)$ parts:

$$H = H_0 + H_1(t) . \tag{20}$$

The anisotropic part is given by

$$H_1(t) = \mu_B \vec{B} \cdot \left[g(t) - g_{\text{iso}} \right] \cdot S + hS \cdot \left[A(t) - A_{\text{iso}} \right] \cdot I . \tag{21}$$

The isotropic parameters are defined as

$$g_{iso} = \frac{1}{3}\mathrm{Tr}\{g\} = \frac{1}{3}\left(g_{xx} + g_{yy} + g_{zz}\right), \tag{22}$$

$$A_{iso} = \frac{1}{3}\mathrm{Tr}\{A\} = \frac{1}{3}\left(A_{xx} + A_{yy} + A_{zz}\right), \tag{23}$$

and anisotropic angular and time-dependent matrices $\mathbf{g}(t)$ and $\mathbf{A}(t)$ are traceless.

In liquid solutions spin-labeled fragments of biomolecules undergo a complex reorientational motion. While the isotropic part of the spin Hamiltonian is invariant to such fluctuations, the anisotropic part, $H_1(t)$, fluctuates in time. Such fluctuations modulate the energy levels and therefore the transition frequencies. Thus, due to stochastic rotational motion the ESR transitions for the same spin should occur at different fields/frequencies, resulting in broadening of the spectral lines.

The magnitude of such a broadening depends upon the magnitude and frequency of the stochastic process. If one would wait for a long time, statistical average of $H_1(t)$ should be equal zero for a random process. However, this is not true for the mean-square interaction: $\langle |H_1(t)|^2 \rangle \neq 0$. Thus, if one measures $H_1(t)$ at moment of time t and then again at $t + \tau$, the results will depend upon how τ is compared with the characteristic time of the fluctuation process. If τ is very short, then the result of measurements should be the same as in moment t. If τ is long, then all "memory" of the preceding event should be lost. It is customary to describe a random process $R(t)$ by a time-dependent function $f(t)$ that has a zero time average and a root-mean-square of unity:

$$R(t) = R^0 f(t), \tag{24}$$

where R^0 is a normalization constant that is equal to the root-mean-square of $R(t)$, i.e., $R^0 = \sqrt{\langle R^2(t) \rangle}$.

The "memory" of a random process could be expressed in terms of autocorrelation function $G(\tau)$, defined as

$$G(\tau) = \overline{f(t)f(t+\tau)}\ , \tag{25}$$

where the horizontal bar indicates an ensemble average. Clearly, autocorrelation function G is independent of time of origin t due to the stationary nature of the random process. Moreover, from the definition of random function $f(t)$, function $G(0) = 1$ and $G(+\infty) = 0$. While the detailed form of $G(\tau)$ is determined by the physical model of the fluctuation process, in general, any statistical model leads to an exponentially decaying $G(\tau)$:

$$G(\tau) = \exp(-|\tau|/\tau_c)\ , \tag{26}$$

where τ_c is the characteristic time of decay. Parameter τ_c is also called the correlation time. For a random tumbling of a spin label in solution, τ_c roughly corresponds to the average time of rotation over one radian. Faster tumbling results in a shorter τ_c and a faster decaying $G(\tau)$.

While the autocorrelation function describes the stochastic fluctuations occurring in the system on the timescale, it is also useful to consider the frequency spectrum of the fluctuations. Indeed, a fluctuating magnetic field might have a component alternating at resonant frequency ω_0 of the spin system. These fluctuations should be more efficient in causing transitions between the states. In the frequency domain the function corresponding to $G(\tau)$ is called the spectral density function, $j(\omega)$, and is related to $G(\tau)$ through a Fourier transform:

$$j(\omega) = \int_{-\infty}^{+\infty} G(\tau)\exp(-i\omega\tau)d\tau . \tag{27}$$

For an exponentially decaying autocorrelation function $G(\tau)$, given by Eq. (26), the corresponding spectral density function is obtained by integration (Eq. (27)). This yields a Lorentzian function:

$$j(\omega) = \frac{2\tau_c}{1+\omega^2\tau_c^2} . \tag{28}$$

Both the frequency and the magnitude of fluctuations of the anisotropic part, $H_1(t)$, of the spin Hamiltonian play important roles in relaxation theory. Indeed, one of the results of such fluctuations is that the electronic spin is experiencing an effective magnetic field $B_1(t)$ that randomly fluctuates. Let us assume for the sake of simplicity that a spin-bearing molecule undergoes an isotropic Brownian rotational diffusion. Then the fluctuating field could be considered as a sum of three orthogonal components $B_{1x}(t)$, $B_{1y}(t)$, and $B_{1z}(t)$ that are characterized by the same correlation time τ_c. Moreover, let us assume for simplicity that all orthogonal components of the fluctuating magnetic field have equal root-mean-square values: $B_{1x}^0 = B_{1y}^0 = B_{1z}^0$. If the random field oscillating in the x-direction has a nonzero density oscillating at spin Larmor frequency ω_0 (in other words, if spectral density $j(\omega_0) \neq 0$), such a field would lead to transitions between spin levels the same way radiofrequency field $B_1\cos(\omega_0)$ causes spin transition in a magnetic resonance experiment. The probability of such a transition is given by time-dependent perturbation theory (Eq. (10)), and therefore corresponding transition rate W_x is given by

$$W_x = \frac{1}{4}\gamma_e^2 \left(B_{1x}^o\right)^2 j(\omega_0) . \tag{29}$$

For isotropic tumbling $B_{1x}^0 = B_{1y}^0$ and therefore the y-component of the fluctuating magnetic field should be equally effective in causing longitudinal spin transitions. Thus, $W_x = W_y$, and because each transition decreases the population difference by 2, overall longitudinal transition rate W becomes

$$W = 2\left(W_x + W_y\right) = \gamma_e^2 \left(B_{1x}^0\right)^2 j(\omega_0)\,. \tag{30}$$

The longitudinal relaxation process is typically described in terms of a characteristic time $T_1 = W^{-1}$. For an exponential approximation of autocorrelation function $G(\tau)$, spectral density function $j(\omega)$ is a Lorentzian (Eq. (28)) and

$$\frac{1}{T_1} = \gamma_e^2 \left(B_{1x}^0\right)^2 \frac{2\tau_c}{1+\omega_0^2\tau_c^2}\,. \tag{31}$$

If rotational correlation time τ_c is short compared to Larmor frequency ω_0, i.e., under "extreme narrowing" condition $\omega_o^2\tau_c^2 << 1$, then the longitudinal relaxation rate becomes proportional to correlation time τ_c:

$$\frac{1}{T_1} \approx 2\gamma_e^2 \left(B_{1x}^0\right)^2 \tau_c\,. \tag{32}$$

In the course of these electronic transitions caused by the fluctuating magnetic field, the spin-system exchanges energy with the reservoir and this constitutes a non-adiabatic thermodynamic process. Thus, the fluctuating magnetic field in both the x- and y-directions leads to non-adiabatic (also called non-secular) contributions to the relaxation process.

Some further insight into relaxation processes can be obtained using a macroscopic model of the magnetic resonance phenomena that considers magnetization of the bulk sample arising from a population difference in spin levels in a static magnetic field. Such a model was initially proposed by Felix Bloch in 1946 and was based mainly on phenomenological arguments. In equilibrium, the magnetization vector is parallel to external magnetic field and is proportional to Δn, given by a Bolztman distribution (Eq. (8)). Thus, for a magnetic field aligned with the z-axis $M_z = M_0$ while $M_x = M_y = 0$. Moreover, if a magnetization vector $\vec{M} = (M_x, M_y, M_z)$ was initially out of equilibrium, it is expected to return to equilibrium through first-order relaxation processes:

$$\begin{aligned} \frac{dM_z}{dt} &= \frac{M_0 - M_z}{T_1}, \\ \frac{dM_y}{dt} &= -\frac{M_y}{T_2}, \\ \frac{dM_x}{dt} &= -\frac{M_x}{T_2}. \end{aligned} \tag{33}$$

These equations represent a simplified case of the famous Bloch equations that describe the macroscopic behavior of a magnetization vector in the presence of an external magnetic field. The Bloch equations are obtained by adding a classical equation of motion of magnetic moment $\vec{M}$ in external magnetic field $\vec{B}$:

$$\frac{d\vec{M}}{dt} = -\gamma_e \vec{B} \times \vec{M} \tag{34}$$

to relaxation Eqs. (33).

According to this macroscopic model, external static magnetic field B_0 induces precession of the magnetization vector around B_0 at Larmor frequency $\omega_0 = -\gamma_e B_0$. If the phase of Larmor precession of all spins is the same, the magnetization vector is expected to have a component that would oscillate in the x-y plane with the same Larmor frequency, ω_0. Then relaxation in the x-y plane could be viewed as a dephasing of the individual magnetization vectors.

By solving full Bloch equations for a magnetic resonance experiment carried out under steady-state (CW) conditions, one obtains a Lorentzian shape for the absorption ESR signal:

$$g(\nu) = \frac{2T_2}{1 + T_1 T_2 \gamma_e^2 B_1^2 + 4\pi^2 T_2^2 (\nu_0 - \nu)^2}, \tag{35}$$

where B_1 is the amplitude of the radiofrequency field of resonance frequency ν_0 applied perpendicular to static field B_0. Under non-saturating conditions, i.e., when $B_1 << \gamma_e^{-1}(T_1 T_2)^{-1/2}$, the shape of the absorption signal is a Lorentzian function:

$$g(\nu) = \frac{2T_2}{1 + 4\pi^2 T_2^2 (\nu_0 - \nu)^2}, \tag{36}$$

with the width at half height:

$$\Delta\nu_{1/2} = \frac{1}{\pi T_2}. \tag{37}$$

It should be noted here that the changes in the magnetization vector along the z-direction could only occur through energy exchange between the spin system and the lattice. This constitutes a non-adiabatic process for the spin system and is also called spin-lattice or longitudinal relaxation. In general, the probability of such a process depends on the quantum states of both the spin system and the lattice. The physical origin of the longitudinal relaxation characterized by relaxation time T_1 lays in spin transitions caused by an oscillating field of molecular origin, as we already discussed in this section (see Eqs. (29)–(32)).

Transverse relaxation of either the M_x or M_y components of the magnetization vector does not involve energy transfer from the spin system to the lattice, and this constitutes an adiabatic process. Such a process could, in principle, occur without any participation from the lattice and therefore should be characterized by a different relaxation time, i.e., $T_1 \neq T_2$.

Several physical mechanisms contribute to T_2. From a quantum-mechanical prospective, any spin transitions between the $|\alpha\rangle$ and $|\beta\rangle$ states should also lead to a loss of phase in the x-y plane. Thus, the same non-secular terms (i.e., spectral density $j(\omega_0)$) contributing to longitudinal relaxation should also be involved in the transverse relaxation. Spins could also exchange energy adiabatically. For example, for spins coupled via spin–spin interaction, a downward transition in energy for the first spin could be accompanied by an upward transition for the second spin (this is also called the flip-flop mechanism). The frequency of such flip-flops depends on secular spectral density $j(0)$. More detailed considerations lead to the following expression for the transverse relaxation time:

$$\frac{1}{T_2} = \frac{1}{2T_1} + \frac{1}{2}\gamma_e^2 \left(B_{1z}^0\right)^2 j(0) = \frac{1}{2}\gamma_e^2 \left(B_{1x}^0\right)^2 \{j(\omega_0) + j(0)\} , \tag{38}$$

where $B_{1x}^0 = B_{1z}^0$. Under extreme narrowing conditions, $\omega_o^2\tau_c^2 << 1$ and $j(\omega_0) = j(0)$, the relaxation times are becoming equal, $T_1 = T_2$, and transverse relaxation time becomes inversely proportional to correlation time τ_c the same way the longitudinal relaxation time does (Eq. (32)). Then, by taking into consideration Eq. (37), we conclude that the half width of the resonance line observed in the CW experiment is proportional to the correlation time of the fluctuating field.

It should be noted that Eqs. (31) and (39) were derived for our simplified model in which the fluctuating field has the same root-mean-square averages in all directions. Depending on the molecular origin of this field, particular forms of spectral density functions will be different, and readers should consult more comprehensive treatments of the relaxation effects in magnetic resonance (Abragam 1961).

2.4. X-Band and High-Field ESR Spectra of Spin Labels: Fast Molecular Motion

The fast-motional regime for a spin system is defined as a stochastic process for which the anisotropic part of the spin Hamiltonian, $H_1(t)$, is effectively averaged out. This condition can be defined as

$$\left|h^{-1}H_1(t)\right|\tau_c \leq 1. \tag{39}$$

In this regime the ESR spectrum of a ^{14}N ($I = 1$) nitroxide spin label consists of three equally spaced lines, and the only spectral manifestation of stochastic tumbling of the nitroxide is in broadening of the ESR lines. Note that at magnetic fields above 0.3 T (i.e., X-band ESR and higher frequencies), the nitroxide spectra can be considered in the high-field limit, so second-order shifts of nitrogen hyperfine lines are negligible.

When a nitroxide molecule reorients in the external static magnetic field, the electronic spin will experience a fluctuating magnetic field because of the anisotropy of both nitrogen hyperfine and Zeeman interactions. It is easy to see that the broadening of each of the nitrogen hyperfine transitions should be different and also be affected by the magnetic field of the ESR experiment from the following simple but approximate considerations. First, let us note that during allowed electronic spin transitions nuclear quantum number m_I should remain the same. Now, let us consider the spins that have $m_I = 0$. Electronic transitions for these spins would give raise to the middle nitroxide hyperfine component. For each of the instantaneous orientations of the nitroxide moiety in the magnetic field the position of this component is only determined by $g(\theta, \phi)$ given by Eq. (18). If we trace the position of the middle nitrogen hyperfine component for the x-, y-, and z-axis crystal orientations of X-band spectra shown in Figure 1, we notice that this component has the smallest spread in magnetic field position compared to the two other components. Thus, an effective fluctuating magnetic field for the spin system with m_I= 0 would be the smallest, and so the line width that is proportional to T_2^{-1}. With increase in the magnetic field to ca. 3.4 T (W-band, Figure 1), the single-crystal spectra become much further separated due to the enhanced contribution of the Zeeman anisotropy, and this should increase the magnitude of relaxation effects and lead to further line broadening.

Without going into further details, the homogeneous line width $\Delta H_{p\text{-}p}(m_I)$ of each of nitrogen hyperfine transitions m_I in the fast-motional regime can be expressed in terms of spectral parameters A, B, and C:

$$\Delta H_{\text{p-p}}(m_I) = (A + A') + Bm_I + Cm_I^2, \tag{40}$$

where all frequency-independent contributions to the line width, such as, for example, those arising from spin-exchange, are included in the A' term. Spin exchange could take place between two electrons that have overlapping orbitals. The Pauli exclusion principle dictates that no two electrons in a system can have identical quantum numbers. Thus, during collision the electronic spins exchange (i.e., hop simultaneously). Such a flip-flop mechanism is independent of magnetic field and would lead to transverse relaxation through dephasing magnetization in the x-y plane. Comprehensive treatment of spin exchange effects has been presented by Molin and coworkers (1980). These spectral parameters contain contributions of several modified spectral densities that are measurable under conditions of ESR experiments (Budil et al. 1993):

$$\begin{gathered} A = 2J_1^{DD}(\omega_n) + \frac{2}{3}J_0^{DD}(\omega_0) + 4J_2^{DD}(\omega_0) + \frac{8}{3}J_0^{GG}(0) + 2J_1^{GG}(\omega_0), \\ B = \frac{16}{3}J_o^{DG}(0) + 4J_1^{DG}(\omega_0), \\ C = \frac{8}{3}J_0^{DD}(0) - J_1^{DD}(\omega_n) + 2J_1^{DD}(\omega_0), \end{gathered} \tag{41}$$

In these equations the spectral densities are modified to include the magnitude of anisotropic magnetic interactions, and therefore we use a capital J instead of a lower case j as in Eq. (27). The modified spectral densities in Eq. (41) arise from rotational modulation of the anisotropic part of nitrogen hyperfine tensor $\boldsymbol{A}$ (the anisotropic part arises from an electron–nuclear dipolar interaction, and therefore this tensor is also called END, and corresponding spectral densities are J^{DD}), the electronic **g**-matrix (J^{GG}), and the correlated effect of both tensors (J^{DG}). In these equations ω_0 and ω_n are electronic and nuclear resonant frequencies corresponding to magnetic field B_0, and the subscript identifies the net change in ($m_S + m_I$), where m_S and m_I are the z component quantum numbers of the electronic spin and nuclear spin, respectively. These spectral densities are defined precisely in the Appendix to (Budil et al. 1993).

One important observation is that the spectral densities arising from the electronic **g**-matrix (J^{GG}) contribute only to spectral parameter A. These densities increase quadratically with the static magnetic field and become dominating at magnetic fields above 1.2 T. Secular spectral density $J_o^{DG}(0)$ arising from rotational modulation of both the **g**-matrix and hyperfine **A**-tensor contributes to parameter B in a correlated way, resulting in an approximate proportionality of this parameter to the magnetic field of ESR experiment (if nonsecular spectral density $J_1^{DG}(\omega_0)$ is neglected). Parameter C remains approximately unchanged, because the contributing secular spectral densities (J^{DD}) arise from rotational modulation of the nitrogen hyperfine tensor and are frequency independent.

The best way to determine the values of line width parameters is by least-squares simulations (Smirnova et al. 1995). However, some estimates of line width can also be obtained from measurements of the peak-to-peak line heights of the first-derivative ESR spectra because for the same line shape the peak height is inversely proportional to the square of the line width.

Characteristic ESR spectra of a nitroxide in a fast-motion regime are shown in Figure 2 (top two spectra recorded at $T = 291$ and 189 K) as an example of a 0.05-mM solution of a small molecule, perdeuterated 2,2′,6,6′-tetramethyl-4-piperidone-1-oxyl (perdeuterated Tempone or PDT), in toluene d_8 at the X- and W-bands. PDT represents a six-membered ring nitroxide without a resolved superhyperfine structure. Both samples were prepared in quartz tubes, carefully deoxygenated by a freeze-pump-thaw technique, and then flame-sealed with a torch.

It is apparent that at room temperature the X-band ESR spectrum of PDT falls into the extreme narrowing regime, in which all three nitrogen hyperfine lines have almost equal line widths and so amplitudes (Figure 2, 291 K). The width of those lines (ca. 0.02 mT peak-to-peak) is determined primarily by inhomogeneous broadening (ca. 0.01 mT peak-to-peak) caused by unresolved hyperfine interactions with deuterons, and by homogeneous contributions mainly from the Heisenberg nitroxide–nitroxide spin exchange and spin–rotation coupling. The homogeneous line width arising from rotational modulation of the **g**-matrix and **A**-tensor is small and difficult to measure under these conditions.

Although the 94-GHz spectrum at the same temperature (Figure 2, 291 K) also falls within the line narrowing regime, the lines are significantly broader primarily due to the contribution of the A term, which increases quadratically with the magnetic field. The magnitude of parameter B, which is equal to $(\Delta H_{p\text{-}p}(m_I = 1) - \Delta H_{p\text{-}p}(m_I = -1))/2$, increases by a factor of 10 from 9.0 to 94 GHz, and this results in a larger difference in width between the high- and low-field nitrogen hyperfine component as compared with fast-motion X-band ESR spectra. While ratio C/B is

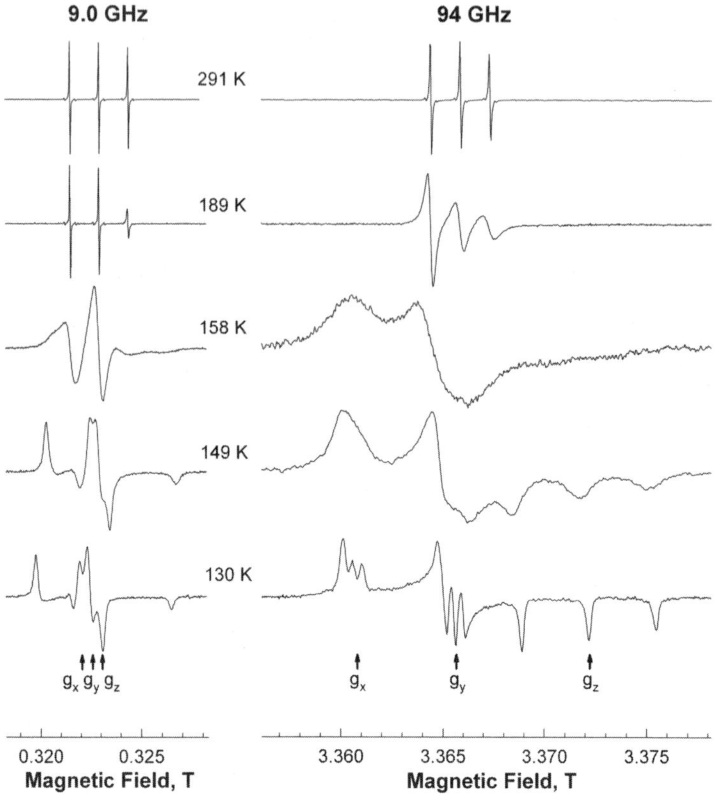

Figure 2. Experimental 9.0- and 95-GHz ESR spectra from 0.05 mM solution of perdeuterated 2,2′,6,6′-tetramethyl-4-piperidone-1-oxyl (Tempone or PDT) in toluene-d_8 recorded at several temperatures. Both samples were carefully deoxygenated. The approximate position of the principal axis orientations of the *g*-matrix is indicated by arrows in the spectra at 130 K that approach the rigid limit. Reproduced with permission from Smirnov et al. (1998). Copyright © 1998, Springer.

affected by the relative magnitude of the components of rotational diffusion tensor (Budil et al. 1993), for many nitroxides at the X-band $(-B) \leq C$. Therefore, the contributions of B and C to the low-field $m_I = 1$ line are nearly cancelled, resulting in almost equal intensity for the first two lines of the nitroxide X-band ESR spectra. Typically, at magnetic fields of ca. 0.3 T the magnitude of the C term is somewhat greater than $(-B)$, and therefore the middle $m_I = 0$ nitrogen hyperfine component appears sharpest in X-band ESR spectra. At the W-band the absolute value of parameter B increases about tenfold while parameter C, which is independent of magnetic field, remains the same. Because parameter B is negative, at the W-band (and higher frequencies) the low-field $m_I = 1$ component is always narrowest, resulting in a very characteristic shape of the fast-motion spectrum (Figure 2, T = 189 K).

Comparison of 9.0- and 94-GHz ESR spectra at different temperatures (Figure 2) also illustrates that at the W-band the onset of slow-motional effects for the same nitroxide–solvent system occurs at a higher temperature than it does at the X-band. This is a direct consequence of an increase in the magnitude of the anisotropic component, $|H_1(t)|$, of the spin Hamiltonian with the magnetic field.

2.5. X-Band and High-Field ESR Spectra of Spin Labels: Slow and Intermediate Motion Regime

While general treatment of anisotropic relaxation effects and modeling of slow-motional ESR spectra are commonly carried out using quantum mechanical matrix formalism, here we would like to use a much simpler two-site jump model to explain some of the basic features of slow-motional high-field ESR spectra. This jump model was initially developed in the early fifties to explain rate-dependent phenomena in NMR spectra for molecules that could coexist in several distinct conformations (Gutowsky et al. 1953; Gutowsky and Saika 1953). One of the simplest examples of rate-dependent NMR spectra for a molecule coexisting of two equally populated conformations is given by Gutowsky and Holm (1956). At low temperatures, the molecule is effectively rigid. Let us assume that the difference in structure for these two conformations gives rise to two distinct magnetic resonance lines separated by $\delta\omega$, such as lines 4 shown in Figure 3. The spectra in Figure 3 are all plotted as absorption signals following the original graphs presented by Gutowsky and Saika (1953). With an increase in temperature the energy of the molecule becomes sufficient to overcome the energy barrier separating these two conformations, and this makes it possible for some of the molecules to "jump" between the two states. With further increase in temperature the jumps should become more frequent. When the jump frequency becomes sufficiently high, the spin is expected to experience only an average environment. Then the information on different conformations is lost and the magnetic resonance signal is expected to appear as a single line centered between the original lines.

Indeed, such behavior is often observed in magnetic resonance (e.g., see Gutowsky and Saika (1953)). The explanation of this effect was first given by Gutowsky and coworkers (1953) and reported independently by Hahn and Maxwell (1952), as well as by several other authors. In the simplest model, two equal lines A and B are separated by $\delta\omega$ and centered at $\omega_0 - \delta\omega/2$ and $\omega_0 + \delta\omega/2$ such as lines 4 in Figure 3. When there are no jumps, the spins obey Bloch equations at each of the sites (for detailed deviations, see Appendix F in Slichter (1992)). Solutions of these equations yield two Lorentzian shapes (lines 4, Figure 3). When a molecule jumps instantaneously between sites A and B (this process is also called spin exchange because it can be viewed as instantaneous swapping of spins A and

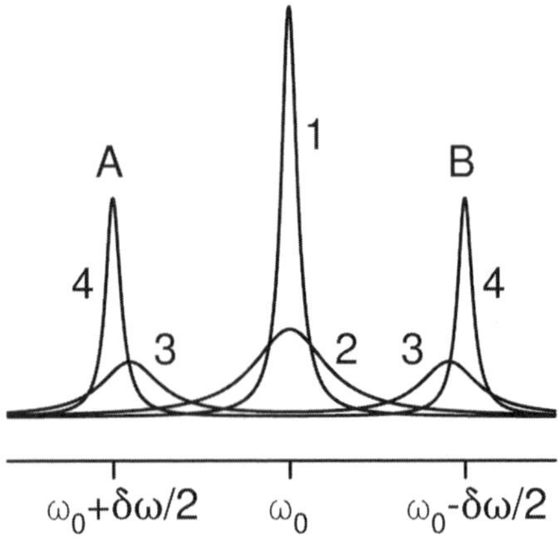

Figure 3. Effect of instantaneous jump diffusion on magnetic resonance spectra from a molecule coexisting in two distinct equally populated conformations. In the absence of jumps, lines A and B are separated by $\delta\omega$. Lines 1–4 are drawn for the following values of parameter $2\tau\delta\omega$, where τ is the time between jumps: (1) -10^{-2}, (2) -10, (3) -1, (4) -10^2. See the text for further details. This figure was redrawn based on results and discussions from Gutowsky and Saika (1953).

B), the differential equations should be solved simultaneously by taking into account the changes in magnetization due to jumps of frequency $F = 1/\tau$ (where τ is the interval between jumps). Solution of these equations yields the line shapes that are plotted in Figure 3 for various products of τ and $\delta\omega$.

When jump frequency $F= 1/\tau$ is small compared to line separation $\delta\omega$, the separate Lorentzian lines are essentially undistorted (for example, lines 4 in Figure 3 correspond to $2\tau\delta\omega = 10^2$). With increasing F (and, thus, decreasing τ) the lines are broadened first (lines 3 in Figure 3; $2\tau\delta\omega = 10$) and then coalesce into a single broad line (line 2, Figure 3; $2\tau\delta\omega = 1$), which becomes even narrower with a further increase in F (line 1, Figure 3; $2\tau\delta\omega = 10^{-2}$).

The instantaneous jump diffusion model is also valid for rapid jumps between different orientations of a spin-labeled molecule in a magnetic field because the resonance conditions depend upon the orientation of the molecular frame with respect to the magnetic field (see Eqs. (17)–(29)). Although for many cases of rotational diffusion in liquids the jump model is only an approximation, the results summarized in Figure 3 are very illustrative for understanding the shapes of ESR spectra of nitroxide spin labels in various motional regimes. Let us assume that the time between jumps can be approximated by rotational correlation time τ_c. The physical meaning of rotational correlation time τ_c is the average period of rotation over one radian. Upon rotating by about one radian, the ESR spectrum of a nitroxide single crystal could change significantly. For example, the orientation of the molecular frame in a magnetic field could change the x-axis to the z-axis. Thus, frequency separation $\delta\omega$ in the jump diffusion model could be approximated by the spin Hamiltonian anisotropy expressed in frequency units, i.e., $\left|h^{-1}H_1(t)\right|$. Then the fast-motion condition given by Eq. (26) corresponds to the $2\tau\delta\omega \leq 1$ spin jump regime under which the lines coalesce into a single resonance. This is illustrated in Figure 3 (top), which shows nitroxide spectra in the fast-motion regime: the lines corresponding to different orientations of a nitroxide molecule in a magnetic field coalesce into a single line for each of the three projections of nitrogen nuclear spin $I = 1$. This gives rise to characteristic three-line spectra with isotropic magnetic parameters given by Eqs. (22) and (23). The fast-motional regime was discussed in the previous section of this chapter.

The jump model is also useful for qualitative characterization of spin-label ESR spectra corresponding to an intermediate motion regime. Let us consider the $m_I = 0$ nitrogen hyperfine transition first. The only anisotropic contribution to this transition is from the $\boldsymbol{g}$-matrix. At the X-band this anisotropy is significantly smaller than that of the $\boldsymbol{A}$-tensor. This is also easy to see from the X-band single-crystal spectra shown in Figure 1 (top): the magnetic field shifts between the middle nitrogen hyperfine components of the spectra for the x-, y-, and z-orientations are much smaller than those for the outer lines. Thus, the fast-motion condition of jump model $2\tau\delta\omega \leq 1$ for the "central" $m_I = 0$ transitions is satisfied for longer τ than for the $m_I = \pm 1$ "outer" transitions. Therefore, the slow-motional effects in X-band ESR spectra appear first as asymmetric shapes and field shifts of the outer hyperfine components, while the electronic transitions for the central $m_I = 0$ line (such as in the X-band spectrum at 158 K in Figure 2) still satisfy fast-motional condition (26). Thus, the width of the middle component of the X-band nitroxide spectra, $\Delta H_{p\text{-}p}(m_I = 0)$, could be used for estimating rotational correlation time even for spectra that have intermediate and even slow-motion features. Experimental studies have confirmed that. For example, a systematic study of thirty spin-labeled cysteine mutants of T4 lysozyme showed that parameter $\Delta H_{p\text{-}p}(m_I = 0)$ is useful as an approximate indicator of the spin-labeled chain mobility (Mchaourab et al. 1996). This experimental line width parameter is now routinely used in site-directed spin-labeling ESR studies to differentiate various protein structural sites, such as surfaces of the helices, interhelical loops, buried sites, and sites involved in tertiary contacts (Mchaourab et al. 1996). It is also becoming customary in spin-labeling literature to report not $\Delta H_{p\text{-}p}(m_I = 0)$ but $1/\Delta H_{p\text{-}p}(m_I = 0)$ which is often called a "mobility" parameter.

Increasing the magnetic field/resonance frequency to 3.4 T/94 GHz (W-band), the anisotropy of the electronic Zeeman term becomes the dominating part of $H_1(t)$. The W-band single-crystal spectra corresponding to the x-, y-, and z-orientations are now well resolved, and the shifts in magnetic field positions (and therefore $\delta\omega$ in the jump model) for the nitrogen hyperfine components with the same m_I also increase. These shifts also become similar in magnitude. Now let us consider instantaneous jumps between the x-, y-, and z-orientations of a spin label. It is easy to conclude that for each of the nitrogen hyperfine transitions the fast-motion condition for such jumps would break down at about the same τ. (Also, because of increased separation of lines the critical τ should also be shorter than that at the X-band.) Then, none of the components of W-band ESR spectra can be used to estimate "mobility" the same way as can be done for X-band spectra. We illustrate this with the example of a small nearly spherical molecule such as PDT. Such a molecule should tumble in solution almost isotropically. Then upon decreasing viscosity the slow-motion features for such a molecule would appear across the entire W-band ESR spectrum, as illustrated in Figure 2 ($T = 158$ K).

Another observation is that with increasing rotational correlation time (this corresponds to lower temperatures in Figure 2) the overall width of both the W- and X-band spectra increases significantly. As the overall width of a magnetic resonance spectrum increases, so does the second moment of the integrated spectrum. The latter is a very useful indicator of molecular motion, as was shown by Van Vleck (1948) in his famous paper. The second moment was previously applied to characterize various spin-labeled sites of T4 lysozyme from X-band ESR spectra (Mchaourab et al. 1996). Figure 2 illustrates that the width of the W-band spectra for nitroxides in an intermediate and/or slow-motional regime exceeds that of corresponding X-band spectra. Thus, for such motional regime the second moment of high-field ESR spectra could be a simple but yet very useful indicator of nitroxide chain dynamics.

It should be noted that some experimental high-field ESR spectra might be distorted by baseline drift and/or admixture of a dispersive component. The latter is rather typical for liquid aqueous samples. For such spectra the second moment might be difficult to compute accurately. However, the second moment could also be approximated as the maximum splitting squared. Then this splitting could be used as an empirical parameter to characterize nitroxide dynamics from HF ESR spectra. Indeed, the plots of this splitting as well as the apparent field positions of the principal axis components were employed by Lebedev (1990) to characterize molecular motion and also by other authors in more recent studies.

Superior angular resolution of nitroxide ESR spectra at high magnetic fields compared to conventional X-band facilitates characterization of anistropic motion such as rotational diffusion of phospholipids and peptides in lipid bilayers. As an illustration, let us consider a fatty acid labeled at acyl chain position 5 with a doxyl nitroxide (5DSA, 5-doxyl stearic acid, Figure 4A). In a lipid bilayer such a molecule aligns along the phospholipids. We doped a lipid bilayer formed from DPPC (1,2-dipalmitoyl-*sn*-glycero-3-phosphatidylcholine) with 5DSA at 1 mol%. When the DPPC bilayer is in a fluid phase, the lipids and stearic acids undergo rapid rotational diffusion about the long axis of the lipid. For 5DSA this axis coincides approximately with the nitroxide z-axis (Figure 4A). Thus, if the rotational reorientation about the z-axis is fast, one could expect that the x- and y-features of high-field ESR spectra will be averaged out while the z-component will not. This is indeed observed in experiment. Figure 4B shows an experimental W-band ESR spectrum from such a system at 314 K. For comparison, a rigid-limit spectrum of such a system taken at 100 K is also shown (Figure 4C). Comparison of these two spectra demonstrates that while the z-axis principal axis component for the "fluid"

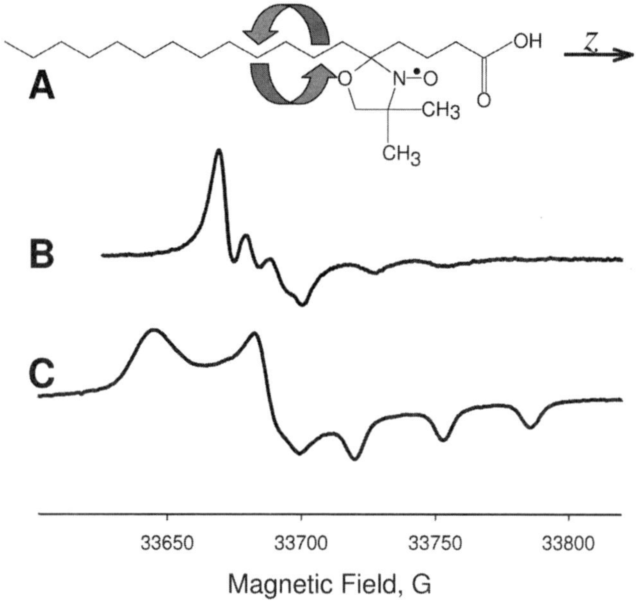

Figure 4. (**A**) Chemical structure of 5-doxyl stearic acid (5DSA) and a cartoon illustrating a predominantly axial rotation of this spin label in a phospholipid bilayer. The doxyl nitroxide plane (z-direction of the nitroxide molecular frame) is approximately perpendicular to the long axis of this molecule. (**B**) Experimental W-band spectrum of 5DSA in DPPC bilayer (1:100 molar ratio) at 314 K exhibits well-resolved z- and pre-averaged x- and y-components. Such a spectrum is characteristic of a fast axial rotation of a nitroxide around its z-axis. For comparison, a rigid limit spectrum (T = 100 K) from the same system is shown at the bottom (**C**).

spectrum (314 K) is similar to that of the rigid-limit powder average (100 K), the x- and y-features are not. Specifically, at 314 K the latter two features coalesce into a single three-line component centered at approximately $(g_x + g_y)/2$.

There are several conclusions that can be drawn from our qualitative discussion of intermediate and slow-motion high-field ESR spectra. All of these conclusions are related to the dominance of g-factor anisotropy over the hyperfine term at ESR frequencies of 95–250 GHz and higher:

(a) The sensitivity to molecular motion in both the fast- and slow-motion regimes is greatly enhanced.

(b) The fast-motion condition breaks down at shorter correlation times. Therefore, for almost all spin-labeled protein and membrane systems, perhaps with the exception of small spin-labeled peptides, one would expect high-field ESR spectra to fall into slow or intermediate motion regimes.

(c) High-field ESR spectra of nitroxides have superior angular resolution compared to conventional X-band ESR. Thus, high-field ESR methods are very advantageous in studying angular orientations of spin-labeled fragments of complex biomolecules with greater accuracy and, in particular, to deduce the static and dynamic behavior along the nitroxide *x*- and *y*- molecular frame axes.

2.5. Introduction to Stochastic Liouville Theory of Slow Motion ESR Spectra

While the qualitative discussion of intermediate and slow motion of high-field ESR spectra of nitroxide spin labels we have presented here is useful for demonstrating the basics, for detailed characterization of molecular dynamics of spin-labeled molecules the reader is directed to the slow-motion theory based on solving the stochastic Liouville equation (SLE). This approach has been the research focus of the group of Professor Freed (Cornell) and has been proven to be an important means of analyzing rather complex spectra that can be observed both in biological and model systems (Schneider and Freed 1989; Polimeno and Freed 1990; Polimeno and Freed 1995; Budil et al. 1996; Earle et al. 1996; Barnes et al. 1999; Pilař et al. 2000; Lou et al. 2001).

In brief, the slow-motion ESR line shape theory employs density matrix formalism to describe the coupled spin and spatial degrees of freedom. The equation of motion for density matrix ρ is governed by the Liouville-von Neumann equation:

$$\frac{d}{dt}\rho = \frac{1}{i\hbar}[H_0 + H_1(t), \rho] \ , \tag{42}$$

where H_0 and $H_1(t)$ are given by Eqs. (20) and (21). While for short correlation times τ_c (fast-motion regime) the effects of $H_1(t)$ may be treated using perturbation theory (motional narrowing theory), more general methods of analysis should be used for longer τ_c. By performing an ensemble average over the density matrix, it is possible to show that the resulting equation of motion for the density matrix is

$$\frac{d}{dt}\langle\rho\rangle = \frac{1}{i\hbar}[H, \langle\rho\rangle] + \Gamma\langle\rho\rangle, \tag{43}$$

where Γ is a stochastic operator describing stochastic processes giving rise to relaxation in the system, and operator *H* is the full quantum-mechanical Hamiltonian of the system. Then the main strategy for determining the slow-motion ESR line shape is:

1. choose the proper model for the stochastic process and derive the analytical form of operator Γ
2. solve Eq. (43) digitally in order to derive the ESR line shape

Several simplest forms of operator Γ, such as, for example, corresponding to isotropic Brownian diffusion, are discussed by Earle and Smirnov (2004). However, local dynamics of spin-labeled protein chains and/or membrane proteins can be rather complicated. For example, for spin-labeled proteins the rotational modes of the tether that links the nitroxide to the protein can also be affected by a slowly fluctuating protein and/or membrane environment as well as the overall tumbling modes of the protein. This complex reorientational dynamics of the nitroxide can be described by the slowly relaxing local structure (SRLS) model introduced by Polimeno and Freed (1995). In the SRLS model, the spin probe is assumed to be reorienting in a local environment that itself is relaxing on a longer timescale. In applications to macromolecular systems, the faster motion describes the internal dynamics, while the slower motions account for the global rotation of the macromolecule. This model has been successfully applied in studies of spin-labeled proteins and other model systems (Earle et al. 1996; Barnes et al. 1999; Pilař et al. 2000; Lou et al. 2001).

The studies of the Cornell group have shown that one of the principal advantages of the multifrequency ESR approach is an ability to "tune" the spectral response of the ESR line shapes to a specific timescale of spin-label motion by choosing the appropriate magnetic field/resonant frequency for the experiment. For example, X-band ESR spectra of spin-labeled proteins are generally sensitive to the overall protein tumbling modes, whereas these modes are often "frozen out" in high-field ESR spectra because of the larger magnitude of the anisotropic Hamiltonian that is subject to averaging. For the same reasons, the diffusion modes corresponding to rapid internal motions may be studied in great detail by analyzing the high-field ESR line shapes. To summarize, the fast tumbling modes are much easier to analyze from high-field ESR spectra while the slower modes have the greatest effect on line shape at lower frequencies. This is a manifestation of the so-called "snapshot" effect (Borbat et al. 2001).

3. HIGH-FIELD ESR OF SPIN-LABELED AQUEOUS SAMPLES: EXPERIMENTAL CONSIDERATIONS

3.1. Resonators for High-Field ESR and Aqueous Sample Configuration

We believe that concentration sensitivity is one of the most important considerations in planning spin-labeling studies of protein and membranes. Usually, the concentration requirements of spin-labeling ESR fall in between those of NMR and fluorescence: concentrations of at least the micromolar level are required for ESR of aqueous samples at the X-band. This is somewhat less than for a typical proton NMR experiment but more than that required for fluorescence.

Liquid aqueous samples create an additional problem for ESR because of the high dielectric losses occurring in water at microwave frequencies of 2–3 GHz and above. These losses are caused by interactions of electrical dipoles of water molecules with electrical component of the microwave field. This results in nonresonant (i.e., magnetic-field-independent) energy absorption and a rather short penetration depth of the electric field component into an aqueous sample. For many liquid aqueous samples the penetration depth of X-band microwaves is <1 mm, and this dictates special arrangements for aqueous samples and special sample holders. The basic idea for constructing such holders is to configure lossy aqueous samples in a way to minimize the Eddy currents through the sample. For example, in a rectangular TE_{102}-mode X-band ESR resonator the central plane of the cavity contains only an electrical field component and is suitable for accommodating a thin sandwich of an aqueous sample with minimal dielectric losses (Poole, 1997). This is accomplished with a special sample holder called a "flat cell" that is made of quartz. Another way to break Eddy currents is to employ a bundle of small capillaries (http://www.bruker-biospin.de/ESR/elexsys/accessory/aquax.html).

With an increase in microwave frequency the wavelength of the microwave field decreases, and so are the linear dimensions of both the ESR resonator and acceptable sample. The wavelength corresponding to X-band ESR is about 3 cm, and this sets the linear scale for resonators and sample holders. Non-lossy samples for X-band ESR are typically 3–4 mm to 10 mm in outside diameter, while capillaries for aqueous samples are only 0.8–1.0 mm i.d. Any such samples are easy to handle in a biophysical laboratory. The W-band wavelength is only 3 mm, and therefore dimensions of both the resonator structures and the samples should be approximately one-tenth those of the conventional X-band. From these simple considerations the largest tube diameter for W-band ESR should be about 1 mm o.d. for non-lossy samples, and aqueous capillaries should have about 0.1 mm i.d. (assuming the same type of resonator and a similar tangent of dielectric losses at both frequencies). Clearly, such fine capillaries are more difficult to handle.

These very simple concerns regarding tiny dimensions of the high-field ESR single-mode resonators and corresponding sample sizes were some of the reasons why the first probeheads of many high-field ESR spectrometers employed not the single-mode cavities but Fabri-Perot resonators (e.g., see Lebedev 1990; Grinberg and Dubinski 2004; Budil et al. 1989). Such resonators are often employed in optical spectroscopy and are based on forming a standing wave between two mirrors, one of which is semitransparent. With further developments in HF ESR instrumentation it was found that single-mode cylindrical resonators often outperform the Fabri-Perot instruments in terms of sensitivity (Lebedev 1990). Theoretical considerations show that under optimal conditions Fabri-Perot resonators should provide better sensitivity than the single-mode cavities (Lebedev 1990; Budil and Earle 2004), but this remains to be demonstrated experimentally.

Nowadays high-field ESR at 95 and 130–140 GHz (W- and D-band, respectively) typically employs single-mode or slightly oversized cylindrical cavities with TE_{011} or TE_{012} microwave modes, while probeheads for the experiments at higher frequencies employ Fabri-Perot resonators (see Poole 1997 for a detailed description of various ESR resonators and microwave modes). The three most common resonator types and sample configurations employed in high-field ESR today are shown in Figure 5.

In a semiconfocal Fabri-Perot resonator (Figure 5A) the mm wave forms a standing wave pattern between the bottom mirror (4) and the flat polarizing mirror (2). In this design the detection circuitry is effectively isolated from the incident

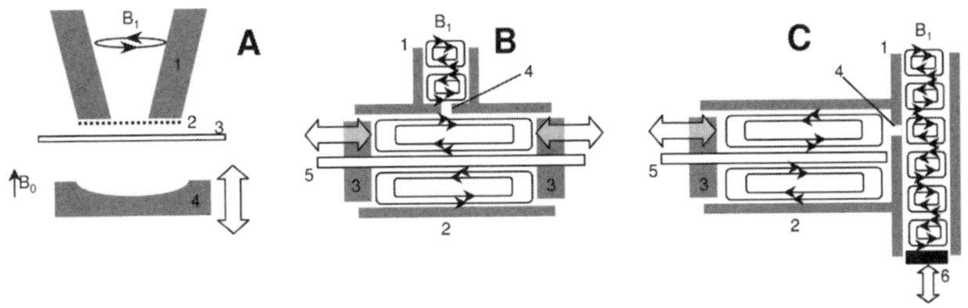

Figure 5. Typical resonators and sample arrangements employed in HF ESR. (**A**) Fabri-Perot resonator (semiconfocal arrangement): 1 = microwave conical taper; 2 = flat polarizing mirror; 3 = thin disc containing the sample; 4 = bottom (adjustable) mirror. (**B**) Cylindrical single TE_{011}-mode cavity coupled from the cylinder side: 1 = mm-waveguide; 2 = cylinder walls; 3 = adjustable plungers to vary resonant frequency; 4 = coupling hole; 5 = capillary sample. (**C**) Cylindrical single TE_{011}-mode cavity coupled from a plunger side: 1 = mm-waveguide; 2 = cylinder walls; 3 = adjustable plungers to vary resonant frequency; 4 = coupling hole; 5 = capillary sample; 6 = adjustable short. A detailed description of resonator modes was given by Poole (1997).

mm-wave radiation. This is a very useful feature for carrying out time-domain high-field ESR experiments (van Tol et al. 2005). In order to minimize non-resonant (dielectric) losses the aqueous sample should be placed in a plane of the magnetic field component of the mm-wave field (i.e., at the electrical field node). This is accomplished by sandwiching an aqueous sample between thin quartz or mylar discs (Barnes and Freed 1997). In order to achieve maximum sensitivity, the thickness of the sample should be optimized by taking into account the frequency dependence of the complex dielectric constant (Budil and Earle 2004), e.g., for water at 20°C and 250 GHz optimal sample thickness should be about 18 μm (Budil and Earle 2004). Such a sample could be configured by putting together two thin quartz discs with one of the discs having a etched groove. However, it might be difficult to maintain the 18-μm thickness uniformly throughout the sample. Nevertheless, this sample arrangement might be very beneficial for spin-labeled samples with flat geometry such as substrate-supported and aligned bilayers and also muscle fibers. For example, Barnes and Freed described a special "shunt" Fabri-Perot resonator to accommodate a thin disc-shaped sample of substrate-supported bilayer with director vector perpendicular to the polarizing magnetic field (Barnes and Freed 1998a). With the help of this resonator, they have carried out a study of macroscopically aligned model membranes composed of mixtures of 1,2-dimyristoyl-*sn*-glycero-3-phosphocholine (DMPC) and dimyristoylphosphatidylserine (DMPS) and a cholestane spin probe, which served as a cholesterol analog (Barnes and Freed 1998b). The use of a shunt Fabry-Perot resonator allowed these authors to examine two different orientations of membrane samples.

It was also shown that in Fabri-Perot cavities the ESR sample tubes could be mounted transversely. For example, Budil and coworkers (1993) described placing a quartz tube of 4-5 mm o.d. and a ≈1 mm wall thickness in a Fabri-Perot resonator. The tube contained essentially non-lossy solution of a spin label in toluene and had dimensions greater than the wavelength of this ESR experiment ($\lambda \approx 1$ mm). Lossy aqueous samples also can be mounted in Fabri-Perot resonators in a transverse orientation. However, in contrast to non-lossy samples, the capillary diameter must be much smaller than the wavelength. For example, recently, van Tol and coworkers (2005) reported good sensitivity for liquid aqueous solutions of spin labels placed in 0.1 mm i.d. capillaries that were mounted transversely in a semiconfocal Fabri-Perot resonator.

While Fabri-Perot resonators offer larger sample dimensions and a flexibility to access and, for example, illuminate the sample, the single-mode cavities provide the best sensitivity at 95–140 GHz over a broad range of temperatures. Figure 5B,C depicts two coupling schemes that are currently employed in such resonators. The scheme in Figure 5B was initially developed by Krymov and colleagues (Krymov et al. 1982). This resonator was successfully employed by Lebedev and coworkers, who pioneered high-field ESR in studies of spin-labeled molecules (see Lebedev 1990 and the references therein). Later the same resonator design was adopted by many other groups working in the field. In brief, this resonator has the same cylindrical TE_{011} microwave mode as the cylindrical single-mode X-band ESR cavities described by Poole (1997). The principal feature of Krymov's design is that the resonant frequency is adjusted by synchronous repositioning of the two plungers (Figure 5B) at both ends of the cylindrical cavity. This frequency adjustment is essential for tuning the cavity resonant frequency to that of the mm-wave source (in conventional X-band ESR the frequency of the microwave source is adjusted instead). Microwave is coupled to the cavity through a miniature hole at the center of the cylindrical sidewall. By rotating the cavity cylinder with respect to the waveguide, a wide range of coupling conditions can be achieved. All parts of this resonator are machined from a non-magnetic alloy. Note that in this design the cavity is disconnected from the waveguide every time the resonator is taken out of the bore of the magnet for sample replacement. In addition, one should be able to rotate the cavity with respect to the waveguide to vary the coupling. Thus, the end of the waveguide should only touch or be very close to the cylindrical wall, and this requires very precise machining.

In such a cavity the sample is placed into a capillary that is centered through the holes in both plungers in order to achieve the maximum B_1 field on the sample and to minimize dielectric losses. The optimum size of liquid aqueous samples depends on the resonant frequency and the sample dielectric constant. Typically, capillaries of 0.1 mm i.d. are about optimal for experiments with lossy aqueous samples at 130 GHz (D-band) while 0.15 or even 0.2 mm i.d. capillaries could be used at 95 GHz (W-band).

An alternative coupling scheme for a cylindrical mode resonator has been developed and employed at the Illinois EPR Research Center for W-band ESR (Figure 5C) (Wang et al. 1994; Nilges et al. 1999). In the Illinois design a cylindrical resonator is coupled to a waveguide through a small hole in a flat and thin wall that separates one side of the cylinder from the waveguide (Figure 5C). The standing wave pattern in the waveguide is tuned by an adjustable short, and this varies the coupling. The advantages of this coupling scheme versus that of Figure 5B are in simplified mechanics and the absence of any parts that should be disconnected mechanically. The Illinois coupling scheme can also be employed in oversized cylindrical resonators with TE_{01n}, $n = 2, 3, ..$, that are useful for accommodating larger samples. The disadvantage is in a somewhat limited range of coupling. The frequency of such a resonator is adjusted by repositioning the single plunger.

The resonator shown in Figure 5C could be built to accommodate the capillary samples through the holes in both the plunger and the opposite wall, allowing one to position the sample precisely through the center of the cavity (not shown). Such a design was initially employed in the Illinois W-band ESR spectrometer (Wang et al. 1994; Nilges et al. 1999). The optimal diameter for liquid aqueous sample was shown to be 0.15–0.20 mm i.d. (depending on the resonator Q-factor). A similar design was later adapted for the commercial Bruker W-band ESR spectrometer. However, in the Bruker design the sample hole is only from the plunger side and the sample cannot be inserted further than the bottom wall (Figure 5C). This breaks the symmetry of the microwave mode in the cavity, and it can be disadvantageous for some samples. Indeed, the sensitivity region of the single-mode W-band cylindrical resonator is only ca. 2 mm long. Then, in the Bruker design the sealed capillary bottom inevitably occupies some part of that volume. This could be particularly problematic for aqueous samples that should be filled into capillaries of only 0.10–0.15 mm in internal diameter. Such small capillaries are very difficult to seal reliably and to keep the length of the seal at a minimum. Accurate positioning of such a microcapillary exactly through the center of the Bruker cavity represents another experimental problem. Nevertheless, good quality room temperature W-band ESR spectra from aqueous solution of spin-labeled azurin have been recorded with a Bruker EleXsys 680 spectrometer (Finiguerra et al. 2003). The liquid sample was drawn into custom thick-wall suprasil quartz capillaries with 0.1 and 0.5 mm i.d. (Wilmad Labglass, Buena, NJ) and the ends of such capillaries were sealed with a Bruker X-Sealant® (Bruker Biospin, Germany). This capillary was then inserted into a standard suprasil quartz Bruker W-band tube (0.60 mm i.d. and 0.84 mm o.d.), pre-sealed at one end, and then into a standard Bruker W-band sample holder. Our experience is that one should follow this procedure with a caution: when a solution of a spin-labeled sample is in direct contact with the Bruker X-Sealant®, some of the spin labels and spin probes could diffuse into the seal, causing erroneous signals. Also, for W-band experiments that do not require in situ sample illumination, the suprasil (synthetic quartz) tubes can often be replaced by capillaries made from less expensive clear-fused quartz. However, the latter should be checked for background ESR signals.

3.2. Sample Deoxygenation and Oxygen Accessibility High-Field ESR Experiments

Molecular accessibility of protein residues including functional sites is of fundamental importance for understanding the mechanisms of molecular recognition, function, and signaling. Static crystallographic structures often offer little or no clues regarding accessibility of the buried sites because of the unknown magnitude of protein structural fluctuations. Accessibility of buried paramagnetic metal ion sites to small paramagnetic molecules such as O_2 and NO is essential for understanding the mechanisms of oxygen transport, redox processes occurring at the metal sites, and nitric oxide signaling. One point of view is that in tightly packed folded proteins there is very little molecular oxygen diffusing through. On the other hand, fluctuations in the protein side chains may be responsible for passing small molecules such as dioxygen and nitric oxide to the myoglobin binding site with little or no barrier (Meuwly et al. 2002).

The ESR oxygen accessibility method is based on magnetic interactions of a dioxygen molecule that serves as a probe with a nitroxide radical that reports such an interaction. The uniqueness of molecular oxygen as a biophysical probe lies in its small size and an appropriate level of hydrophobicity that allows this molecule to enter even the smallest vacant pockets transiently formed in proteins or membranes (Altenbach et al. 1989; Hyde and Subczynski 1989; Hubbell and Altenbach 1994; Ashikawa et al. 1994). During bimolecular collisions with oxygen both T_1 and T_2 electronic relaxation times of the radicals are shortened. This effect could be observed from continuous wave ESR as broadening of the spectral lines or effects on the rollover saturations curves. A time-domain saturation-recovery method could be used as well (Hyde and Subczynski 1989). The effect of oxygen on nitroxide ESR spectra became the foundation of ESR oximetry, which has found numerous biochemical and biophysical applications (Lai et al. 1982; Rosen et al. 1988; Hyde and Subczynski 1989; Subczynski et al. 1989; Altenbach et al. 1990; Likhtenshtein 1993; Smirnov et al. 1996). It also led to remarkable site-directed spin labeling experiments for elucidating protein structures and transmembrane protein location based on the oxygen accessibility of protein residues selectively labeled with nitroxides (Altenbach et al. 1990).

In order to observe the effects of molecular oxygen on ESR spectra, one should be able to manipulate the gas content in a sample. For organic solutions of nitroxide radicals it is customary to remove oxygen as well as other gases by applying multiple freeze-pump-thaw cycles. However, the design of high-field ESR resonators often imposes severe restrictions on the diameter and, sometimes, the length of the sample tube. Even for non-lossy solutions of spin labels and open Fabri-Perot cavities, the dimensions of the sample tube are rather limited. For example, Budil and coworkers (1993) used a short quartz tube (4–5 mm o.d., $\approx$1 mm wall thickness) for deoxygenating a toluene solution of PDT (perdeuterated 2,2′,6,6′-tetramethyl-4-piperidone-1-oxyl). The solution was degassed by a freeze-pump-thaw method and sealed with an oxygen torch, producing an ampoule only 20 mm long. This is perhaps the shortest conventional ESR quartz tube that can be flame-sealed. It should be noted that special care should be taken in flame-sealing such short ESR tubes containing such highly flammable chemicals as toluene.

Surprisingly, high-field ESR cylindrical cavities may offer more options for sample deoxygenation and manipulating sample gas content. Most W-band resonators will accommodate quartz capillaries up to 0.9 mm in outside diameter. In the

Bruker W-band spectrometer equipped with a split-coil superconductive magnet, the axis of the cavity is directed along the vertical bore of the magnet. Thus, the length of the sample tube is not limited by the size of the cavity nor the magnet bore diameter as with solenoid magnets. Then liquid non-lossy samples can be drawn into quartz capillaries pre-sealed at one end (such as o.d. = 0.89, i.d. = 0.70 mm, clear-fused or synthetic quartz tubes from VitroCom, Mountain Lakes, NJ, sold unsealed). Such capillaries are supplied 10 cm long and can be connected to a vacuum line using a simple adapter made out of a disposable plastic pipette tip. The liquid inside such a capillary can then be deoxygenated by applying several freeze-pump-thaw cycles. Note that it is important to keep the sample "head space" to a minimum in order to minimize solvent loss during cycles. With minimal practice one can seal such a microampoule with an oxygen torch while keeping the end of the capillary containing the sample under liquid nitrogen. For solenoid-type magnets the sample length is usually limited by the diameter of the variable-temperature cryostat or the room-temperature bore, and is typically ≈3–5 cm. For such a configuration, the quartz capillaries can be custom bent to form an L-shape and sealed at the short ends, as described by Smirnova et al. (1995). Spin-label solutions can be drawn into such capillaries (primarily into the short end beyond the bend), deoxygenated by a freeze-pump-thaw method, and sealed with a torch while the solution is kept away from the heat under liquid nitrogen.

It should be noted that aqueous solutions of biological interest such as spin-labeled proteins and membranes usually cannot be deoxygenated for high-field ESR experiments the same way for two reasons: (i) freeze-thaw cycles may destroy the sample, and (ii) the optimal diameter of the capillary might be too small for degassing viscous and/or foam-forming solutions. For example, at 95 GHz the optimal diameter for liquid aqueous samples is only 0.15–0.2 mm, and this makes freeze-pump-thawing of viscous solutions of lipid dispersions and/or proteins rather difficult due to excessive foaming. In order to resolve this problem we have adapted a degassing method of conventional X-band ESR that is based on gas-permeable plastic capillaries (Windrem and Plachy 1980; Hyde and Subczynski 1989). The only difference is that high-field ESR plastic capillaries are much smaller in diameter than at the X-band. For example, for W-band ESR we utilized a microbore Teflon (poly(tetrafluoroethylene) or TFE) tubing (i.d. = 0.152 ± 0.025 mm; wall thickness 0.127 ± 0.025; Cole-Parmer Instrument Company, Vernon Hills, IL). An aqueous sample was drawn into such a capillary, and the ends were sealed with a Critoseal sealant (Sherwood Medical, purchased from Fisher Scientific, Pittsburgh, PA). The capillary was positioned through the center of a cylindrical W-band TE_{012} resonator so that the sealed ends remained outside the cavity. This provides uniform distribution of the sample along the cavity center and does not bring the sealant material that could contain paramagnetic impurities inside. Note that this is an easy task if the cavity has two access openings at both ends, such as in the Illinois design. In the Bruker W-band cavity at least a part of the seal will be located within the resonator sensitivity region. A gas flow with the desired oxygen content was directed to the cryostat that enclosed the cavity and, through the waveguide, directly to the sample. At room temperature, re-equilibration of an aqueous PDT solution drawn inside such a capillary occurred with a time constant of $t_{1/2} \approx 4$ min. This is acceptable for many experiments. With increasing temperature the oxygen permeability coefficient for TFE increases, reducing sample re-equilibration time (Smirnov et al. 1994).

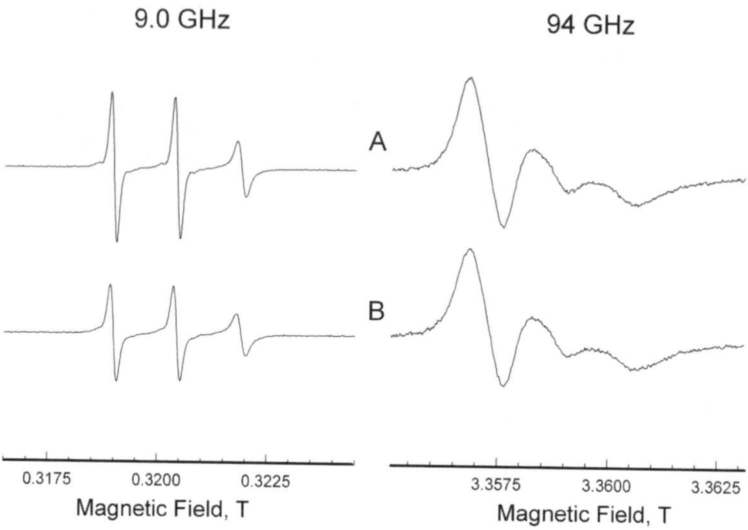

Figure 6. Comparative experimental X-band (9.0 GHz) and W-band (94 GHz) ESR spectra from a multibilayer aqueous dispersion of 1,2-dimyristoyl-*sn*-glycero-3-phosphocholine (DMPC; 20% v/v; 1.1 M borate buffer at pH 9.5) labeled with 16-doxyl stearic acid (16DSA, label-to-lipid ratio 1:100). Sample was drawn into a microbore poly(tetrafluoroethylene) tubing (i.d. = 0.152 ± 0.025 mm; wall thickness 0.127 ± 0.025) and equilibrated at 41°C with nitrogen (A) and then with air (B). Reproduced with permission from Smirnov et al. (1998). Copyright © 1998, Springer.

Figure 6 shows the effect of deoxygenation on 9.0- and 94-GHz ESR spectra taken at 41°C from a multilamellar aqueous dispersion of 1,2-dimyristoyl-*sn*-glycero-3-phosphocholine (DMPC; 20% v/v; 1.1 M borate buffer at pH 9.5) labeled with 16-doxyl stearic acid (16DSA; label-to-lipid ratio 1:100). Under these conditions the X-band ESR spectra of 16DSA fall into a regime of motional narrowing. The nitroxide moiety of 16DSA is located within the center of the bilayer where the oxygen permeability is maximal (e.g., see Smirnov et al. 1996). The oxygen broadening effect is easily seen from the changes in peak-to-peak amplitude of the double-integral-normalized X-band spectra. Oxygen-induced broadening for this sample, as measured by least-squares convolution-based fitting, was

0.0267 ± 0.0005 mT (Smirnov et al. 1996), while the peak-to-peak linewidth of the middle nitrogen hyperfine component was ≈0.136 mT. A W-band spectrum from the same carefully deoxygenated sample has much broader lines (ca. 1 mT), and therefore the oxygen effect on the overall shape of the spectrum is much smaller than at the X-band (Figure 6).

As discussed in the previous section, a significant increase in contribution from the *g*-factor anisotropy to the anisotropic part of the spin Hamiltonian at high magnetic fields results in a line width increase for spin-label high-field ESR spectra regardless of the motional regime. The oxygen-induced broadening occurs primarily through a spin-exchange mechanism and is frequency independent (Hyde and Subczynski 1989). Thus, at high magnetic fields it will be more difficult to determine the oxygen broadening of the same magnitude than from conventional X-band ESR spectra. Deriving the oxygen accessibility parameter from power saturation curves could also be more difficult. Indeed, enhanced contribution of *g*-factor anisotropy to electronic relaxation of spin labels in liquid solutions shortens both the electronic T_1 and T_2. Thus, ESR spectra from spin labels are more difficult to saturate at high magnetic fields assuming comparable B_1 on the sample. Moreover, critical coupling of high-field ESR resonators containing aqueous samples could also be a problem. Thus, it appears that conventional X-band ESR is more suitable than high-field ESR for deriving an oxygen accessibility parameter, with perhaps a few exceptions.

The first exception is the samples containing multiple components that are unresolved at conventional X-band ESR but become resolved at high magnetic fields. We speculate that a spin-labeled protein could exist in several conformations that could have different oxygen accessibility. If the lifetime of these states is longer than the spin-label relaxation time, this different accessibility could, in principle, be observed in an ESR experiment. Enhanced resolution of high-field ESR would assist these measurements. We believe that the multi-quantum ESR method developed by Klug and colleagues (King et al. 2005) might be particularly suitable for such measurements when extended to high magnetic fields.

The second exception is comprised of paramagnetic species that ESR spectra become narrower with an increase in magnetic field. One example of such species is the Mn^{2+} ion, which plays an important role in enzyme regulation (Sigel and Signel 2000). Increasing the magnetic field above 2–3 T, the Mn^{2+} ESR spectrum becomes narrower and first order. Depending on ion coordination, the peak-to-peak line width of each of the six nuclear hyperfine components of the ESR spectra corresponding to the $\Delta m_S = \pm 1/2$ electronic transition could be as narrow as 0.7–1.5 mT at the W-band versus 2.5–6.0 mT at the X-band. The electronic spins of Mn^{2+} are less shielded than, for example Gd^{3+} (another ion that has an exceptionally narrow ESR line at the W-band (Smirnova et al. 1998) and, thus, one could expect some magnetic interactions in the course of molecular collisions with oxygen.

We recently reported initial experiments to directly observe dynamic magnetic interactions between Mn^{2+} ions and molecular oxygen in solutions from the line width effects on high-field ESR (95 GHz, 3.4 T, W-band) spectra (Smirnova and Smirnov 2003). Re-equilibration of aqueous solutions of various Mn^{2+} salts and protein solutions with different gases was carried out inside a W-band ESR cavity using the same microbore polytetrafluoroethylene tubing as we used for PDT (see above). In order to facilitate observation of small oxygen broadening effects on the Mn^{2+} ESR line, we carried out ESR measurements at elevated oxygen pressures by pressurizing the entire variable-temperature cryostat (Oxford Instruments, Ltd., Concord, MA) containing the W-band probe head with a desired compressed gas. To separate the oxygen broadening effect from the intrinsic line width of the Mn^{2+} ion, we applied a one-line-width-parameter fitting model previously developed for fitting inhomogeneously broadened spin-label ESR spectra (Smirnov and Belford 1995; Smirnov et al. 1996). This model is based on simulating CW ESR spectra with a convolution equation:

$$F(B) = \int_{-\infty}^{+\infty} F_0(B')m(B-B')dB', \tag{44}$$

where $F_0(B)$ is the spectrum taken in the absence and $F(B)$ in the presence of oxygen; $m(B)$ is Lorentzian broadening, measured as $\Delta B_{p\text{-}p}$, peak-to-peak line width. We have previously shown that this model is applicable for deriving oxygen-induced line width broadening from CW ESR spectra from nitroxide membrane probes in phospholipid bilayers at temperatures corresponding to ripple and fluid bilayer phases (Smirnov et al. 1996).

The least-squares simulations were carried out in automatic mode for all sequential ESR spectra collected upon oxygenation or deoxygenation of solutions inside a gas-permeable capillary fixed in a W-band ESR cavity. Figure 7 shows the kinetics of oxygen broadening observed for $Mn(NO_3)_2$ and MnEDTA at the W-band (please note that the timescale shown has a different starting point for switching the bathing gas to dioxygen in these two experiments). The curves illustrate that, despite a small magnitude of the effect, our method produced repeatable and consistent results, i.e., the curves are well approximated by first-order kinetics, which is a good model for the final stages of oxygen diffusion to the limited volume of a gas-permeable capillary (Smirnov et al. 1994). Fitting these curves to first-order kinetics and averaging the results of multiple measurements yield greater accuracy for measurements of oxygen-induced broadening.

Experiments at elevated oxygen pressures described by Smirnova and Smirnov (2003) demonstrated that the molecular oxygen effect on CW ESR spectra from aqueous Mn^{2+} complexes can be described as an additional Lorentzian broadening whose magnitude is independent of the frequency of the ESR experiment but changes with the oxygen permeability coefficient of the solvent. The observed effect is caused by Heisenberg exchange between electronic spins of oxygen and Mn^{2+} during bimolecular collisions.

Another interesting observation was that the dioxygen effect on Mn^{2+} CW ESR spectra depends upon the immediate microenvironment of the ion (Smirnova and

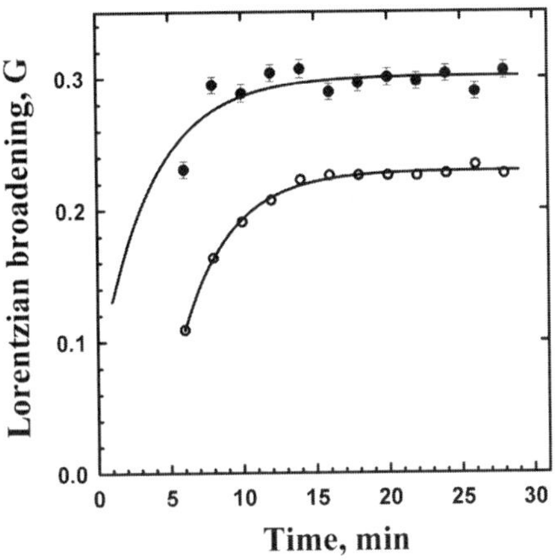

Figure 7. Kinetics of Lorentzian broadening upon re-equilibration of initially deoxygenated solutions of $Mn(NO_3)_2$ (open circles) and MnEDTA (filled circles) with oxygen. Reproduced with permission from Smirnova and Smirnov (2003). Copyright © 2003, American Chemical Society.

Smirnov 2003). It should be noted here that high-field ESR spectra of Mn^{2+} ions bound to proteins and RNA are easily observed at room temperature, and therefore the oxygen accessibility method might be useful for characterizing the transient environment of these ions. As a first step to demonstrating this effect for characterization of oxygen accessibility to a metalloprotein site, the Mn^{2+} center in Concanavalin A (Con-A) from *Canavalia ensiformis* was examined (Smirnova and Smirnov 2003). Figure 8 (squares) shows the pressure dependence of oxygen broadening for the Con-A Mn^{2+} center. While oxygen broadening at 1 atm is only 53 ± 30 mG (5.3 ± 3.0 μT, which is the limit of detection), the pressure dependence shows that at elevated pressures the effect is measured easily and reliably.

In order to demonstrate that oxygen accessibility to the Mn^{2+} center in Con A changes with modifications in metal coordination, we fully denatured Con-A in 4 M guanidine hydrochloride (GndHCl). The room-temperature 95 GHz ESR spectrum from the Mn^{2+} center changed drastically, reflecting release of Mn^{2+} ion into the guanidine hydrochloride solution. Oxygen broadening observed at 1 atm pressure for the Mn^{2+} center in denatured protein was 18 ± 3 μT, which is very close to that of the aqua ion (Smirnova and Smirnov 2003).

While experimental results reviewed here provided the first demonstration of a high-field ESR study of dynamic molecular oxygen accessibility to Mn^{2+} sites in biological macromolecules, more experiments are necessary in order to determine how the magnitude of the effect is related to the structure of the ion complex in solution. Once this method is developed further, it would complement the rich but static structural information obtained from ESR/ESEEM/ENDOR experiments with frozen samples of Mn^{2+}-containing proteins.

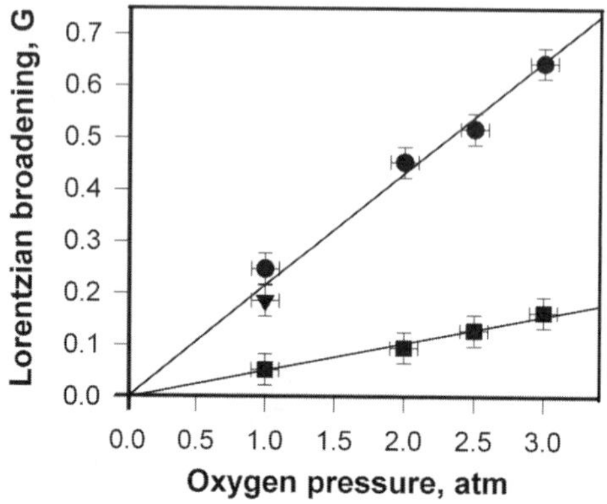

Figure 8. Oxygen-induced Lorentzian line width broadening as a function of oxygen pressure: • = 2 mM aqueous solution of $MnCl_2$; ■ = 0.75 mM solution of Concanavalin A (Con-A) in pH 6.4; HEPES; ▼ = 0.75 mM solution of denatured Con-A in 4 M GndHCl. Reproduced with permission from Smirnova and Smirnov (2003). Copyright © 2003, American Chemical Society.

It should also be noted here that high-field ESR measurements of oxygen accessibility to Mn^{2+} sites should not be viewed as an ESR tool to establish a protein residue periodicity pattern and/or transmembrane location of a side-chain of interest as in site-directed spin-labeling studies. Rather, the method offers a way to directly measure dioxygen accessibility to catalytic paramagnetic metal ion sites containing native Mn^{2+} or Mg^{2+}. While Mg^{2+} is diamagnetic, Mn^{2+} could easily substitute it because of the similar size.

4. HIGH-FIELD ESR IN STUDIES OF MOLECULAR DYNAMICS

Although a rather small fraction of recent site-directed spin-labeling literature is devoted to applications of high-field/high-frequency methods, many of the principal advantages of high-field ESR have been already demonstrated for a variety of systems ranging from small nitroxides in simple fluids to complex multimode reorientations of spin-labeled protein side chains. Several reviews of spin-labeling high-field ESR have also been published (Smirnov et al. 1998; Freed 2000; Smirnov 2002; Marsh et al. 2002; Earle and Smirnov 2004; Livshits and Marsh 2004).

In this section we review the applications of high-field ESR to studying molecular dynamics of biological molecules with nitroxide spin labels. Specifically, we would like to illustrate the kind of new information high-field ESR is capable of providing on the dynamics of complex and heterogeneous molecular structures such as membranes and proteins. We will also emphasize the underlying physical aspects as well as direct the reader to emerging applications that, in our opinion, may be worthy of future explorations.

From a very general prospective, an experimentalist becomes interested in a new spectroscopic method only if it will offer higher sensitivity and better resolution over a traditional one. While high-field ESR holds both advantages over conventional ESR of spin labels across a wide range of rotational correlation times, the advances are most striking when the spectra fall into fast- or slow-motion limits. This is illustrated by the variable-temperature 95-GHz ESR spectra of a nitroxide PDT shown in Figure 2: essentially all the spectra have sharp and well-defined features except the one measured at $T = 158$ K. The latter spectrum is close to an intermediate motion regime where the electronic T_2 reaches the minimum and the spectral features are excessively broad. Thus, we will primarily focus on fast- and slow-motion regimes.

4.1. High-Field ESR in Spin-Labeling Studies of Biological Systems: Fast Motion Limit

The fast-motion condition defined by (39) dictates that with an increase in the anisotropic part of the spin Hamiltonian, $H_1(t)$, the onset of slow-motional effects will occur at shorter rotational correlation times τ_c. With an increase in magnetic field, the contribution from the anisotropic Zeeman term to $H_1(t)$ grows proportionally, and, therefore, fewer spin-labeled systems will satisfy the fast-motion condition at, for example, the W-band, than at the conventional X-band frequency. Nevertheless, at the W-band the fast-motion condition still could be fulfilled for small nitroxides and spin adducts in simple and complex fluids, including fluid phospholipid bilayers. In addition, it has been found that high-field ESR spectra from short spin-labeled peptides, DNA monomers, and flexible sections of proteins might also fall into a fast-motion regime when studied at room temperature and above. The fast-motion regime is very advantageous for separating overlapping species since the ESR line widths can be very narrow. These narrow signals are also easier to detect than the broader slow-motion CW ESR spectra for the same spin label concentrations. We illustrate these considerations on several examples ranging from free radical spin adduct molecules to partially unfolded proteins.

We start by illustrating the principal advantage of high-field ESR in resolving multiple spin-labeled species from a mixture of spin adducts formed during spin-trapping experiments. These experiments are typically carried out at conventional X-band frequencies because of the good concentration sensitivity and ready availability of conventional ESR. The trapped radicals are then characterized by comparing isotropic hyperfine coupling constants with those measured for the reference spin adducts. These experiments encounter two major challenges: (*i*) spin adducts from many carbon-centered free radicals have *g*-factors that are nearly identical (resulting in strongly overlapping spectra at 9.5 GHz), and (*ii*) measurable hyperfine couplings correspond to interactions of the electron spin with just the nearest nuclei. Therefore, very little or no information is obtained on the overall structure of the spin adduct molecule.

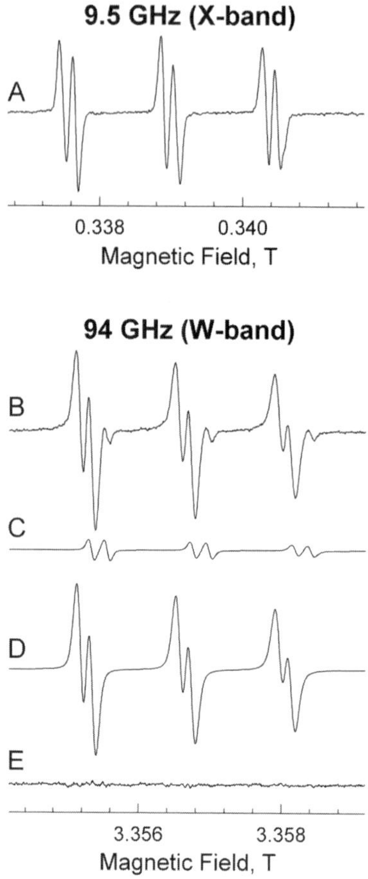

Figure 9. (**A**) X-band (9.5 GHz) ESR spectrum from a mixture of phenyl-PBN and trichloromethyl-PBN spin adducts in benzene at 24°C. Arrow points to some extra broadening on the line shoulder that is caused by the presence of two species. (**B–D**) Experimental W-band (94 GHz) spectrum of the mixture and the corresponding least-squares-simulated spectra of phenyl-PBN and trichloromethyl-PBN spin adducts, respectively. (**E**) Difference between the experimental (B) and simulated spectra (C,D). Reproduced with permission from Smirnova et al. (1997). Copyright © 1997, American Chemical Society.

Previously, we have shown that some of these difficulties can be overcome by carrying out ESR characterization of spin adducts at high magnetic fields (Smirnova et al. 1997). Figure 9A shows an X-band (9.5 GHz) spectrum from a mixture of two very similar spin adducts: phenyl- and trichloromethyl-PBN (PBN, phenyl-*tert*-butylnitrone). The spectrum was measured at 24°C from a benzene solution that was carefully deoxygenated. Figure 9A shows that only the high-field line (m_I = –1 nitrogen hyperfine component) has some small extra broadening on the line shoulder (shown by an arrow) that might hint at the presence of two species. This strong overlap of the X-band ESR spectra makes an accurate determination of the number of species as well as nitrogen and β-hydrogen hyperfine coupling parameters a difficult task. Figure 9B shows the 94-GHz ESR spectrum from the same mixture of spin adducts at the same temperature. The W-band sample was also deoxygenated using the freeze-pump-thaw procedure described in the previous section. Because spectral separation ΔB is proportional to resonance frequency ν at the same Δg (Eq. (6)), the presence of two radical adducts becomes clear at the W-band. For this better-resolved ESR spectrum least-squares fitting was employed to extract isotropic g-factors as well as nitrogen and hydrogen hyperfine coupling parameters of these two different spin adducts. The best-fit results and the corresponding ESR spectra from individual spin adducts are presented in Figure 9B–D (simulated mixture spectrum, phenyl- and trichloromethyl-PBN adducts, respectively). Residual of the fit (difference between the experimental and simulated spectra) is shown in Figure 8E.

It should be noted that the 94-GHz ESR spectra from spin adducts are broader than that at the conventional X-band. The reason for that resides in enhanced contributions from the spectral densities arising from rotational modulation of the electronic g-matrix and is discussed in detail in §2.4. However, as can be seen from the spectrum, the line broadening at 94 GHz is much too small to defeat the gain in resolution due to Δg, which for these two spin adducts is a mere $\Delta g_{iso} = 1.2\times10^{-4}$. The main reason for this is that even at the W-band the ESR line width of these spin adducts is dominated by unresolved proton hyperfine couplings and therefore increases with the resonant frequency only slightly (Smirnova et al. 1997). Thus, we can conclude that enhanced g-factor resolution of high-field ESR could be very useful for resolving ESR spectra of various spin adducts from a mixture and that the g-factor could be a new parameter for spin adduct characterization.

Identification of radical precursors by ESR becomes even more challenging when the formed spin adducts have a similar or identical structure in the close vicinity of the unpaired electron. Examples of this include PBN spin adducts formed from alkyl radicals with variable chain length. We have measured magnetic parameters for two very different alkyl PBN spin adducts — methyl and tetradecyl in a nonpolar solvent toluene — and found these parameters are almost identical (Table 2). Note that the isotropic g-factors of these two PBN adducts are identical with the accuracy of our W-band experiment. Thus, magnetic parameters alone might be insufficient for radical identification. (Of course, the methyl spin adduct could be identified from the three protons of the methyl group, but detection of more remote protons for other alkyl spin adducts is complicated.)

We have shown that spin adducts with identical magnetic parameters but different structures, such as PBN adducts formed by alkyl radicals with variable chain length, can still be characterized and in some cases differentiated by measuring the parameters of rotational diffusion of these molecules (Smirnova et al. 1997). These parameters can be measured more accurately at the W-band, assisting one in characterizing the spin adduct molecule as a whole. As an example, line width parameters (Eq. (40)) of methyl-, ethyl-, and tetradecyl-PBN were measured at the W-band for toluene solutions (Smirnova et al. 1997). Figure 10 shows line width parameter B (on a logarithmic scale) as a function of the logarithm of the viscosity-to-temperature ratio (η/T) as measured at the W-band for methyl-, ethyl-, and tetradecyl-PBN spin adducts. For the same magnetic and anisotropy parameters, the value of parameter B is proportional to rotational correlation time τ if the contributions from nonsecular and pseudosecular spectral densities can be neglected. Figure 10 demonstrates that these rotational dynamics parameters measured from W-band spectra were strikingly different for all three spin adducts studied. This indicates a very different rotational correlation times and rotational radii even for ethyl- and methyl-PBN, which have rather similar structures.

Table 2. Magnetic Parameters of Methyl and Tetradecyl Adducts of PBN in Toluene

	Methyl-PBN	Tetradecyl-PBN
g_x	$2.00934 \pm 3\times10^{-5}$	$2.00933 \pm 3\times10^{-5}$
g_y	$2.00619 \pm 3\times10^{-5}$	$2.00617 \pm 3\times10^{-5}$
g_z	$2.00234 \pm 3\times10^{-5}$	$2.00237 \pm 3\times10^{-5}$
$\langle g\rangle^a$	$2.00596 \pm 5\times10^{-5}$	$2.00596 \pm 5\times10^{-5}$
g_{iso}^b	$2.00589 \pm 1\times10^{-5}$	$2.00588 \pm 1\times10^{-5}$
$A_x\times(2\pi/\gamma_e)$	4.5 ± 0.2	4.1 ± 0.2
$A_y\times(2\pi/\gamma_e)$	6.6 ± 0.2	5.8 ± 0.2
$A_z\times(2\pi/\gamma_e)$	33.4 ± 0.2	33.7 ± 0.2
$\langle A\rangle\times(2\pi/\gamma_e)^c$	14.8 ± 0.3	14.5 ± 0.3
$A_{iso}\times(2\pi/\gamma_e)$	14.83 ± 0.03	14.54 ± 0.03

[a] $\langle g\rangle=(g_x+g_y+g_z)/3$.
[b] Measured from fast-motion ESR spectra at 94 GHz.
[c] $\langle A\rangle=(A_x+A_y+A_z)/3$.
Reproduced with permission from Smirnova et al. (1997). Copyright © 1997, American Chemical Society.

Overall, the experiments described by Smirnova et al. (1997) demonstrated that high-field ESR can be a valuable supplement to conventional (9.5 GHz) ESR for spin adduct characterization. Specifically, it has been shown that at 94 GHz the gain in g-factor resolution is sufficient to offset the line broadening caused by enhanced contributions from rotational modulation of the g-matrix at high magnetic fields. Line width parameter B and the rotational diffusion tensor characterize a molecule as a whole, providing data on configuration of the molecule; this can be used as supplemental information for spin adduct characterization. In addition, we have already discussed that high-field ESR offers superior absolute point sensitivity. In application to nitroxides many experiments can be carried out with picomolar quantities of the sample — essentially the same sample requirement as for mass spectroscopy. One could take advantage of this by concentrating the spin adducts from a biological sample for further characterization with high-field ESR. Thus, we expect to see more spin-trapping experiments with high-field ESR in the future.

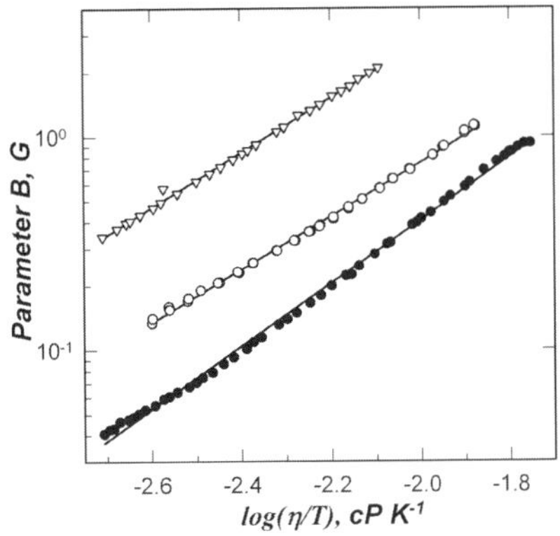

Figure 10. A log–log plot of W-band (94.5 GHz) line width parameter B for 0.1 mM solutions of methyl-PBN (closed circles), ethyl-PBN (open circles), and tetradecyl-PBN (open triangles) in toluene measured as functions of viscosity-to-temperature ratio. Abscissa is $\log(\eta/T)$, where η/T is in cP × K^{-1}. Results of first-order regressions of the linear portions of the curves are shown as solid lines. Reproduced with permission from Smirnova et al. (1997). Copyright © 1997, American Chemical Society.

Similar to spin adducts, many other small spin-labeled molecules in liquids may satisfy fast-motion condition Eq. (39) at magnetic fields of high-field ESR. Some examples include membrane spin probes di-*tert*-butyl-nitroxide (DTBN) and 2,2′,6,6′-tetramethyl-4-piperidone-1-nitroxide (Tempone) when partitioned in phospholipid bilayers (Smirnov et al. 1995; Smirnov and Smirnova 2001). We will postpone discussion of the use of these probes in high-field ESR studies of lipid bilayers until §6 of this chapter. While at high magnetic fields the motion of spin-labeled protein side chains and doxyl-labeled lipids usually falls into an intermediate or even slow-motion regime, it was found that some short spin-labeled peptides, DNA monomers, and flexible sections of proteins when studied at room temperature and above may still fall into a fast-motion regime. The narrow line width associated with motion regime is advantageous for separating overlapping nitroxide species. The signal-to-noise ratio of these narrow signals is superior to those observed for broader slow-motion CW ESR spectra at the same spin label concentrations.

One example of application of high-field ESR to studying side-chain dynamics in a helix-forming peptide — Ac-$(AAAAK)_2$CAAAKA-NH_2 labeled with 1-oxyl-2,2,5,5-tetramethyl-3-pyrroline-3-methylthiosulfonate (MTSL) — is given by Bennati et al. (1999). It was found that experimental ESR spectra recorded at 140 GHz over the temperature interval from 277 to 306 K were amenable to analysis with simple fast-motion theory. The analysis revealed anisotropic rotational dynamics of the side chain with $\tau_\perp/\tau_\parallel$ approaching 2.6 at 306 K. Another example is a W-band

ESR titration experiment with a synthetic P11 peptide fragment of the laminin B1 chain labeled with methanethiosulfonate derivative of a pH-sensitive imidazolidine nitroxide (Smirnov et al. 2004). Room-temperature ESR spectra from this spin-labeled peptide were found to satisfy the fast-motion condition. Another example of a nitroxide in a fast-motion regime was provided by Budil et al. (2000), who reported at 220 GHz ESR an aqueous solution of an oligonucleotide labeled with a newly synthesized cytosine nitroxide. The anisotropic diffusion tensor estimated from this fast-motion spectrum was consistent with the expected rapid motion of the probe molecule around its tether.

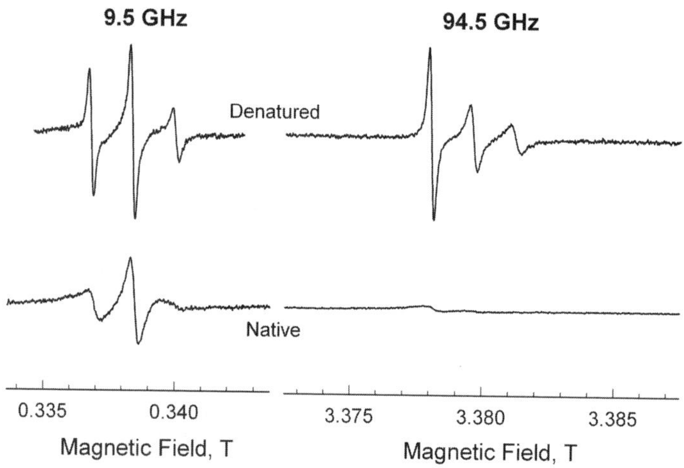

Figure 11. Comparison of 9.5- and 94.4-GHz ESR spectra of phosphoglycerate kinase labeled at the native cysteine 97 with a maleimide proxyl spin label illustrates an enhanced sensitivity of W-band ESR line shape to changes in local protein dynamics upon protein denaturing. The spectra at each of the frequencies were taken at room temperature and normalized by the value of the double integral. The protein was denatured by addition of 2 M of guanidine hydrocholoride (Pierce, IL). Protein concentration was 0.2 mM in a HEPES buffer at pH 6.5 (Smirnov et al., unpublished).

While the fast-motion regime for high-field ESR spectra from short peptides and oligonucleotides should not be surprising, local tumbling of flexible protein chains might also satisfy Eq. (39) in certain experiments. One example of this is the W-band ESR spectra of a recombinant human growth hormone, labeled at the NH_2 terminus with a succinimidyl 2,2,5,5-tetramethyl-3-pyrroline-1-oxyl-3-carboxylate (Smirnov et al. 1998). Although fast-motion high-field ESR spectra of spin-labeled proteins and peptides are relatively uncommon because they are indicative of rather flexible and/or disordered parts, they have the advantage of being well resolved from otherwise broad high-field ESR spectra characteristic of slow or intermediate motion. This could be useful for differentiating the protein fractions that are disordered or partially unfolded. Figure 11 shows comparative W-band and X-band ESR spectra from phosphoglycerate kinase (PGK) labeled at the native cysteine 97 with a maleimide proxyl spin label and illustrates an enhanced sensitivity of W-band ESR line shape to changes in local protein dynamics upon protein denaturing. PGK is a two-domain protein that frequently serves as a model for studying folding and interdomain interactions (Ballew et al. 1996; Lillo et al. 1997a,b). Previous studies have shown that two domains of PGK unfold sequentially (Lillo et al. 1997b) and indicated that continuity of the domains is not required for correct folding (Lillo et al. 1997a). The ESR spectra from spin-labeled PGK were measured at room temperature at both ESR frequencies (X- and W-band) and were normalized by the value of the double integral (Figure 11). The protein was denatured by addition of 2 M of guanidine hydrocholoride (Pierce, IL). The protein concentration was 0.2 mM in a HEPES buffer at pH 6.5. From comparisons of the double-integral normalized ESR spectra at the X- and W-band (Figure 11), it is clear that even for cysteine 97, which is located in the mobile protein loop, high-field ESR is more sensitive to changes in local dynamics upon protein unfolding.

4.2. High-Field ESR in Spin-Labeling Studies of Biological Systems: Intermediate and Slow Motion Limit

One of the most important features of the SDSL ESR method is that it reveals the protein tertiary fold through differential mobility of nitroxide side chains. The ESR spectra of such spin-labeled chains are affected by tertiary interactions and backbone dynamics. Systematic studies of the relationship between nitroxide side-chain mobility and protein structure of various sites of T4 lysozyme have shown that essentially all ESR spectra of surface-exposed chains fall into an intermediate motion regime at the X-band (Mchaourab et al. 1996). This should not be surprising because typically for a protein in a solution the side chains fluctuate with a correlation time of a few nanoseconds. At the X-band the fast-motion condition of Eq. (39) breaks down at about $\tau_c \geq 1$ ns, and therefore ESR spectra undergo large and characteristic changes within a relatively narrow range of rotational correlation times $\tau_c \approx 1$–3 ns.

One of the conclusions of our analysis of spin-label ESR spectra at high magnetic fields was that the fast-motion condition breaks down at shorter correlation times because of the greatly enhanced contribution from *g*-factor anisotropy to the anistropic part of the time-dependent Hamiltonian, $H_1(t)$. Thus, one could expect that some rotational modes that manifest at the X-band in intermediate-motion regime ESR spectra would appear as characteristic slow-motion spectra at frequencies of W-band ESR and higher. Also, fast librations of protein side chains start to contribute significantly to the ESR line shapes observed at high magnetic fields. An example of this was provided by Poluektov et al. (2003), who utilized spin labels and ESR at 9.5 and 130 GHz to study the local dynamics of a photosynthetic

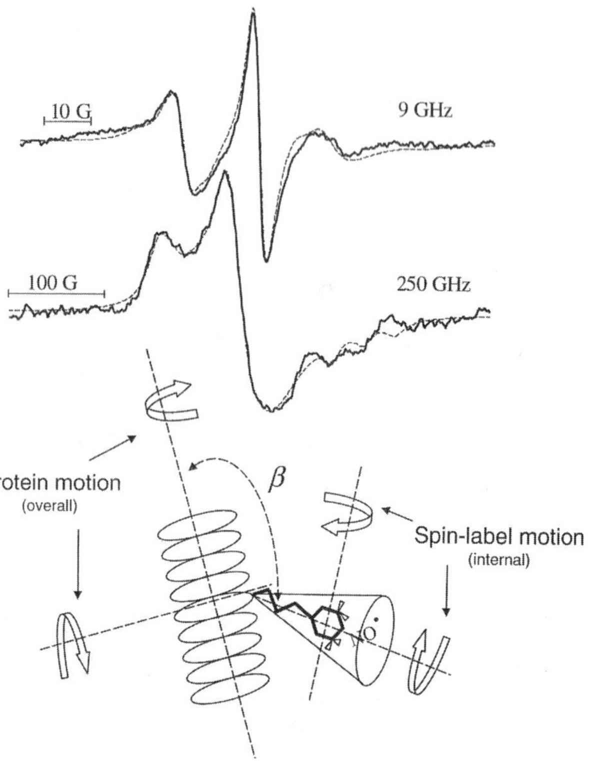

Figure 12. Results of a multifrequency ESR study of T4 lysozyme labeled with a nitroxide at residue 44. The 9- and 250-GHz ESR spectra were measured for aqueous protein solutions at 10°C. The illustration at the bottom shows the relevant modes of motion. The protein tumbles slowly about its principal axes (this motion can be analyzed in the framework of the slow-relaxing cage model of Polimeno and Freed (1995)). The motion of the spin label moiety is restricted by its tether and its surroundings to be within a cone whose main axis makes angle β with the protein main axis. The spatial extent of the internal rotational diffusion modes and their rates are distinguished from the overall protein tumbling rates. Reproduced with permission from Barnes et al. (1999). Copyright © 1999, The Biophysical Society.

bacterial reaction center. Thus, one could say that the high-field ESR spectra provide us with a faster "snapshot" of the protein dynamics. Then for high-field ESR spectra of many spin-labeled proteins and membranes global tumbling modes will be effectively frozen and the fast internal (librational) modes are expected to have the most effect. This has been demonstrated in a multifrequency ESR study of several spin-labeled T4 lysozyme mutants (Barnes et al. 1999). Specifically, it was found that the overall rotation of this small water-soluble protein was too slow to significantly affect the ESR spectrum at 250 GHz. (A similar observation was made by Hustedt et al. (1997) for 95-GHz ESR spectra from a larger tetrameric glyceraldehyde-3-phosphate dehydrogenase.) Thus, the 250-GHz ESR spectra from spin-labeled T4 lysozyme (Figure 12) were analyzed to deduce internal dynamic behavior (Barnes et al. 1999). These internal dynamic parameters were then fixed and slower "snapshot" X-band ESR spectra from the same spin-labeled mutant were fitted to yield the rates of the global dynamics. Overall, Barnes and coworkers (1997) demonstrated that these two types of motion can be effectively separated and studied using a consistent multifrequency ESR approach.

Another observation that emerged from analysis of multifrequency ESR spectra from spin-labeled proteins is that the spectra obtained and analyzed at a single resonant frequency often provide somewhat limited information on protein dynamics. For example, in a study of several spin-labeled azurin mutants in aqueous solutions it was found that, while X-band ESR spectra can be well modeled as an isotropic tumbling, the spectra at W-band are clearly anisotropic (Finiguerra et al. 2003). It is very straightforward to qualitatively explain these results using the "snapshot" concept we already discussed. Indeed, azurin is a rather small protein, and therefore its rotational diffusion in aqueous solutions at room temperature should occur with a correlation time of a few ns. At the X-band this corresponds to an intermediate motion regime when the fast-motion condition breaks down and slow-motional spectral features appear. As discussed, such CW ESR spectra are extremely sensitive to changes in the rotational correlation time and therefore are dominated by global protein tumbling. The azurin molecule could be well approximated by a sphere, so that rotational diffusion of such a protein should be isotropic. Thus, X-band ESR spectra should appear isotropic, as was indeed reported by Finiguerra and coworkers (2003). Librational modes of spin-labeled tether and backbone fluctuations should have only minor effects on X-band spectra because of the much shorter timescale. However, at high magnetic fields these modes can move into an intermediate motion regime while the overall protein tumbling mode should be approaching the rigid limit. Thus, a very different mode of motion should dominate the W-band spectra. Indeed, Finiguerra and coworkers (2003) found that W-band ESR spectra of 2 out of 3 mutants could only be modeled by highly anisotropic rotational diffusion.

A faster "snapshot" feature of high-field ESR could be especially useful in differentiating the global and local dynamics of DNA. Clearly, the faster timescale of high-field ESR, where the overall tumbling modes of diffusion are frozen out, and the unique sensitivity to non-axial (i.e., *x*- *y*-) modes of motion, would be very beneficial in providing new insights on position-dependent DNA dynamics. An initial example was provided by Okonogi and colleagues (2000), who reported a W-band rigid-limit ESR spectrum of a DNA to which a perdeuterated and ^{15}N-substituted nitroxide was attached in a rigid manner. The W-band spectrum demonstrated excellent resolution and negligible strain effects, thus indicating the great potential of such labels in DNA research. In separate efforts, Budil and coworkers (2000) and Kolaczkowski et al. (2001) reported a different approach to labeling of oligonucleotides using a newly synthesized cytosine label that has a flexible attachment. Variable temperature 220-GHz ESR studies of spin-labeled monomers and pentamers in aqueous buffers with 75 and 35 vol% of glycerol, respectively, showed good agreement of the slow-motional line shapes with the microscopic-order macroscopic disorder model (MOMD) (Budil et al. 2000). The viscosity of the DNA solution was adjusted by adding glycerol in order to bring the motion of an otherwise rather mobile label, due to its flexible attachment, into the slow-motion regime. Least-squares simulations of 220-GHz ESR spectra demonstrated that for the monomer species the rotational rates around all three molecular axes of the label could be measured independently because the principal axis orientations are well resolved in 220-GHz ESR spectra.

Different modes of protein dynamics, such as rotation of the protein as a whole, backbone fluctuations, as well as internal motions of the spin-labeled side chains, could be fully accounted for by the slow relaxing local structure model (SRLS) we briefly discussed in §2.5. An initial testing of this model was provided by Liang et al. (2004) in a variable-temperature study of two spin-labeled T4 lysozyme mutants with CW ESR at 9.5 and 250 GHz. One of the conclusions of their study was that the deconvolution of various modes of protein motion from multifrequency ESR spectra still remains a challenging task, especially if several conformers are suspected of contributing to ESR spectra. However, the multifrequency ESR approach opens new timescale windows that allow one to study these modes selectively and more accurately than ESR at any particular frequency. The authors are expecting further progress in developing specialized theoretical and least-squares simulation tools tailored to multifrequency ESR of proteins.

Another principal feature of slow-motion ESR spectra at high magnetic field is the full resolution of all three principal axis orientations illustrated in Figure 1B. Such a significant resolution enhancement simplifies analysis of ESR spectra for anisotropic rotational diffusion. This greatly enhanced angular resolution is particularly useful in studying anisotropic motion and local molecular ordering, such as one would expect to occur for biomolecules in phospholipid bilayers. One example of such a study was provided by Gaffney and Marsh (1998), who investigated anisotropic motion and, in particular, non-axial lipid ordering in cholesterol-rich phospholipid membranes by W-band ESR. Using fully hydrated bilayer membranes of dimyristoyl phosphatidylcholine containing 40 mol% cholesterol, Gaffney and Marsh (1998) obtained an indication of the way in which lateral ordering of lipid chains in membranes is induced by cholesterol. More recently, Livshits and Marsh (2000) undertook a theoretical study to identify the dynamic origin of the residual ($g_x - g_y$) anisotropy observed in non-axial high-field ESR spectra of cholesterol-containing membranes. It was concluded that this anisotropy arises from in-plane ordering of the lipid chains by cholesterol. This hypothesis was further tested and confirmed by Kurad et al. (2001) by studying cholesterol-rich membranes with W-band ESR. This result could be a significant step toward understanding domain formation in cellular membranes. Timely reviews of high-field ESR of spin labels in membranes by Marsh and coworkers have now appeared (Marsh et al. 2002; Livshits and Marsh 2004). It should be noted, though, that while the model of Marsh and colleagues is useful for obtaining qualitative insights into the data trends, quantitative analysis by rigorous slow-motional theory would be necessary to extract all available information from the spectrum. We believe that these studies of model bilayer systems will also be very useful in providing a solid background for understanding the dynamics of transmembrane proteins in the future.

5. HIGH-FIELD ESR IN STUDIES OF MOLECULAR STRUCTURE

One way of using spin-labeling ESR for structural studies of membranes and proteins is in distance determination between interacting unpaired electronic spins. Usually, a pair of nitroxides is introduced into a system using site-directed spin-labeling methods, although nitroxide–metal ion pairs can also be used. It should be noted that, depending on the ESR experiment, the range of measurable distances spreads from ca. 0.6 to 8 nm (Pannier et al. 2000), and therefore both intramolecular (such as within protein) and intermolecular distances (such as in protein and/or protein membrane complexes) can be measured. A detailed review of various ESR methodologies are now used for interspin distance measurements has recently been edited by Eaton and colleagues (2001). Most notable progress in this field was achieved with development of both ESR hardware and software for double-quantum coherence (DQC) pulsed-ESR experiments with nitroxides (Borbat et al. 2004) and extending this method to Cu^{2+}–Cu^{2+} ion pairs (Becker and Saxena 2005). With these new developments, both shorter and longer distances as well as the distribution of distances can be evaluated with much better accuracy.

Although electronic spin–spin interactions (e.g., spin exchange, dipolar, or a combination thereof) are magnetic field independent, applications of high magnetic fields both simplify and improve the accuracy of distance measurements that are based on analysis of such interactions. Perhaps the first studies of nitroxide pairs with high-field ESR were carried out by Lebedev and coworkers in Moscow (Ondar et al. 1981; Lebedev 1990). These authors noted that, although the number of spectral components of the rigid-limit 140-GHz ESR spectra of a nitroxide biradical is greater than at the conventional X-band, the spectrum is still easier to interpret because of an excellent angular resolution that allows one to sort out the spectral components and the corresponding dipolar coupling. Basically, for well-resolved spectra, the components of the dipolar tensor can be estimated directly from the spectrum.

Another principal advantage of high-field ESR for spin labeling studies lies in exceptional angular resolution. We believe that one of the most efficient ways to utilize this advantage in studies of phospholipid bilayers and membrane proteins is by carrying out high-field ESR with macroscopically aligned samples. We will illustrate this with an example of several different bilayer alignment techniques. We will also show that in some cases the excellent angular resolution of high-field ESR would allow one to evaluate mutual orientations of interacting nitroxides even for samples that are not macroscopically aligned.

5.1. Distance and Mutual Orientation Measurements in Disordered Systems

Perhaps one of the best demonstrations of advantages of dipolar spin labeling high-field ESR has been provided by Hustedt and coworkers (1997) in a study of spin-labeled coenzyme NAD^+ (SL-NAD^+) bound to a microcrytalline, tetrameric glyceraldehyde-3-phosphate dehydrogenase (GAPDH). The structure of the tetramer was such that for the purpose of dipolar ESR the four spins can be considered two independent pairs. The work was carried out with an ammonium sulfate-precipitated form of SL–NAD^+–GAPDH that was re-suspended in a small volume of saturated ammonium sulfate as a slurry at a concentration of about 200 mM. For such a complex the spin labels populate unique conformations and remain virtually immobilized with respect to the protein. ESR spectra were recorded at 4°C at the X-, Q-, and W-bands and fit simultaneously with a least-squares software package developed at Vanderbilt University (Hustedt et al. 1997; Hustedt and Beth 1999, 2001). The results, shown in Figure 13, illustrate excellent agreement between the experimental spectra and the fits. Based on the results of this study, several conclusions regarding the prospects of high-field ESR in such experiments can be drawn:

(i) The spectral features corresponding to all principal axis components of the nuclear hyperfine and dipolar coupling interactions are clearly resolved, and at the conditions of the experiment no strain broadening of the lines was observed. The line assignment is significantly simplified at the W-band vs. the lower ESR frequencies.

(ii) Not only the z- but also the x- and y-axes magnetic tensor components are well resolved in the W-band ESR spectrum. This makes the least-squares simulations much more sensitive to the mutual orientations of the spin labels on the proteins and allows one to determine mutual orientations of spin-labeled protein residues with better accuracy.

(iii) The timescale of motional averaging at the W-band is moved toward faster motions, bringing the whole spectrum essentially into the rigid limit at the conditions of the experiment. We have already discussed this "snapshot" effect previously with examples of 250-GHz spin-label ESR spectra.

Another example of the use of CW dipolar ESR for distance determination was given recently by Bottle and colleagues (2000). An initial X-band ESR study of a carboxy-derivative of an anionic isoindoline aminoxyl, CTMIO, revealed the presence of extra broadening in the ESR spectra recorded in frozen toluene-d_8 at 123 K (Bottle et al. 1999). The broadening was interpreted as an indication of dimmer formation, although no resolved dipolar splittings were observed. Consequent reexamination of this system with W-band ESR demonstrated a well-resolved dipolar splitting of the g_z components essentially at the same temperature of the experiment (130 K) (Bottle et al. 2000). Direct observation of dipolar interaction led to a more accurate estimate of the interspin distance of 19.3 Å. This study is a good illustration of the resolution gain achieved with high-field ESR.

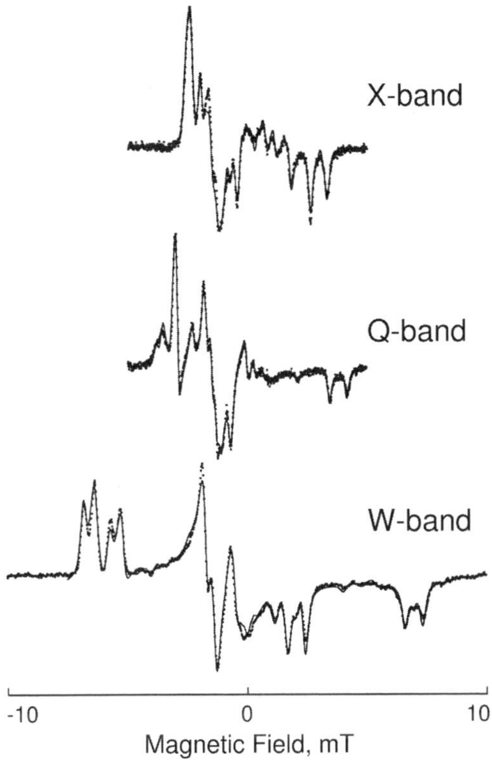

Figure 13. Experimental CW ESR spectra (dots) at the X-, Q-, and W-band of spin-labeled NAD^+ bound to GAPDH at a ratio of 3.4 labels per GAPDH tetramer. Dipolar coupling between the nitroxides results in well-resolved splitting of the spectral components corresponding to principal axis orientations. Note the progressive improvement in spectral resolution and angular selectivity from the X- to the W-band. The simultaneous multifrequency least-squares simulations are shown as solid lines superimposed with the spectra. Reproduced with permission from Hustedt et al. (1997). Copyright © 1997, The Biophysical Society.

We believe that with the growing availability of high-field ESR more double-label experiments will be carried out at high magnetic fields. Clearly, the excellent angular and *g*-factor resolution achieved at high magnetic fields would facilitate accurate determination of distances and angles between the interacting spins from CW and time-domain ESR experiments. These methods are awaiting detailed exploration in applications to spin-labeled molecules.

5.2. Measurement of Transmembrane Orientations from Macroscopically Aligned Samples

As discussed in §2 of this chapter, an increase in magnetic field/resonant frequency makes slow-motion ESR spectra of spin labels extremely sensitive to the orientation of the nitroxide molecular frame with respect to the external magnetic field because of the dominance of the *g*-factor anisotropy of a rhombic character. This sensitivity can be exploited in studies of microscopic ordering and mutual spin orientations for disordered samples, as we discussed in the preceding section. When carried out at high magnetic fields, spin-labeling high-field ESR provides even more information on studies of samples that are macroscopically aligned even if this alignment is only partial. If the macroscopic order is relatively high, then the orientation of the spin-labeled fragment, at least in first approximation, can be read directly from the spectrum. This puts spin-labeling high-field ESR in a unique position among other spectroscopic techniques because only a few of these techniques are suitable for elucidating molecular orientations in (partially) aligned samples. Thus, spin-labeling ESR at high magnetic fields should be particularly useful for studying orientations of biological molecules such as peptides and proteins with respect to membrane surfaces. Additionally, there is a growing interest in inferring the orientations of proteins and peptides on solid interfaces — and biochip surfaces in particular.

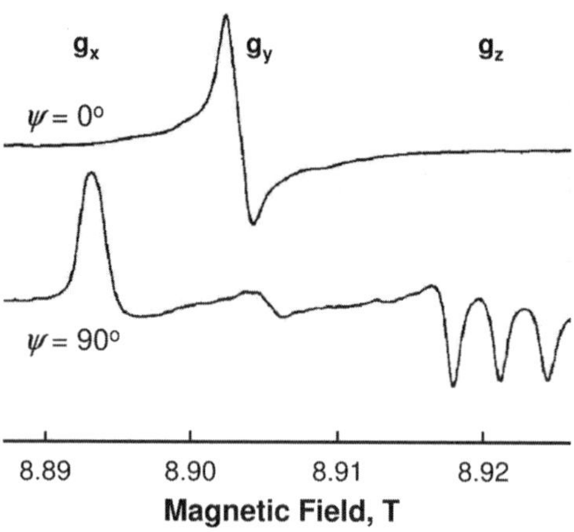

Figure 14. Experimental room-temperature 250-GHz ESR spectra of planar substrate-supported DMPC bilayers doped at 1 mol% with a spin-labeled cholesterol analog cholestane. The spectra were measured at two different orientations of the bilayer director vector with respect to the magnetic field: (**top**) parallel to the field ($\psi = 0°$) and (**bottom**) perpendicular to the field ($\psi = 90°$). Reproduced with permission from Barnes and Freed (1998b). Copyright © 1998, The Biophysical Society.

Macroscopic alignment of spin-labeled bilayers and/or membrane proteins for high-field ESR can be achieved in several ways. Typically, such samples are prepared by forming aligned lipid bilayers on planar solid substrates. One of the methods to produce such a sample is isopotential spin-dry ultracentrifugation (ISDU) (Ge et al. 1994). It involves sedimentation of the membrane fragments (in the gel phase) with simultaneous evaporation of the water phase in a vacuum ultracentrifuge. This method has been utilized by Barnes and Freed (1998) to prepare a macroscopically aligned bilayer composed of mixtures of DMPC and 1,2-dimyristoyl-*sn*-glycero-3-[phospho-*L*-serine] (DMPS) and doped with a cholestane spin probe that served as a cholesterol analog [47]. In order to study these samples with 250-GHz ESR at two different orientations of the bilayer director vector (i.e., membrane normal) with respect to the magnetic field (Figure 14), these authors employed a shunt Fabry-Perot resonator (discussed in §3.1; see also Barnes and Freed 1998a). Cholestane is a so-called "*y*-axis" membrane probe because the nitroxide molecular *y*-axis approximately coincides with the long axis of this molecule. In bilayers composed of phosphatidylcholine lipids (such as DMPC) this molecule aligns with the lipids and the nitroxide *y*-axis becomes approximately parallel to the bilayer director. Then the 250-GHz ESR spectrum from macroscopically aligned lipid bilayers with a director vector parallel to the magnetic field should resemble that of a single crystal when the field is directed along the nitroxide *y*-axis. According to the bottom part of Figure 1, such a spectrum should be centered at a magnetic field corresponding to the g_y principal axis component, and this is indeed what was observed in the experiment (Figure 14, top). At a perpendicular bilayer orientation (Figure 14, bottom), the cholestane *y*-molecular axis will always be perpendicular to the magnetic field, so no spectral component centered at g_y should be observed. This remarkable angular resolution of 250-GHz ESR permitted detailed study of the effects of bilayer composition on the orientation of this spin-labeled cholesterol analog. Specifically, by increasing the fraction of DMPS in DMPC Barnes and Freed (1998b) observed significant changes in 250-GHz ESR spectra that can only be interpreted in terms of a strong local biaxial environment. This biaxiality was initially predicted from molecular dynamics simulations, but spin-labeling 250-GHz ESR provided the first experimental evidence for such cholesterol behavior. It should be noted here that orientational resolution of conventional X-band ESR was insufficient for unambiguous data interpretation.

Another way to macroscopically align phospholipid bilayers is to make use of spontaneous alignment of certain bilayer structures in high magnetic fields. Magnetic field alignment is a well-established technique in solid-state NMR of proteins (Sanders and Schwonek 1992; Sanders et al. 1994; Sanders and Landis 1995; Tjandra and Bax 1997; Katsaras et al. 1997; Prosser et al. 1998; Prosser et al. 1999). The method is based on a spontaneous alignment of discoids formed from a mixture of long- and short-chain phospholipids at magnetic fields above 2.5 T and temperatures of 37–40°C (Katsaras et al. 1997). Typically, these discoids — also called bicelles — are made from about a 3:1 mixture of a long-chain 1,2-dimyristoyl-*sn*-glycero-3-phosphocholine (DMPC) and a short-chain 1,2-dihexanoyl-*sn*-glycero-3-phosphocholine (DHPC). Because the resonant magnetic field for W-band ESR of nitroxides is about 3.4-3.5 T, it is sufficient to align the bicelles. This was first demonstrated by Mangels and colleagues (2001) in a W-band study of the order parameter of a cholestane spin probe in mixed DMPC: bicelles. These studies demonstrated that the order parameter for cholestane in DMPC/DHPC bicelles containing 10% of cholesterol is $S = 0.64 \pm 0.04$, which is very similar to that determined for mechanically aligned DMPC bilayers.

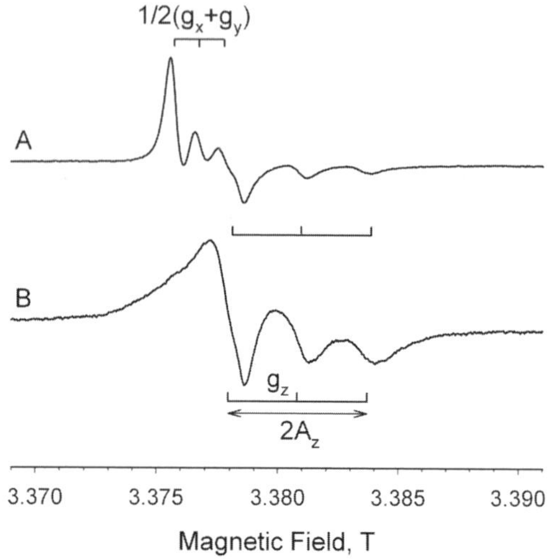

Figure 15. Experimental W-band (94.4 GHz) ESR spectra from mixed DMPC/DHPC bicelles labeled at 1 mol% with 5-doxyl stearic acid. At the temperature of the experiment of 40°C the bicelles were found to spontaneously align in the external magnetic field of ca. 3.4 T. The top spectrum (**A**) is illustrative of the bicelle orientation in which the bilayer vector is perpendicular to the external magnetic field. Upon addition of lanthanide Tm^{3+} ions that bind to the bilayer in the polar head region and change the sign of the magnetic susceptibility anisotropy, the bicelles are flipped by 90°, resulting in parallel orientation of the director vector vs. the magnetic field (bottom spectrum **B**). Reproduced with permission from Smirnov (2002). Copyright © 2002, RSC Publishing.

The magnetic alignment of bicelles is the consequence of the anisotropy of magnetic susceptibility tensor $\Delta\chi$. It was demonstrated in NMR studies that the latter could be manipulated by addition of paramagnetic lanthanide ions. It has been discovered that certain lanthanides shift $\Delta\chi$ to a positive value, causing the bicelles to flip by 90° such that the membrane normal becomes parallel to the external magnetic field (Prosser et al. 1998, 1999). This is illustrated in Figure14 with another probe — 5-doxyl stearic aids (5DSA). As already discussed in §2.4, this spin probe (a mimic of a fatty acid) has a nitroxide moiety that is positioned close the bilayer polar lipid head and aligned with the nitroxide *z*-axis along the vector of the bilayer director. At the temperature of the experiment (40°C) the probe is undergoing a rapid highly anisotropic rotation around the probe molecular *z*-axis. Spectrum A in Figure 15 corresponds to the bicelle orientation with the director vector perpendicular to the external magnetic field. The rapid motion of the probe affectively averages the *x*- and *y*-components of the *g*-matrix, resulting in maximum spectral intensity in the vicinity of the $½(g_x + g_y)$ region. Addition of a small amount of lanthanide ions (in this case Tm^{3+}, which binds to the bilayer in the polar head region and changes the sign of the magnetic susceptibility anisotropy) flips the bilayer director with respect to the magnetic field by 90°. At the latter orientation the *z*-axis of the probe becomes parallel to the magnetic fields, and this results in shifting the spectral intensity to the g_z-region (Figure 15B). Thus, the method of magnetic field alignment may be useful in combination with high-field ESR because the sample preparations are much easier than for substrate-supported samples and no special resonators are required.

Although the magnetic field alignment of bicelles has been also demonstrated in X-band experiments (Mangels et al. 2000; Cardon et al. 2001), the use of ESR at magnetic fields >3.3 T is clearly advantageous for at least two reasons: (*i*) the concentration of lanthanides, such as Yb^{3+}, can be dropped to 1 mol% of lipids (vs. 10 mol% at the X-band (Mangels et al. 2000; Cardon et al. 2001) and (*ii*) the bicelle alignment is spontaneous even at magnetic fields corresponding to the W-band, while field cycling is known to be a necessity for X-band studies (Mangels et al. 2000).

Recently, Smirnov and Poluektov described another approach for aligning phospholipid bilayers for magnetic resonance studies (2003). In brief, they showed that the lipids would assemble into nanotubular bilayers when placed inside nanoporous anodic aluminum oxide membranes (AAO) (Figure 16). These structures — which they called lipid nanotube arrays — have a high density of the nanoporous channels, providing at least a 600-fold gain in the bilayer surface area for the same size as the planar substrate chips. Subsequent studies have shown that these new substrate-supported bilayers retain many biophysical properties of unsupported bilayers (Alaouie and Smirnov 2005) and are suitable for aligning membrane proteins for high-resolution solid-state NMR studies (Chekmenev et al. 2005). Recent NMR studies provide detailed characterization of lipid properties in such nanotubular bilayers (Lorigan et al. 2004; Gaede et al. 2004).

The initial demonstration of self-assembly of phospholipids into cylindrical structures when confined to nanoporous AAO channels was obtained with spin-labeling ESR at 95 GHz (Smirnov and Poluektov 2003). Because local order parameter *S* of the lipid fatty acid chains is known to decrease progressively toward the bilayer center, 1-palmitoyl-2-stearoyl-(5-doxyl)-*sn*-glycero-3-phosphocholine (5PC) lipids containing nitroxide moiety at position 5 were chosen for the experiments. Dynamic lipid disorder and partial averaging of spectral anisotropies were further reduced by taking ESR spectra at low temperature (150 K), at which dynamics of the phospholipids is approaching the rigid limit. The largest changes in the 5PC spectra upon reorientation of the AAO substrate in the magnetic field were observed at 95 GHz (W-band) because of an enhanced angular resolution of high-field ESR over conventional X-band. The relative intensities of characteristic peaks of these spectra (Figure 16, left) are clearly different for two orientations of the

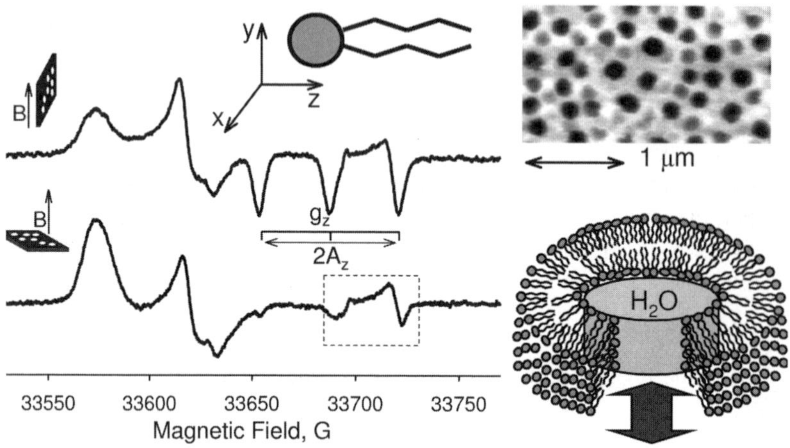

Figure 16. (**Left**) Experimental rigid-limit (T = 150 K) high-resolution 94.4-GHz (W-band) ESR spectra of AAO substrate with deposited DMPC:5PC (100:1 molar ratio) at two orientations of the substrate surface in the magnetic field. Cartoon at the top shows orientations of the magnetic axes with respect to the phospholipid. Note that the bottom ESR spectrum has a low intensity in the g_z region (the feature in the dashed box is due to a paramagnetic AAO impurity), indicating that at this substrate orientation the lipid chains are perpendicular to the magnetic field. (**Right**) Scanning electron microscopy (SEM) image of the substrate after heat treatment and a cartoon of a lipid nanotube formed inside the substrate nanopores.

substrate in the magnetic field. Particularly noticeable are the changes in the g_z region (i.e., high-field component that spreads from ca. 3.3850 to 3.3940 T): when the surface of the AAO substrate is perpendicular to the magnetic field (bottom spectrum), the z-component almost completely disappears (signals inside the dashed box are mainly due to paramagnetic impurities in the AAO substrate). This means that at this substrate orientation only a very small fraction of molecules has the z-axis of the N–O frame aligned with the external magnetic field. The orientation of the nitroxide magnetic axes of 5PC is such that the z-axis is approximately aligned along the phospholipid chain (Figure 16, cartoon in top right corner). Thus, it must be concluded that a majority of the phospholipids inside the nanopores (since the surface phospholipids were mechanically removed during sample preparation) are positioned with their long axis perpendicular to the magnetic field and therefore perpendicular to the direction of pores. The AAO surface is known to be hydrophilic (Liew et al. 1997), and therefore in fully hydrated samples the lipids are organized in bilayers rather than in monolayers. The bilayer lipid organization was further confirmed by experiments with the lipids labeled at the end of the acyl chain (16PC; not shown): these samples produced ESR spectra essentially identical to those from unsupported liposomes (not shown). This is consistent with the lipid nanotube geometry shown in Figure 16 (bottom right). The static order parameter of lipids in such a nanotube was determined by a "center of mass" method (Grinberg et al. 1990) and was found to be exceptionally high: $S_{static} \approx 0.9$.

It appears that one of the most important advantages of the nanoporous AAO substrate is its excellent accessibility to the bilayer surface for exposure to solute molecules. This makes it possible to expose membrane protein samples to different solution media and to repeat this procedure multiple times with the sample. In addition, fully hydrated membrane proteins can be exposed, for example, to buffers at various pH, ion, and drug concentrations, facilitating a wider range of structural and functional studies in the native-like environment by means of magnetic resonance and spin-labeling ESR.

6. CHARACTERIZATION OF THE NITROXIDE MICROENVIRONMENT BY HIGH-FIELD ESR

Magnetic parameters of nitroxides are known to be sensitive to intermolecular interactions and, in particular, to hydrogen bonding and local solvent polarity (Kawamura et al. 1967; Griffith et al. 1974; Ondar et al. 1985). In general, when a spin-labeled molecule is transferred from an aqueous (polar) to a hydrocarbon (non-polar) environment, isotropic nitrogen hyperfine constant A_{iso} decreases by up to 0.1–0.2 mT, while the isotropic g-factor, g_{iso}, increases only slightly (by $\approx$0.0004). At the X-band this change in g_{iso} corresponds to a $\approx$0.068 mT shift of the ESR line, which is smaller than the typical nitroxide line width (ca. 0.1 mT). Therefore, X-band ESR spectra from spin labels residing in different environments are only partially resolved for non-perdeuterated nitroxides with line widths of one tenth of an mT or more. As discussed in the first section of this chapter, the shift in the resonance line positions due to g-factor differences grows proportionally with increasing magnetic field/resonant frequency of the experiment. Even at the W-band the shifts in the nitroxide ESR specra due to the g-factor are typically about threefold greater than those due to A_{iso}. This difference increases even further with an increasing magnetic field. The sensitivity on nitroxide ESR spectra to the hydrophobicity of the immediate molecular environment could be assessed from both fast-motion and rigid-limit ESR spectra. Here we will discuss the basics of solvent effects on the nitroxide high-field ESR spectra, review the use of this method for probing local polarity and proticity (i.e., the ability of a complex solvent to form a hydrogen bond with a solute) of membrane and proteins, and the emerging applications of high-field ESR in site-directed pH-sensitive spin-labeling.

6.1. Probing Local Polarity and Proticity of Membrane and Proteins with High-Field ESR

Solvent effects on magnetic parameters in organic free radicals including nitroxides have been described since the early sixties (see Kawamura et al. 1967 and the references therein). At that time, Kawamura and coworkers (1967) presented the first theoretical interpretation of solvent effects and hydrogen bonding on the nitroxide electronic g-factor. The theoretical treatment was based on the theory of

Stone (1963) for the g-matrix of π-radicals. However, as discussed in this chapter, the g-factor resolution of conventional ESR at the X-band was clearly insufficient to measure solvent and hydrogen bonding line shifts accurately and unambiguously. Thus, over the years researchers focused on evaluating polarity effects solely from measurements of nitrogen hyperfine coupling constants. For example, Griffith and coworkers (1974) deduced the shape of the hydrophobic barrier of phospholipid bilayers and presented evidence for water penetration into biological membranes based on A_{iso} measurements.

With the development of high-field ESR it became possible to measure the g-factor of nitroxides much more accurately. Also, the superior resolution of high-field ESR allowed researchers to accurately determine the principal axis components of the $\boldsymbol{g}$-matrix and $\boldsymbol{A}$-tensor from powder pattern spectra, in many cases eliminating the need for preparing and studying single crystals (see Figure 1 and related discussion). For example, an early high-field ESR study of the magnetic parameters of 43 different nitroxide radicals in common solvents demonstrated a great potential of the high-field ESR method to elucidate intermolecular interactions and local solvent polarity (Ondar et al. 1985). Specifically, it was shown that the g_x component of the $\boldsymbol{g}$-factor matrix is the most sensitive to polarity effects. More recently, high-resolution ESR at 250 GHz was employed to derive polarity profiles in phospholipid membranes doped with a series of doxyl-labeled stearic acids (Earle et al. 1994). The studies of bilayer polarity profile were revisited by Marsh and coworkers, who employed 95-GHz ESR and a larger series of spin-labeled phospholipids (Kurad et al. 2003). High-field ESR studies of polarity effects were also extended to proteins. Specifically, several studies of membrane proteins such as bacteriorhodopsin (Steinhoff et al. 2000; Wegener et al. 2001) and pore-forming colicin a toxin (Savitsky et al. 2004) with SDSL ESR demonstrated that accurate determination of nitroxide magnetic parameters at high fields might assist in determination of transmembrane protein location.

Semiquantitative interpretation of polarity effects on the electronic g-factor of nitroxides could be derived from Stone's (1963) theory of the g-factor of aromatic free radicals. According to this theory (see also Möbius et al. 2005),

$$g_x \cong g_e + \frac{2\zeta(O)\rho_\pi^O c_{ny}^2}{\Delta E_{n\pi^*}}, \tag{45}$$

where $g_e \cong 2.0023$ is the free-electron g-factor, $\zeta(O)$ is the oxygen spin–orbit coupling parameter, ρ_π^O is the π spin density on oxygen $2p_z$ atomic orbital, c_{ny} is a coefficient accounting for the $2p_y$ atomic orbital contributing to the oxygen lone-pair orbital n, and $\Delta E_{n\pi^*}$ is the $n \to \pi^*$ electronic excitation energy; $\Delta E_{n\pi^*} = E_n - E_{\pi^*}$. Local electric fields and formation of hydrogen bonds may affect many of these parameters, perhaps with exception of $\zeta(O)$. Then the relative change in $\Delta g_x = g_x - g_e$ is given by

$$\frac{\delta(\Delta g_x)}{\Delta g_x} = \frac{\delta\rho_\pi^O}{\rho_\pi^O} + \frac{\delta c_{ny}^2}{c_{ny}^2} - \frac{\delta(\Delta E_{n\pi^*})}{\Delta E_{n\pi^*}}. \tag{46}$$

The deviation in Eq. (46) is based on the properties of differentials. Indeed, if *f* is the following function of three independent variables *a*, *b*, and *c*:

$$f = abc^{-1}, \tag{47}$$

then the full differential, *df*, is

$$df = \left(bc^{-1}\right)da + \left(ac^{-1}\right)db - \left(ab\right)c^{-2}dc\,. \tag{48}$$

The relative change in function *f* can be easily found:

$$\frac{df}{f} = \frac{df}{abc^{-1}} = \frac{\left(bc^{-1}\right)da + \left(ac^{-1}\right)db - \left(ab\right)c^{-2}dc}{abc^{-1}} = \frac{da}{a} + \frac{db}{b} - \frac{dc}{c}. \tag{49}$$

For planar structures, nitrogen hyperfine component A_z is known to be proportional to the π spin density on oxygen, $A_z = Q_\pi^N \rho_\pi^N$, and therefore:

$$\frac{\delta A_z}{A_z} = \frac{\delta \rho_\pi^N}{\rho_\pi^N}. \tag{50}$$

For nitroxide radicals the electronic spin density delocalization from the NO fragment is known to be no more than 1–2%; therefore, $\rho_\pi^O + \rho_\pi^N \cong 1$, and

$$\delta \rho_\pi^N = -\delta \rho_\pi^O\,. \tag{51}$$

From Eqs. (50) and (51):

$$\frac{\delta \rho_\pi^O}{\rho_\pi^O} = -\frac{\delta \rho_\pi^N}{\rho_\pi^O} = -\frac{\delta \rho_\pi^N}{\rho_\pi^N}\cdot\frac{\rho_\pi^N}{\rho_\pi^O} = -\frac{\rho_\pi^N}{\rho_\pi^O}\cdot\frac{\delta A_z}{A_z}, \tag{52}$$

and Eq. (46) can be written as

$$\frac{\delta(\Delta g_x)}{\Delta g_x} = -\frac{\rho_\pi^N}{\rho_\pi^O}\cdot\frac{\delta A_z}{A_z} - \frac{\delta(\Delta E_{n\pi^*})}{\Delta E_{n\pi^*}} + \frac{\delta c_{ny}^2}{c_{ny}^2}. \tag{53}$$

Equation (53) predicts a negative slope for $\delta(\Delta g_x)/\Delta g_x$ vs. $\delta(\Delta A_z/Az)$, and this was indeed observed in the experiments (Steinhoff et al. 2000; Wegener et al. 2001; Savitsky et al. 2004) (see also Figure 17, bottom, for an illustration).

In order to make use of Eq. (53) to evaluate the magnitude of molecular interactions on ESR spectra and the magnetic parameters of nitroxides, one has to include solvent molecules in considerations. One of the first steps in this direction has been made by Griffith and coworkers (1974), who estimated the effects of electric field along the NO bond on A_z. The electric field effects on isotropic nitrogen hyperfine coupling constant A_{iso} were further studied by Schwartz and colleagues (1997). These authors reported a calibration between the shift in isotropic nitrogen

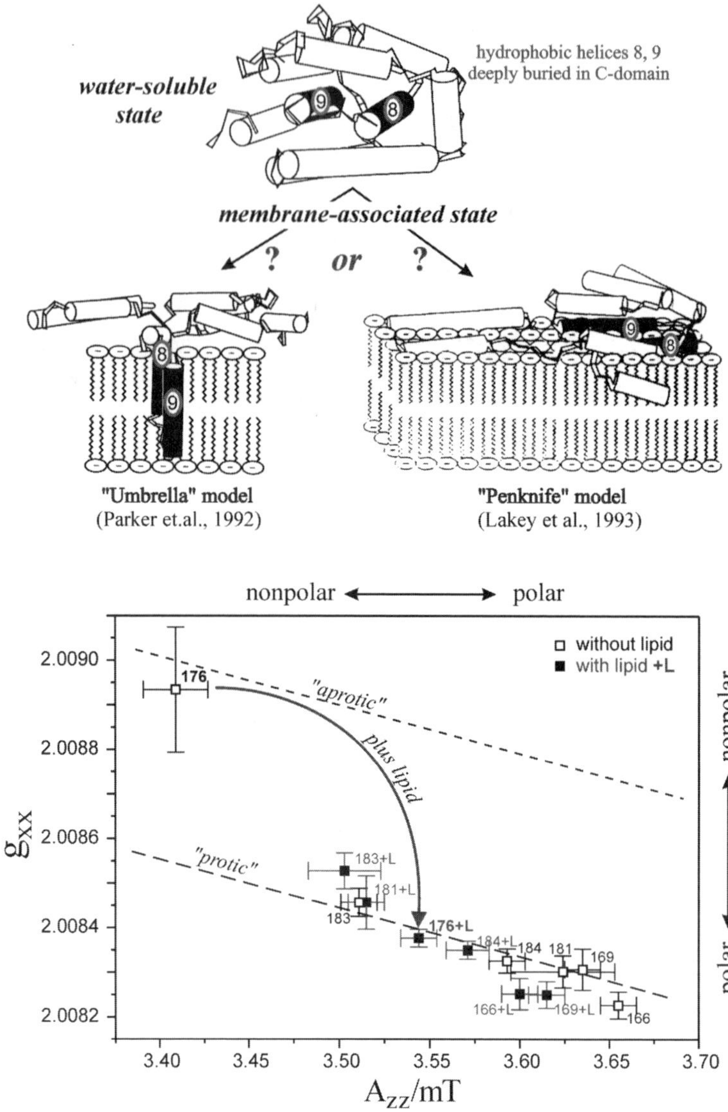

Figure 17. (**Top**) Cartoon of the "umbrella'' (Parker et al. 1992) and the ''penknife'' (Lakey et al. 1993) models of the membrane-associated state of colicin *a*. (**Bottom**) Polarity plot of g_x vs. A_z for the various nitroxide-labeled mutants of helix 9 of colicin *a* A mutants. The data points from samples prepared with and without lipid vesicles are marked by filled and open squares, respectively. The dashed lines define the limits between the non-hydrogen-bonded (short dashes) and the fully-hydrogen-bonded (long dashes) nitroxides. Reproduced with permission from Möbius et al. (2005). Copyright © 2005, RSC Publishing.

hyperfine coupling constant A_{iso} and the local charge on the nitroxide. In their study the aqueous solvent was approximated with a continuum model and the effect on A_{iso} calculated within the framework of standard Hückel theory. Gullá and Budil (2001) extended the calibration of electric field effects to **g**-matrix calculations, and the results were compared with the anisotropic *g*-factors of nitroxides containing internal groups that can be ionized. Using Stone's theory they have shown that the electric fields of the magnitude typically found in proteins may lead to *g*-factor shifts measurable by high-field ESR. They also suggested that the **g**-matrix might be sensitive not only to the magnitude of the electric field but also to the electric field orientation with respect to the nitroxide frame. Ding et al. (2001) presented the results of ab initio calculations of the electric field effect on the **g**-matrix of nitroxide radicals. Their approach utilized an intermediate level Rayleigh-Schrödinger perturbation theory based on unrestricted Hartree-Fock ab initio calculations. It was found that the calculated **g**-matrices agree well with the experimentally measured data.

Hydrogen bonding is another factor that one should consider when evaluating the effects of intermolecular interactions on the magnetic parameters of nitroxide radicals. Results from previous studies suggested that nitroxides might form hydrogen-bonded complexes when immobilized (Griffith et al. 1974; Hwang et al. 1975; Johnson 1981; Hustedt et al. 1995). Hydrogen bond formation and its lifetime are thought to be modulated by the chemical environment and temperature, and thereby may complicate polarity assignment. Hydrogen bonding effects on nitroxide ESR spectra were revisited by Engström and cowoerkers (2001), who applied the restricted open-shell Hartree-Fock (ROHF) linear response method with the atomic mean field approximation (AMFI) to calculate the effect of hydrogen bond formation on the **g**-matrix of MTSL. Their results showed that g_x is reduced most among all three *g*-factor principal components. The decrease in g_x is traced to a higher excitation energy and lower spin–orbit coupling and angular momentum matrix elements for $n \rightarrow \pi^*$ excitation. Subsequently, a density functional theory (DFT) calculation of **g** and **A** for MTSL and MTSL covalently attached to β-mercaptoethanol were performed under conditions of a varying dielectric constant of the medium and the number of hydrogen bonds formed with the nitroxide oxygen (Owenius et al. 2001). These theoretical and experimental high-field ESR studies should be continued to further assist in interpretation of the rigid-limit data from spin-labeled peptides and proteins in terms of environmental properties such as polarity, proticity, and local electric field.

A demonstration of the utility of high-field ESR in characterizing the structural details of protein organization in biological membranes has been provided by W-band spin-labeling ESR studies of membrane proteins bacteriorhodopsin and colicin *a* (Steinhoff et al. 2000; Wegener et al. 2001; Savitsky et al. 2004). A study of membrane insertion of colicin *a* toxin provides a particularly illustrative example (Savitsky et al. 2004). In brief, the C domain of colicin *a* can exist in either a water-soluble or transmembrane form. In order to explain the massive structural transformation of this protein that leads to formation of a transmembrane pore, two models — the "umbrella" and the "penknife" models — have been discussed in the literature (see Figure 17, top) (Parker et al. 1992; Lakey et al. 1993; Savitsky et al. 2004; Möbius et al. 2005). According to the "umbrella" model, after docking to the membrane surface, the hydrophobic 8,9-hairpin spontaneously traverses the membrane, whereas in the "penknife" model the refolding initially leaves the 8,9-hairpin close to the membrane surface. Only then a change in the electric transmembrane potential can trigger an insertion of the 8,9-hairpin into the membrane.

In order to differentiate between these two models, several different spin-labeled mutants of helix 9 of colicin *a* were studied with 95-GHz ESR. Low-temperature rigid-limit ESR spectra were recorded and magnetic parameters measured in order to evaluate the local polarity of the nitroxide side chains. The results are summarized in Figure 17 (bottom) as a g_x vs. A_z plot. The first observation was that all but one of the data points followed a linear dependence with a negative slope, as predicted by Eq. (53) (Möbius et al. 2005). All these data points that followed the bottom dashed line were attributed to nitroxide labels that formed hydrogen bonds with (presumably) solvent molecules but experienced different local polarities. In the absence of lipid bilayers, only one labeled site (176) of colicin *a* had a much higher g_x than the rest. This is typical for non-hydrogen-bonded nitroxides. Upon mixing with aqueous solution of small unilamellar vesicles formed from DMPG (1,2-dimyristoyl-*sn*-glycero-3-phospho-*rac*-(1-glycerol), the local polarity of this site increased dramatically (arrow in Figure 16, bottom). In the "umbrella" model, helices 8 and 9 should penetrate the membrane spontaneously, and the central part of helix 9, where the nitroxide side chain 176 is located, should be placed into the membrane interior (Möbius et al. 2005). Thus, upon addition of lipids nitroxide chain 176 should experience a very nonpolar environment. This contradicted the experimental observation of an increase in the local polarity of this chain upon addition of DMPG. Thus, Möbius and coworkers (2005) concluded that the W-band ESR polarity data disagreed with the umbrella model. It was found that the experimental results were in agreement with the penknife model, in which helix 9 should remain for some time in a transient state at the bilayer interface and therefore experience relatively high polarity.

Assessment of local polarity from the magnetic parameters of nitroxides can be carried out from accurate measurements of isotropic magnetic parameters from HF ESR spectra approaching fast-motional limit. This could be very useful for characterization of biological systems with multiple compartments. One example of this was given by Smirnov and Smirnova (2001) in W-band studies of lipid bilayer interdigitation by ethanol. In order to report on the lipid bilayer state, these authors employed the small nitroxide Tempo, which is known to partition between the lipid and aqueous phases of phospholipid bilayers (Shimshick and McConnell 1973). Figure 18 shows typical experimental X- and W-band ESR spectra from this system at temperatures below and above the main phase transition ($T_c \approx 40.3$°C).

While at 9.0 GHz (X-band) only the high-field component of the Tempo spectra is partially resolved, all the spectral components are fully separated at 94.4 GHz (W-band). This is another example of the superior *g*-factor resolution of high-field ESR in spin-labeling studies (see other examples in §4.1). These high-resolution

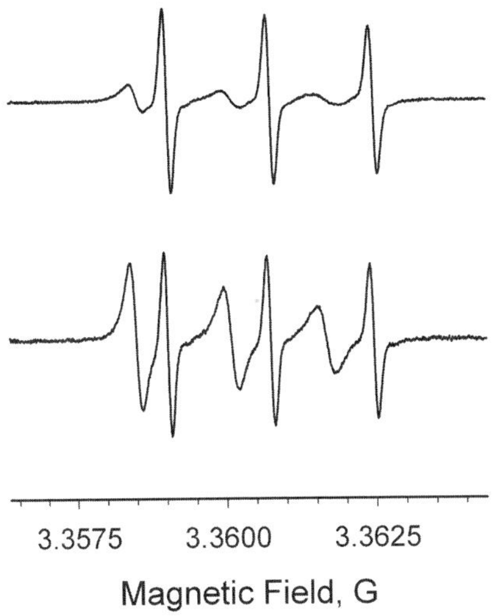

Figure 18. Representative experimental 9.0- and 94.4-GHz ESR spectra of multibilayer vesicles prepared from 1,2-dipalmitoyl-*sn*-glycero-3-phosphatidylcholine (DPPC) and labeled with the nitroxide Tempo (2,2,6,6-tetramethylpiperedine-1-oxyl). Fhe final concentration of DPPC in aqueous media was 200 mg/ml; Tempo 620 μM. The effect of partitioning of Tempo between the aqueous and lipid phases of the membrane is indicated by arrows in the spectra taken at 42°C and is discussed in the text. Reprinted with permission from Smirnov (2002). Copyright © 2002, RSC Publishing.

W-band ESR spectra were least-squares simulated to yield data on the rotational dynamics of this small probe in the lipid bilayer, the actual partition coefficients, and, also, very accurate values for g_{iso} and A_{iso} (Smirnov et al. 1995; Smirnov and Smirnova 2001). Based on g_{iso} vs. A_{iso} data, Smirnov and Smirnova were able to determine the local polarity of the Tempo probe partitioned in DPPC bilayers. The assignment was carried out using an empirical relationship between g_{iso} and A_{iso} for Tempo established for a series of protic and aprotic solvents and their mixtures (Figure 19, open circles). The isotropic g-factor of Tempo in water was taken as a reference point (i.e., $\Delta g_{iso} = 0$). The plot shows an exceptional linear correlation between g_{iso} and A_{iso} for this nitroxide. The data for Tempo partitioned in the lipid phase of DPPC are shown as filled squares in Figure 19: A corresponds to the average magnetic parameters of Tempo just before the phase transition and B was measured at ca. 45°C (above the phase transition). It is clear that points A and B fit closely our calibration curve. The closest calibration points are from acetone (Figure 19, point 7) and olive oil/ethanol (9:1, v/v) mixture (Figure 19, point 8). It is also clear that the observed values deviate significantly from that for Tempo in the bulk hydrocarbon phase (Figure 19, point 10, olive oil) and from protic polar solvents such as ethanol and methanol (Figure 19, points 4 and 5 respectively). Thus, it is quite likely that the Tempo molecules in the lipid phase are in a rapid exchange between the polar head and the central membrane regions and/or travel with aqueous defects through the bilayer. Figure 19 also shows two data points (C and D) that correspond to two different membrane compartments of Tempo observed for DPPC bilayers interdigitated (i.e., mispacked into a single leaflet of a mixed head-and-tail configuration) by 1.2 M of ethanol. From these data the polarity of the first compartment (point C in Figure 19) corresponds to an approximately pure lipid phase while the second compartment (point D in Figure 19) is close to point 3, which was measured for an ethanol–water (7:3, v/v) mixture. These data demonstrate that the micropolarity of the nitroxide probe partitioned in a multicompartment system can be assigned based on accurate *g*-factor measurements at high magnetic fields.

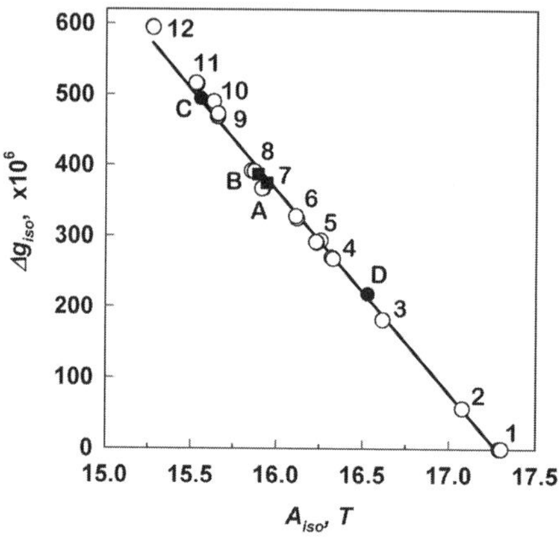

Figure 19. Isotropic magnetic parameters g_{iso} and A_{iso} for Tempo in a series of protic and aprotic solvents and their mixtures: 1 = water; 2 = water/ethanol solution (7:3, v/v); 3 = water/ethanol solution (3:7, v/v); 4 = methanol; 5 = ethanol; 6 = isopropanol; 7 = acetone; 8 = olive oil/ethanol solution (9:1, w/w); 9 = acetonitrile; 10 = olive oil; 11 = toluene; 12 = hexane. The isotropic *g*-factor of Tempo in water was taken as a reference point (i.e., $\Delta g_{iso} = 0$). The estimated errors are within the size of the symbols. The magnetic parameters for Tempo partitioned in the lipid phase of the DPPC bilayer are shown as filled squares. Parameters corresponding to the membrane in the gel phase (before the main phase transition) are marked as A and above the phase transition (g_{iso} at ca. 45°C) marked as B. Parameters for two components of Tempo in non-aqueous phase of DPPC bilayer interdigitated by 1.2 M ethanol are marked as C and D. Reproduced with permission from Smirnov (2002). Copyright © 2002, RSC Publishing.

6.2. Site-Directed pH-Sensitive Spin Labeling: Differentiating Local pK and Polarity Effects by High-Field ESR

As discussed in the preceding section, local electric fields and hydrogen bonding can be observed from changes in magnetic parameters of nitroxide ESR spectra measured with high resolution at high magnetic fields. However, nitroxides that are traditionally employed in spin-labeling studies, such as MTSL, Tempo, and its derivative, cannot be used for probing another important parameter that affects intermolecular interactions — local pH — simply because the pK_a of these nitroxides lays outside the useful range. Currently, there is considerable interest in assessing local proton concentration (reported as pH) because this parameter plays a very important role in many biological processes such as, for example, self-cleavage of viral ribozymes (Luptak et al. 2001), refolding of proteins on membrane interfaces, and protein insertion into phospholipid bilayers (Zhan et al. 1994).

Recently, Smirnov and coworkers (2004) described a general approach to mapping local pK_a values of peptides with high-field ESR using MTS (methanothiosulfonate) derivatives of imidazolidine nitroxides. The tertiary amine nitrogen N3 of this label readily participates in proton exchange reactions, which are monitored through changes in the ESR spectra of nitroxide moiety (Figure 20).

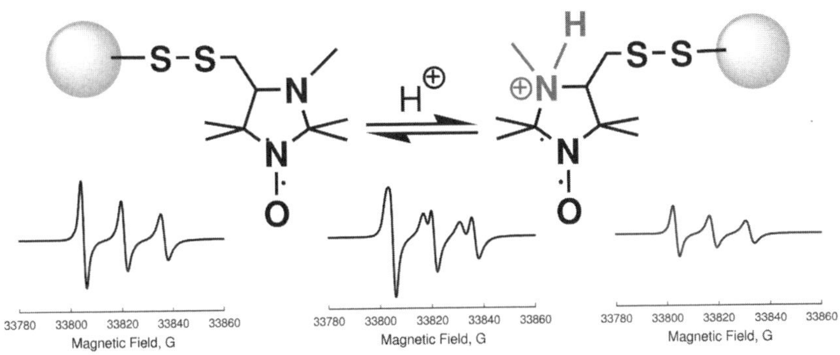

Figure 20. An imidazolidine nitroxide (IMTSL) covalently attached to a peptide through a disulfide bridge shows pH-dependent 95-GHz ESR spectra. See text and Smirnov et al. (2004) for more details. Reproduced with permission from the Smirnov et al. (2004). Copyright © 2004, American Chemical Society.

It is worthwhile to note here that reversible pH effects on ESR spectra of stable nitroxide radicals have been known for more than 20 years (Khramtsov et al. 1982; Keana et al. 1982; Khramtsov et al. 1985). One of the first steps in specific attachment of pH-sensitive nitroxides to biomolecules was undertaken by Khramtsov et al. (1985) and Khramtsov et al. (1992), who reported studies of surface potentials and polarity of phospholipid bilayers and human serum albumin with several imidazolidine nitroxide derivatives. However, the covalent attachment of pH-sensitive nitroxides employed in those studies was not fully specific. To achieve the high specificity of covalent attachment of a pH-sensitive imidazolidine nitroxide to thiols, the Novosibirsk group of nitroxide chemists synthesized the MTS derivative IMTSL, methanethiosulfonic acid S-(1-oxyl-2,2,3,5,5-pentamethyl-imidazolidin-4-ylmethyl) ester. This is the same attachment group as in spin-label MTSL typically employed in SDSL ESR studies. Moreover, the nitroxide side chain formed upon attachment of IMTSL to a cysteine is very similar to that of MTSL (see also Figure 20). Thus, one could expect that both spin labels should exhibit very similar rotational dynamics and, thus, ESR spectra when attached to proteins.

To calibrate g_{iso} and A_{iso} of IMTSL for pH dependence, a series of room-temperature W-band ESR spectra of aqueous solutions of a free label and a label attached to a cysteine (IMTSL-cys) and glutathione (IMTSL-glu) were measured as a function of pH. At intermediate pH the W-band ESR spectra of these compounds consisted of two partially resolved nitroxide components (bottom middle spectrum shown in Figure 20) characteristic of slow exchange conditions. The spectra were simulated with least-squares program described earlier (Smirnov and Belford 1995). The A_{iso} titration data and corresponding least-squares Henderson-Hasselbalch titration curves are shown in Figure 21.

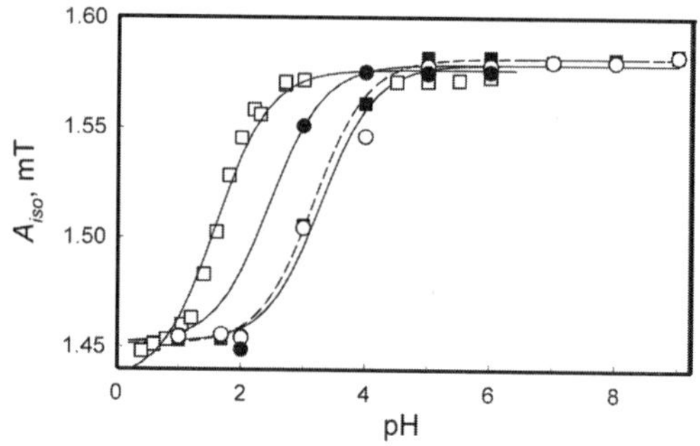

Figure 21. Weighted A_{iso} for two-component W-band (95 GHz) ESR spectra as a function of pH and corresponding Henderson-Hasselbalch titration curves (shown as lines): open squares = IMTSL; filled squares = IMTSL–glutathione; filled circles = IMTSL–P11 (solid lines); open circles = IMTSL–cysteine (dashed line). Reproduced with permission from Smirnov et al. (2004). Copyright © 2004, American Chemical Society.

The ESR titration curves in Figure 21 show that for all compounds studied A_{iso} undergoes large (>0.13 mT) changes upon protonation of the tertiary amine nitrogen N3. Also, the pK_a of IMTSL-cys and IMTSL-glu were respectively shifted to more basic pKa = 3.21 ± 0.04 and 3.15 ± 0.03 from that of the free label (1.58 ± 0.03 units) while A_{iso}(base) and A_{iso}(acid) were not affected. It is likely that after labeling, the electronegativity of the side chain is reduced. Since the side chain is closer to N3 than to the nitroxide moiety, this decrease in the electronegativity shifts the N3 pK_a to more basic values without any measurable effects on the unpaired electron spin density. Thus, for these small spin-labeled peptides that are not expected to acquire any globular structure it was observed that the presence of other ionizable groups in the side chain have no effects on the IMTSL ESR spectra, which showed only one transition caused by proton exchange reaction at the N3 position.

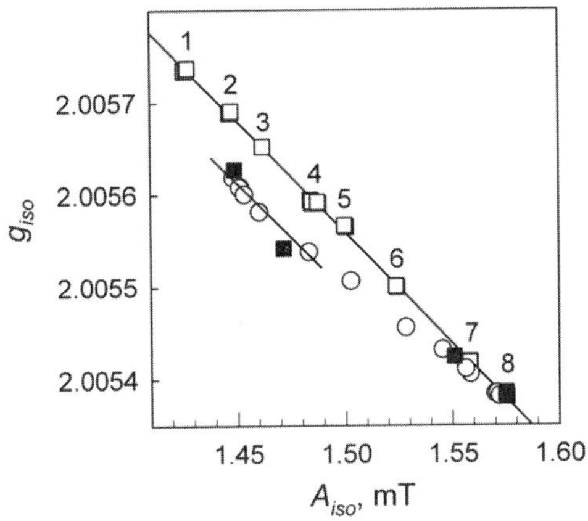

Figure 22. Isotropic magnetic parameters g_{iso} and A_{iso} from solution W-band ESR spectra at room temperature. Open squares: basic form of IMTSL in 1 = toluene, 2 = acetonitrile, 3 = acetone, 4 = isopropanol, 5 = ethanol, 6 = water/ethanol (3:7, v/v) and 7 = (7:3, v/v), 8 = water buffered to pH 6.0. Titration parameters for: IMTSL (open circles) and IMTSL–P11 (filled squares). See text for details. Reproduced with permission from Smirnov et al. (2004). Copyright © 2004, American Chemical Society.

To elucidate solvent effects on IMTSL W-band ESR spectra, a series of measurements in protic and aprotic solvents were carried out. The data, summarized in Figure 22 as g_{iso} versus A_{iso} plots, suggest a linear correlation similar to the polarity plot for Tempo (Figure 19). One interesting question is whether g_{iso} and A_{iso} for titration curves would fall on the same correlation plot. Figure 22 shows that at pH values below the N3 pK_a, magnetic parameters g_{iso} and A_{iso} clearly fall below the correlation line. We speculate that this deviation is related to the effect of an asymmetric charge located chiefly on N3 onto the oxygen lone-pair orbital of the reporter nitroxide moiety. As one would expect, an acidic form of IMTSL would have a somewhat different charge configuration of the nitroxide moiety that cannot be mimicked by a symmetric rearrangement of the solvent molecules with respect the axis of the N–O bond occurring in solvents with different polarity. This would result in different correlations between g_{iso} and A_{iso} that we observed for acidic and basic forms of IMTSL (see also Eq. (53)). This difference could be very useful for determining whether the changes in ESR spectra are caused by N3 protonation or by local polarity effects. It is worthwhile to note here that the accurate g-factors provided by high-field ESR are essential for unambiguous assignment.

To illustrate the utility of IMTSL and high-field ESR in site-directed pKa measurements, we have selectively labeled the thiol group of a synthetic P11 peptide fragment of the laminin B1 chain (Cys–Asp–Pro–Gly–Tyr–Ile–Gly–Ser–Arg). Previous studies have shown that this laminin fragment could inhibit lung tumor colony formation by blocking tumor cell invasion through basement membranes (Iwamoto et al. 1987). W-band titration experiments with IMTSL-P11 have shown that upon a decrease in pH the W-band ESR signal splits into two components, similar to that observed for IMTSL-cys and IMTSL-glu. From pH = 6.0 to 3.0 the effective A_{iso} decreased only slightly, but at pH = 2.0 A_{iso} was essentially equal to the A_{iso}(acid) value for IMTSL (Figure 21). Because at acidic pH one could expect a change in the peptide conformation, only the data from pH = 3 to 6 were fitted to the Henderson-Hasselbalch equation (Figure 21, solid line). Figure 21 demonstrates that for IMTSL-P11 pKa = 2.5 ± 0.1 is less than that of either IMTSL-glu or IMTSL-cys (pKa ≈ 3.18).

It seems possible that local polarity effects arising from the peptide backbone could also explain the A_{iso} value observed in our experiments. For example, if the nitroxide ring aligns with a tyrosine, this would decrease the nitroxide local polarity and cause A_{iso} to decrease. To rule out this possibility, the correlation of the g_{iso} versus A_{iso} plot was analyzed. Figure 22 shows that the g_{iso} versus A_{iso} plot of the magnetic parameters for IMTSL-P11 (filled squares) closely follows the IMTSL titration curve that falls below the polarity calibration line. Thus, it was concluded that the observed changes in g_{iso} and A_{iso} are related to a proton exchange reaction of the nitroxide label but not local polarity changes (Smirnov et al. 2004).

While small spin-labeled peptides such as IMTSL-glu are expected to be unfolded, a larger P11 peptide can adopt a folded or a partially folded state through tertiary interactions. Specifically for P11, an energy optimized structure (not shown) places the nitroxide ring in a close proximity to the carboxylic group of the aspartic acid. Because the pK_a of that group in peptides ranges from 2.0 to 4.93 units (Lohse et al. 1997; Noble et al. 2000), it is feasible that the observed shift of the IMTSL-P11 titration curve from that of, e.g., IMTSL-glu reflects the close proximity of the carboxilic group.

Overall, Smirnov et al. (2004) demonstrated an extension of SDSL ESR to protein site-specific pH measurements. We anticipate that the next applications of this method could be in studies of ion channels. The method will also be very useful for studying pH-triggered conformations in biomolecules and for better understanding of the role of hydrophobic and electrostatic interactions in membrane protein structure and function.

7. PERSPECTIVES

The field of spin labeling high-field ESR is rapidly expanding as more research groups are getting directly involved in site-directed spin-labeling studies of membranes and proteins and acquiring/developing high-field ESR instrumentation.

The method already produces valuable and recognizable results for complex biological systems, ranging from large water-soluble protein complexes and DNA to membrane proteins, the systems that are rather difficult to approach by other spectroscopic means. The new developments and, specifically, time-domain experiments and double resonance techniques, such as DEER (double electron–electron resonance) or PELDOR (pulsed electron double resonance), and DQC (double quantum coherence) should be extended to high-field ESR and fully explored in studies of membrane protein structure and dynamics. The work on synthesis of new nitroxide labels, as illustrated here in an example of pH-sensitive probes, is going to continue with the goals to tailor these new labels for high-field applications. Rapid development of computer hardware and software will lead to better understanding of the complex dynamics and structural organization of spin labeled systems. One way to pursue these studies would be to take advantage of modern molecular dynamics software packages to derive local potentials experienced by spin labels in protein systems as well as further developing direct simulations of ESR spectra from molecular trajectories.

Another very promising prospect in employing spin-labeling ESR at high fields for membrane protein studies lies in electrostatic mapping of membrane proteins using nitroxide tags that are specifically tailored for such studies. Although handling of miniature high-field ESR samples and optimizing signals from lossy aqueous samples at high fields remain very tricky, it is our expectation that these problems will be solved in a timely fashion. We also expect studies involving determination of orientations of spin-labeled molecules and distance-angle measurements to become increasingly important in the study of membrane proteins.

8. ACKNOWLEDGEMENTS

A.I.S. would like to acknowledge DOE Contract DE-FG02-02ER15354, NIH 1R01GM072897, and NSF ECS 0420775 for providing partial support for this work. Acknowledgment is made to the Donors of the American Chemical Society Petroleum Research Fund for partial support of this research through ACS PRF# 40232-G7 (A.I.S.). T.I.S. would like to thank the National Science Foundation for their support through Grant No. MCB-0451510. Acknowledgment is made to the Donors of the American Chemical Society Petroleum Research Fund for partial support of this research through ACS PRF# 40771-G4 (T.I.S.). We would also like to express our gratitude to T. Gray Chadwick (NCSU) for his expert help with chemical structures and references.

9. REFERENCES

Abragam A. 1961. *The principles of nuclear magnetism*. Oxford: Clarendon.

Alaouie AM, Smirnov AI. 2005. Cooperativity and kinetics of phase transitions in nanopore-confined bilayers studied by differential scanning calorimetry. *Biophys J* **88**:L11–L13.

Altenbach C, Flitsch SL, Khorana HB, Hubbell WL. 1989. Structural studies on transmembrane proteins, 2: spin labeling of bacteriorhodopsin mutants at unique cysteines. *Biochemistry* **28**:7806–7812.

Altenbach C, Marti T, Khorana HG, Hubbell WL. 1990. Transmembrane protein structure: spin labeling of bacteriorhodopsin mutants. *Science* **248**:1088–1092.

Ashikawa I, Yin J-J, Subczynski WK, Kouyama T, Hyde JS, Kusumi A. 1994. Molecular organization and dynamics in bacteriorhodopsin-rich reconstituted membranes: discrimination of lipid environments by the oxygen transport parameter using a pulse ESR spin-labeling technique. *Biochemistry* **33**:4947–4952.

Ballew RM, Sabelko J, Gruebele M. 1996. Direct observation of fast protein folding: the initial collapse of apomyoglobin. *Proc Natl Acad Sci USA* **93**:5759–5764.

Barnes JP, Freed JH. 1997. Aqueous sample holders for high-frequency ESR. *Rev Sci Instrum* **68**:2838–2846.

Barnes JP, Freed JH. 1998a. A shunt Fabry-Perot resonator for high frequency ESR using a variable coupling scheme. *Rev Sci Instrum* **69**:3022–3027.

Barnes JP, Freed JH. 1998b. Dynamics and ordering in mixed model membranes of DMPC and DMPS: a 250 GHz ESR study. *Biophys J* **75**:2532–2546.

Barnes JP, Liang Z, Mchaourab HS, Freed JH, Hubbell WL. 1999. A multifrequency electron spin resonance study of T4 lysozyme dynamics. *Biophys J*, **76**:3298–3306.

Becker J, Saxena S. 2005. Double quantum coherence electron spin resonance on coupled Cu(II)–Cu(II) spins. *Chem Phys Lett* **414**:248–252.

Bennati M, Gerfen GJ, Martinez GV, Griffin RG, Singel DJ, Millhauser GL. 1999. Nitroxide side-chain dynamics in a spin-labeled helix-forming peptide revealed by high-frequency (139.5-GHz) EPR spectroscopy. *J Magn Reson* **139**:281–286.

Bernhard WA, Fouse GW. 1989. The simulation of powder ESR spectra of organic free radicals. *J Magn Reson* **82**:156–162.

Bloch F. 1946. Nuclear induction. *Phys Rev* **70**:460–474.

Borbat PP, Costa-Filho AJ, Earle KA, Moscicki JK, Freed JH. 2001. Electron spin resonance in studies of membranes and proteins. *Science* **291**:266–269.

Borbat PP, Davis JH, Butcher SE, Freed JH. 2004. Measurement of large distances in biomolecules using double-quantum-filtered refocused electron-spin-echoes. *J Am Chem Soc* **126**:7746–7747.

Bottle SE, Gillies DG, Micallef AS, Reid DA, Sutcliffe LH. 1999. ESR measurements of the partitioning of some new spin probes in n-octanol-water. *Magn Reson Chem* **37**:730–734.

Bottle SE, Gillies DG, Hughes DL, Micallef AS, Smirnov AI, Sutcliffe LH. 2000. Synthesis, characterization, single crystal x-ray structure and W-band EPR spectroscopy of a new anionic isoindoline nitroxide:synthesis and characterisation of some derivatives. *J Chem Soc* **2**:1285–1291.

Brunel L-C, van Tol J, Angerhofer A, Hill S, Krzystek J, Maniero AL. 2004. Modern developments and prospects in multi-frequency high field EMR. *Biol Magn Reson* **22**:466–538.

Budil DE, Earle KA. 2004. Sample resonators for quasioptical EPR: a practical guide for biological applications. *Biol Magn Reson* **22**:353–400.

Budil DE, Earle KA, Lynch WB, Freed JH. 1989. Electron paramagnetic resonance at 1 millimeter wavelengths. In *Advanced EPR applications in biology and biochemistry*, p. 307–340. Ed AJ Hoff. Amsterdam: Elsevier.

Budil DE, Earle KA, Freed JH. 1993. Full determination of the rotational diffusion tensor by electron paramagnetic resonance at 250 GHz. *J Phys Chem* **97**:1294–1303.

Budil DE, Lee S, Saxena S, Freed JH. 1996. Nonlinear-least-squares analysis of slow-motion EPR spectra in one and two dimensions using a modified Levenberg-Marquardt algorithm. *J Magn Reson A* **120**:155–189.

Budil DE, Kolaczkowski SV, Perry A, Varaprasad C, Johnson F, Strauss PR. 2000. Dynamics and ordering in a spin-labeled oligonucleotide observed by 220 GHz electron paramagnetic resonance. *Biophys J* **78**:430–438.

Cardon TB, Tiburu EK, Padmanabhan A, Howard KP, Lorigan GA. 2001. Magnetically aligned phospholipid bilayers at the parallel and perpendicular orientations for X-band spin-label EPR studies. *J Am Chem Soc* **123**:2913–2914.

Chekmenev EY, Hu J, Gor'kov PL, Brey WW, Cross TA, Ruuge A, Smirnov AI. 2005. ^{15}N and ^{31}P solid-state NMR study of transmembrane domain alignment of M2 protein of influenza A virus in hydrated cylindrical lipid bilayers confined to anodic aluminum oxide nanopores. *J Magn Reson* **173**:322–327.

Clarkson RB, Smirnov AI, Smirnova TI, Belford RL. 2001. Multi-frequency EPR of high-spin S-state ions: examples from Gd^{3+} chelates. In *The chemistry of contrast agents in medical magnetic resonance imaging*, pp. 383–415. Ed A Merbach. New York: Wiley.

Ding Z, Gullá AF, Budil DE. 2001. Ab initio calculations of electric field effects on the g-tensor of a nitroxide radical. *J Chem Phys* **115**:10685–10693.

Earle KA, Freed JH. 1999. Quasioptical hardware for a flexible FIR-EPR spectrometer. *Appl Magn Reson* **16**:247–272.

Earle KA, Smirnov AI. 2004. High field ESR: applications to protein structure and dynamics. *Biol Magn Reson* **22**:95–143.

Earle KA, Budil DE, Freed JH. 1993. 250-GHz EPR of nitroxides in the slow motional regime: models of rotational diffusion. *J Phys Chem* **97**:13289–13297.

Earle KA, Moscicki JK, Ge M, Budil D, Freed JH. 1994. 250-GHz electron spin resonance studies of polarity gradients along the aliphatic chains in phospholipid membranes. *Biophys J* **66**:1213–1221.

Earle KA, Moscicki JK, Polimeno A, Freed JH. 1997. A 250-GHz ESR study of o-terphenyl: dynamic cage effects above T_c. *J Chem Phys* **106**:9996–10015.

Eaton GR, Eaton SS, Rinard GA. 1998. Frequency dependence of EPR sensitivity. In *Spatially resolved magnetic resonance: methods, materials, medicine, biology, rheology, geology, ecology, hardware*, pp. 65–74. Ed P Blumler, B Blümich, R Botto, E Fukushima. New York: Wiley.

Eaton SS, Eaton GR, Berliner LJ. eds. 2001. Distance measurements in biological systems by EPR. *Biol Magn Reson* **19**:450.

Engström M, Owenius R, Vahtras O. 2001. Ab initio g-tensor calculations of hydrogen bond effects on a nitroxide spin label. *Chem Phys Lett* **338**:407–413.

Finiguerra MG, van Amsterdam IMC, Alagaratnam S, Ubbink M, Huber M. 2003. Anisotropic spin label mobilities in azurin from 95 GHz electron paramagnetic resonance spectroscopy. *Chem Phys Lett* **382**:528–533.

Freed JH. 2000. New technologies in electron spin resonance. *Annu Rev Phys Chem* **51**:655–689.

Gaede HC, Luckett KM, Polozov IV, Gawrisch K. 2004. Multinuclear NMR studies of single lipid bilayers supported in cylindrical aluminum oxide nanopores. *Langmuir* **20**:7711–7719.

Gaffney BJ, Marsh D. 1998. High-frequency spin-label EPR of nonaxial lipid ordering and motion in cholesterol-containing membranes. *Proc Natl Acad Sci USA* **95**:12940–12943.

Gamliel D, Levanon H. 1995. *Stochastic processes in magnetic resonance*. Singapore: World Scientific.

Ge M, Budil DE, Freed JH. 1994. ESR studies of spin-labeled membranes aligned by isopotential spin-dry ultracentrifugation: lipid–protein interactions. *Biophys J* **67**:2326–2344.

Griffith OH, Dehlinger PJ, Van SP. 1974. Shape of the hydrophobic barrier of phospholipid bilayers (evidence for water penetration in biological membranes). *J Membr Biol* **15**:159–192.

Grinberg OYa, Dubinski AA. 2004. The early years. *Biol Magn Reson* **22**:1–18.

Grinberg OYa, Dubinski AA, Shuvalov VF, Oranskii LG, Kurochkin VI, Lebedev YaS. 1976. Submillimeter ESR spectroscopy of free radicals. *Dokl Phys Chem* [Engl Trans] **230**:923–926.

Grinberg OYa, Dubinski AA, Lebedev YaS. 1983. Electron paramagnetic resonance of free radicals in the two-millimetre wavelength range. *Russ Chem Rev* [Engl Trans] **52**:850–865.

Grinberg OYa, Dubinskii AA, Poluektov OG, Lebedev YaS. 1990. Study of orientational order in liquid crystals by the spin label technique in the 2 mm EPR range. *Sov J Chem Phys* **6**:2685–2704.

Groenen EJJ, Schmidt J. 2004. High-frequency EPR, ESEEM and ENDOR studies of paramagnetic canters in single-crystalline materials. *Biol Magn Reson* **22**:277–304.

Gullá AF, Budil DE. 2001. Orientation dependence of electric field effects on the g factor of nitroxides measured by 220 GHz EPR. *J Phys Chem B* **105**:8056–8063.

Gutowsky HS, Holm C. 1956. Rate processes and nuclear magnetic resonance spectra, II: hindered internal rotation of amides. *Chem Phys* **25**:1228–1234.

Gutowsky HS, Saika A. 1953. Dissociation, chemical exchange, and the proton magnetic resonance in some aqueous electrolytes. *J Chem Phys* **21**:1688–1694.

Gutowsky HS, McCall DW, Slichter CP. 1953. Nuclear magnetic resonance multiplets in liquids. *J Chem Phys* **21**:279–292.

Hahn EL, Maxwell DE. 1952. Spin echo measurements of nuclear spin coupling in molecules. *Phys Rev* **88**:1070–1084.

Hofer P, Kamlowski A, Maresch GG, Schmalbein D, Weber RT. 2004. The Bruker ELEXSYS E600/680 W-band spectrometer series. *Biol Magn Reson* **22**:306–352.

Hubbell WL, Altenbach C. 1994. Investigation of structure and dynamics in membrane proteins using site-directed spin labeling. *Curr Opin Struct Biol* **4**:566–573.

Hustedt EJ, Beth AH. 1999. Nitroxide spin–spin interactions: applications to protein structure and dynamics *Annu Rev Biophys Biomol Struct* **28**:129–153.

Hustedt EJ, Beth AH. 2001. Structural information from CW-EPR spectra of dipolar coupled nitroxide spin labels. *Biol Magn Reson* **19**:155–184.

Hustedt EJ, Kirchner JJ, Spaltenstein A, Hopkins PB, Robinson BH. 1995. Monitoring DNA dynamics using spin-labels with different independent mobilities. *Biochemistry* **34**:4369–4375.

Hustedt EJ, Smirnov AI, Laub CF, Cobb CE, Beth AH. 1997. Measurement of molecular distances using dipolar coupled nitroxide spin-labels: the global analysis of multifrequency CW-EPR data. *Biophys J* **72**:1861–1877.

Hwang JS, Mason RP, Hwang L-P, Freed JH. 1975. Electron spin resonance studies of anisotropic rotational reorientation and slow tumbling in liquid and frozen media, III: perdeuterated 2,2,6,6-tetramethyl-4-piperidone N-oxide and an analysis of fluctuating torques. *J Phys Chem* **79**:489–511.

Hyde JS, Subczynski WK. 1989. Spin-label oximetry. *Biol Magn Reson* **8**:399–425.

Iwamoto Y, Robey FA, Graf J, Sasaki M, Kleinman HK, Yamada Y, Martin GR. 1987. YIGSR, a synthetic laminin pentapeptide, inhibits experimental metastasis formation. *Science* **238**:1132–1134.

Johnson ME. 1981. Apparent hydrogen bonding by strongly immobilized spin-labels. *Biochemistry* **20**:3319–3328.

Katsaras J, Donaberger RL, Swainson IP, Tennant DC, Tun Z, Vold RR, Prosser RS. 1997. Rarely observed phase transitions in a novel lyotropic liquid crystal system. *Phys Rev Lett* **78**:899–902.

Kawamura T, Matsunami S, Yonezawa T. 1967. Solvent effects on the g-value of di-t-butyl nitric oxide. *Bull Chem Soc Jap* **40**:1111–1115.

Khramtsov VV, Weiner LM, Grigoriev IA, Volodarsky LB. 1982. Proton exchange in stable nitroxyl radicals: EPR study of the pH of aqueous solutions. *Phys Chem Lett* **91**:69–72.

Khramtsov VV, Weiner LM, Eremenko SI, Belchenko OI, Schastnev PV, Grigor'ev AI, Reznikov VA. 1985. Proton exchange in stable nitroxyl radicals of imidazoline and imidazolidine series. *J Magn Reson* **61**:397–408.

Khramtsov VV, Marsh D, Weiner L, Reznikov, VA. 1992. The application of pH-sensitive spin labels to studies of surface potential and polarity of phospholipid membranes and proteins. *Biochim Biophys Acta* **1104**:317–324.

Keana JFW, Acarregui MJ, Boyle SLM. 1982. 2,2-disubstituted-4,4-dimethylimidazolidinyl-3-oxy nitroxides: indicators of aqueous acidity through variation of aN with pH. *J Am Chem Soc* **104**:827–830.

Klug CS, Camenisch TG, Hubbell WL, Hyde JS. 2005. Multiquantum EPR spectroscopy of spin-labeled arrestin K267C at 35 GHz. *Biophys J* **88**:3641–3647.

Kochelaev BI, Yablokov YV. 1995. *The beginning of paramagnetic resonance*. Singapore: World Scientific.

Kolaczkowski SV, Perry A, Mckenzie A, Johnson F, Budil DE, Strauss PR. 2001. A spin-labeled abasic DNA substrate for AP endonuclease. *Biochem Biophys Res Comm* **288**:722–726.

Krymov VN, Oranskii LG, Kurochkin VI, Lebedev YS, Grinberg OY. 1982. Sample chamber for EPR spectrometer. *Bull Isobr Otkrit [Rus]*, 13 USSR Patent #918832.

Kurad D, Jeschke G, Marsh D. 2001. Spin-label HF-EPR of lipid ordering in cholesterol containing membranes. *Appl Magn Reson* **21**:469–481.

Kurad D, Jeschke G, Marsh D. 2003. Lipid membrane polarity profiles by high-field EPR. *Biophys J* **85**:1025–1033.

Lai C-H, Hopwood LE, Hyde JS, Lukiewicz S. 1982. ESR studies of O_2 uptake by chinese hamster ovary cells during the cell cycle. *Proc Natl Acad Sci USA* **79**:1166–1170.

Lakey JH, Duché D, González-Manas J-M, Baty D, Pattus F. 1993. Fluorescence energy transfer distance measurements: the hydrophobic helical hairpin of colicin A in the membrane bound state. *J Mol Biol* **230**:1055–1067.

Lebedev YaS. 1990. High-frequency continuous wave electron spin resonance. In *Modern pulsed and continuous-wave electron spin resonance*, pp. 365–404. Ed L Kevan, MK Bowman. New York: Wiley.

Liang Z, Lou Y, Freed JH, Columbus L, Hubbell WL. 2004. A multifrequency ESR study of T4 lysozyme dynamics using the slowly relaxing local structure model. *J Phys Chem* **108**:17649–17659.

Liew MKH, Fane AG, Rogers PL. 1997. Fouling of microfiltration membranes by broth-free antifoam agents. *Biotechnol Bioeng* **56**:89–98.

Likhtenshtein GI. 1993. *Biophysical labeling methods in molecular biology*. New York: Cambridge UP (pp. 70–71).

Lillo MP, Beechem JM, Szpikowska BK, Sherman M, Mas MT. 1997a. Design and characterization of a multisite fluorescence energy-transfer system for protein folding studies: a steady-state and time-resolved study of yeast phosphoglycerate kinase. *Biochemistry* **36**:11261–11272.

Lillo MP, Szpikowska BK, Mas MT, Sutin JD, Beechem JM. 1997b. Real-time measurement of multiple intramolecular distances during protein folding reactions: a multisite stopped-flow fluorescence energy-transfer study of yeast phosphoglycerate kinase. *Biochemistry* **36**:11273–11281.

Livshits VA, Marsh D. 2000. Simulation studies of high-field EPR spectra of spin-labeled lipids in membranes. *J Magn Reson* **147**:59–67.

Livshits VA, Marsh D. 2004. HF EPR spectra of spin labels in membranes. *Biol Magn Reson* **22**:431–464.

Lohse DL, Denu JM, Santoro N, Dixon JE. 1997. Roles of aspartic acid-181 and serine-222 in intermediate formation and hydrolysis of the mammalian protein-tyrosine-phosphatase PTP1. *Biochemistry* **36**:4568–4575.

Lorigan GA, Dave PC, Tiburu EK, Damodaran K, Abu-Baker S, Karp ES, Gibbons WJ, Minto RE. 2004. Solid-state NMR spectroscopic studies of an integral membrane protein inserted into aligned phospholipid bilayer nanotube arrays. *J Am Chem Soc* **126**:9504–9505.

Lou Y, Ge M, Freed JH. 2001. A multifrequency ESR study of the complex dynamics of membranes. *J Phys Chem B* **105**:11053–11056.

Luptak A, Ferre-Drámare AR, Zhou K, Zilm KW, Doudna JA. 2001. Direct pK_a measurement of the active-site cytosine in a genomic hepatitis delta virus ribozyme. *J Am Chem Soc* **123**:8447–8452.

Mangels ML, Cardon TB, Harper AC, Howard KP, Lorigan GA. 2000. Spectroscopic characterization of spin-labeled magnetically oriented phospholipid bilayers by EPR spectroscopy. *J Am Chem Soc* **122**:7052–7058.

Mangels ML, Harper AC, Smirnov AI, Howard KP, Lorigan GA. 2001. Investigation of magnetically aligned phospholipid bilayers with EPR spectroscopy at 94 GHz. *J Magn Reson* **151**:253–259.

Marsh D, Kurad D, Livshits VA. 2002. High-field electron spin resonance of spin labels in membranes. *Chem Phys Lipids* **116**:93–114.

Mchaourab HS, Lietzow MA, Hideg K, Hubbell WL. 1996. Motion of spin-labeled side chains in T4 lysozyme: correlation with protein structure and dynamics. *Biochemistry* **35**:7692–7704.

Meuwly M, Becker OM, Stote R, Karplus M. 2002. NO rebinding to myoglobin: a reactive molecular dynamics study. *Biophys Chem* **98**:183–207.

Molin YuN, Salikhov KM, Zamaraev KI. 1980. *Spin exchange: principles and applications in chemistry and biology*. New York: Springer.

Möbius K, Savitsky A, Schnegg A, Plato M, Fuchs M. 2005. High-field EPR spectroscopy applied to biological systems: characterization of molecular switches for electron and ion transfer. *Phys Chem Chem Phys* **7**:19–42.

Nilges MJ, Smirnov AI, Clarkson RB, Belford RL. 1999. Electron paramagnetic resonance W-band spectrometer with a low-noise amplifier. *Appl Magn Reson* **16**:167–183.

Noble MA, Gul S, Verma CS, Brocklehurst K. 2000. Ionization characteristics and chemical influences of aspartic acid residue 158 of papain and caricain determined by structure-related kinetic and computational techniques: multiple electrostatic modulators of active-centre chemistry. *Biochem J* **351**:723–733.

Okonogi TM, Alley SC, Reese AW, Hopkins PB, Robinson BH. 2000. Sequence-dependent dynamics in duplex DNA. *Biophys J* **78**:2560–2571.

Ondar MA, Grinberg OYa, Dubinskii AA, Grigoriev IA, Volodarkii LB, Lebedev YaS. 1981. Determination of the magnetic parameters and structure of nitroxide biradicals from the electron spin resonance spectra in the 2-mm range. *Zh Strukt Khim* **22**:59–66.

Ondar MA, Grinberg OYa, Dubinskii AA, Lebedev YaS. 1985. Study of the effect of the medium on the magnetic-resonance parameters of nitroxyl radicals by high-resolution EPR spectroscopy. *Sov J Chem Phys* **3**:781–792.

Owenius R, Engström M, Lindgren M, Huber M. 2001. Influence of solvent polarity and hydrogen bonding on the EPR parameters of a nitroxide spin label studied by 9-GHz and 95-GHz EPR spectroscopy and DFT calculations. *J Phys Chem A* **105**:10967–10977.

Pannier M, Veit S, Godt A, Jeschke G, Spiess HW. 2000. Dead-time free measurement of dipole–dipole interactions between electron spins. *J Magn Reson* **142**:331–340.

Parker MW, Postma JPM, Pattus F, Tucker AD, Tsernoglou D. 1992. Refined structure of the pore-forming domain of colicin A at 2.4 Å resolution. *J Mol Biol* **224**:639–657.

Pilar J, Labsky J, Marek A, Budil DE, Earle KA, Freed JH. 2000. Segmental rotational diffusion of spin-labeled polystyrene in dilute toluene solution by 9 and 25 GHz ESR. *Macromolecules* **33**:4438–4444.

Polimeno A, Freed JH. 1990. A many-body stochastic approach to rotational motions in liquids: complex decay times in highly viscous fluids. *Chem Phys Lett* **174**:481–488.

Polimeno A, Freed JH. 1995. Slow motional ESR in complex fluids: the slowly relaxing local structure model of solvent cage effects. *J Chem Phys* **99**:10995–11006.

Poluektov OG, Utschig LM, Dalosto S, Thurnauer MC. 2003. Probing local dynamics of the photosynthetic reaction center with a cysteine specific spin label. *J Phys Chem B* **107**:6239–6244.

Poole ChP. 1997. Electron paramagnetic resonance (a comprehensive treatise on experimental techniques), 2nd ed. New York: Dover.

Prosser RS, Hwang JS, Vold RR. 1998. Magnetically aligned phospholipid bilayers with positive ordering: a new model membrane system. *Biophys J* **74**:2405–2418.

Prosser RS, Bryant H, Bryant RG, Vold RR. 1999. Lanthanide chelates as bilayer alignment tools in NMR studies of membrane-associated peptides. *J Magn Reson* **141**:256–260.

Rinard GA, Quine RW, Harbridge JR, Song R, Eaton GR, Eaton SS. 1999. Frequency dependence of EPR signal-to-noise. *J Magn Reson* **140**:218–227.

Rosen GM, Halpern HJ, Brunsting LA, Spencer DP, Strauss KE, Bowman MK. 1988. Direct measurement of nitroxide pharmacokinetics in isolated hearts situated in a low-frequency electron spin resonance spectrometer: implications for spin trapping and in vivo oxymetry. *Proc Natl Acad Sci USA* **85**:7772–7776.

Sanders CR, Landis GC. 1995. Reconstitution of membrane proteins into lipid-rich bilayered mixed micelles for NMR studies. *Biochemistry* **34**:4030–4040.

Sanders CR, Schwonek JP. 1992. Characterization of magnetically orientable bilayers in mixtures of dihexanoylphosphatidylcholine and dimyristoylphosphatidylcholine by solid-state NMR. *Biochemistry* **31**:8898–8905.

Sanders CR, Hare BJ, Howard KP, Prestegard JH. 1994. Magnetically-oriented phospholipid micelles as a tool for the study of membrane-associated molecules. *Prog Nucl Magn Reson Spectrosc* **26**:421–444.

Savitsky A, Kuhn M, Duche D, Möbius K, Steinhoff H-J. 2004. Spontaneous refolding of the pore-forming colicin A toxin upon membrane association as studied by X-band and W-band high-field electron paramagnetic resonance spectroscopy. *Phys Chem B* **108**:9541–9548.

Schneider DJ, Freed JH. 1989. Calculating slow motional magnetic resonance spectra. *Biol Magn Reson* **8**:1–75.

Schwartz RN, Peric M, Smith SA, Bales BL. 1997. Simple test of the effect of an electric field on the ^{14}N-hyperfine coupling constant in nitroxide spin probes. *J Phys Chem B* **101**:8735–8739.

Shimshick EJ, McConnell HM. 1973. Lateral phase separation in phospholipid membranes. *Biochemistry* **12**:2351–2360.

Sigel A, Sigel H, eds. 2000. *Metal ions in biological systems*. New York: Marcel Dekker.

Slade EF, Ingram DJE. 1969. An ESR spectrometer operating at 4 mm wavelength for the investigation of large zero field splittings. *Proc Roy Soc A* **312**:85–88.

Slichter CP. 1992. *Principles of magnetic resonance*, 3rd ed. New York: Springer.

Smirnov AI. 2002. Spin labeling in high field ESR. *Electron Paramagn Reson* **18**:109–136.

Smirnov AI, Belford RL. 1995. Rapid quantitation from inhomogeneously broadened EPR spectra by a fast convolution algorithm. *J Magn Reson A* **113**:65–73.

Smirnov AI, Poluektov OG. 2003. Substrate-supported lipid nanotube arrays. *J Am Chem Soc* **125**:8434–8435.

Smirnov AI, Smirnova TI. 2001. Resolving domains of interdigitated phospholipid membranes with 95 GHz spin labeling EPR. *Appl Magn Reson* **21**:453–467.

Smirnov AI, Norby S-W, Weyhenmeyer JA, Clarkson RB. 1994. The effect of temperature on respiration of cultured neural cells as studied by a novel electron paramagnetic resonance technique. *Biochim Biophys Acta* **1200**:205–214.

Smirnov AI, Smirnova TI, Morse II PD. 1995. Very high frequency electron paramagnetic resonance of 2,2,6,6-tetramethyl-1-piperidinyloxy in 1,2-dipalmitoyl-sn-glycero-3-phosphatidylcholine liposomes: partitioning and molecular dynamics. *Biophys J* **68**:2350–2360.

Smirnov AI, Clarkson RB, Belford RL. 1996. EPR linewidth (T_2) method to measure oxygen permeability of phospholipid bilayers and its use to study the effects of low ethanol concentrations. *J Magn Reson B*. **111**:149–157.

Smirnov AI, Belford RL, Clarkson RB. 1998. Comparative spin label spectra at X-band (9.5 GHz) and W-band (95 GHz). *Biol Magn Reson* **16**:83–107.

Smirnov AI, Ruuge VA, Reznikov MA, Voinov M, Grigor'ev IA. 2004. Site-directed electrostatic measurements with a thiol-specific pH-sensitive nitroxide: differentiating local pK and polarity effects by high field EPR. *J Am Chem Soc* **126**:8872–8873.

Smirnova TI, Smirnov AI. 2003. Effect of molecular oxygen on continuous wave EPR spectra of Mn^{2+} in solution: a high field EPR experiment to probe ion microenvironment. *J Phys Chem B* **107**:7212–7215.

Smirnova TI, Smirnov AI, Clarkson RB, Belford RL. 1995. W-band (95 GHz) EPR spectroscopy of nitroxide radicals with complex proton hyperfine structures: fast motion. *J Phys Chem* **99**:9008–9016.

Smirnova TI, Smirnov AI, Clarkson RB, Belford RL. 1998. Lipid-MRI contrast agent interactions: a spin-labeling and a multifrequency EPR study. *J Am Chem Soc* **120**:5060–5072.

Smirnova TI, Smirnov AI, Clarkson RB, Belford RL, Kotake Y, Janzen EG. 1997. High-frequency (95 GHz) EPR spectroscopy to characterize spin adducts. *J Phys Chem B* **101**:3877–3885.

Steinhoff H-J, Savitsky A, Wegener C, Pfeiffer M, Plato M, Möbius K. 2000. High-field EPR studies of the structure and conformational changes of site-directed spin labeled bacteriorhodopsin. *Biochim Biophys Acta* **1457**:253–262.

Stone AJ. 1963. g factors of aromatic free radicals. *Mol Phys* **6**:509–515.

Subczynski WK, Hyde JS, Kusumi A. 1989. Oxygen permeability of phosphatidylcholine: cholesterol membranes. *Proc Natl Acad Sci USA* **86**:4474–4478.

Thomas DD, Dalton LR, Hyde JS. 1976. Rotational diffusion studies by passage saturation transfer electron spin resonance. *J Chem Phys* **65**:3006–3024.

Tjandra N, Bax A. 1997. Direct measurement of distances and angles in biomolecules by NMR in a dilute liquid crystalline medium. *Science* **278**:1111–1113.

van Tol J, Brunel L-C, Wylde RJ. 2005. A quasioptical transient electron spin resonance spectrometer operating at 120 and 240 GHz. *Rev Sci Instrum* **76**:074101.

Van Vleck JH. 1948. The dipolar broadening of magnetic resonance lines in crystals. *Phys Rev* **74**:1168–1183.

Wang DM, Hanson GR. 1995. A new method for simulating randomly oriented powder spectra in magnetic resonance: the Sydney Opera House (SOPHE) method. *J Magn Reson* **117**:1–8.

Wang W, Belford RL, Clarkson RB, Davis PH, Forrer J, Nilges MJ, Timken MD, Walczak T, Thurnauer MC, Norris JR, Morris AL, Zhang Y. 1994. Very high frequency EPR, 94 GHz instrument and applications to primary reaction centers from photosynthetic red bacteria and to other disordered systems. *Appl Magn Reson* **6**:195.

Wegener C, Savitsky A, Pfeiffer M, Möbius K, Steinhoff H-J. 2001. High-field EPR-detected shifts of magnetic tensor components of spin label side chains reveal protein conformational changes: the proton entrance channel of bacteriorhodopsin. *Appl Magn Reson* **21**:441–452.

Windrem DA, Plachy WZ. 1980. The diffusion-solubility of oxygen in lipid bilayers. *Biochim Biophys Acta* **600**:655–665.

Zavoisky EK. 1944. Paramagnetic absorption in orthogonal and parallel fields for salts, solutions, and metals [in Russian]. PhD dissertation. Kazan University, Kazan, Russia.

Zhan H, Choe S, Huynh PD, Finkelstein A, Eisenberg D, Collier RJ. 1994. Dynamic transitions of the transmembrane domain of diphtheria toxin: disulfide trapping and fluorescence proximity studies. *Biochemistry* **33**:11254–11263.

10. APPENDIX: CHEMICAL STRUCTURES OF NITROXIDE SPIN LABELS IN MEMBRANE BIOPHYSICS

DTBN: di-tert-butyl-nitroxide

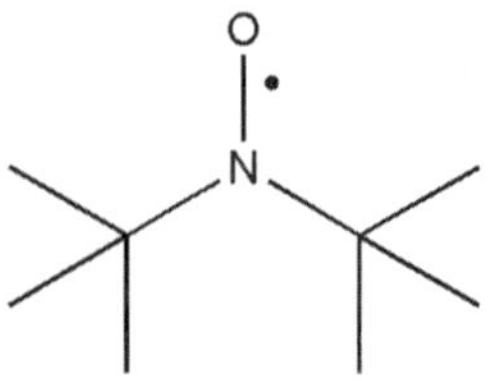

Tempo: 2,2,6,6-tetramethyl piperidone-1-oxyl

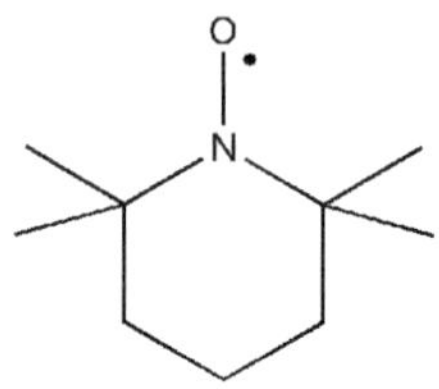

Tempone: 2,2',6,6'-tetramethyl-piperidone-1-nitroxide

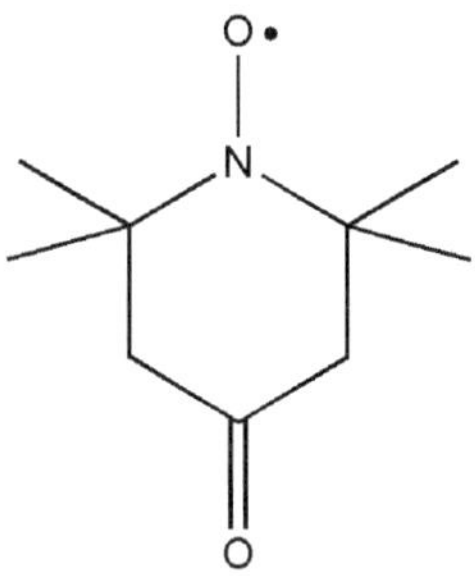

MSL: 2,2,5,5-tetramethylpyrrolidine-1-4-oxyl-3-maleimide

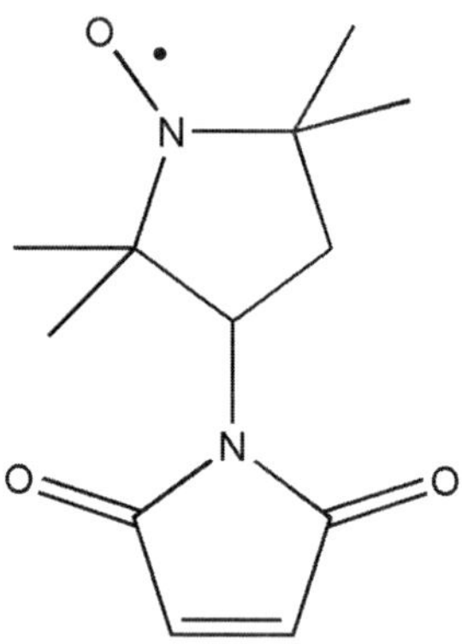

MTSL: 1-oxyl-2,2,5,5-tetramethyl-3-pyrroline-3-methyl

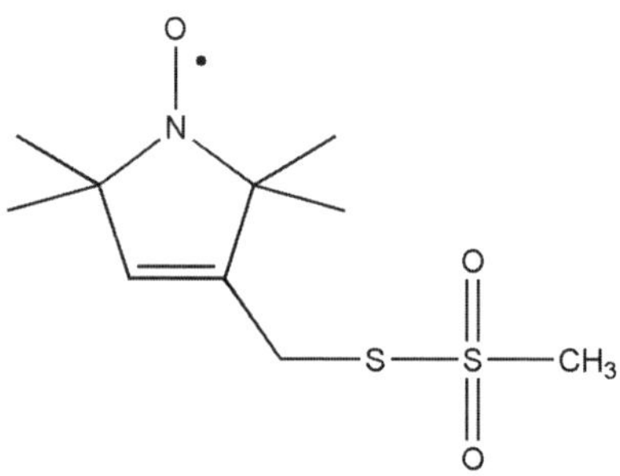

IMTSL: S-(1-oxyl-2,2,3,5,5-pentamethyl-imidezolidin-4-ylmethyl) ester

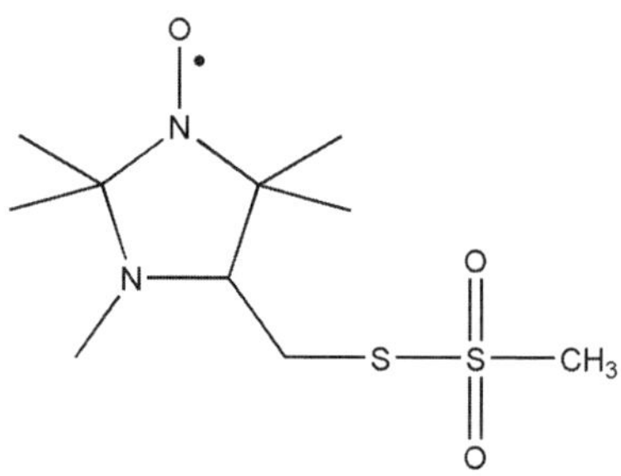

SSL: Succinimidyl 2,2,5,5-tetramethyl-3-pyrroline-1-oxyl-3-carboxylate

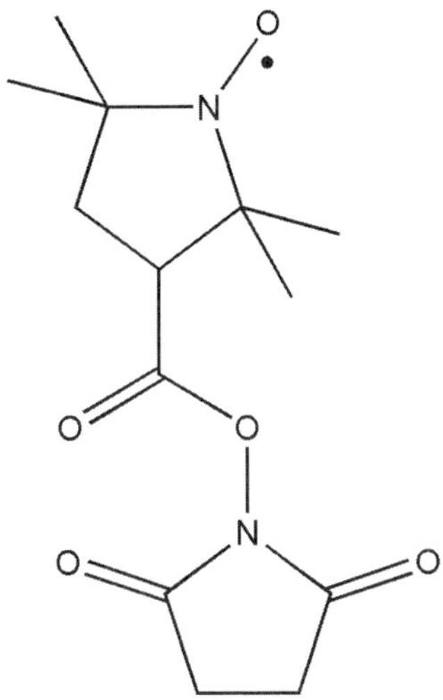

Cytosine: N4-[2,2,6,6-tetramethyl-4-(TAPDC) (2-aminoethyl)-piperidin-1-oxyl]-2'-deoxycytosine

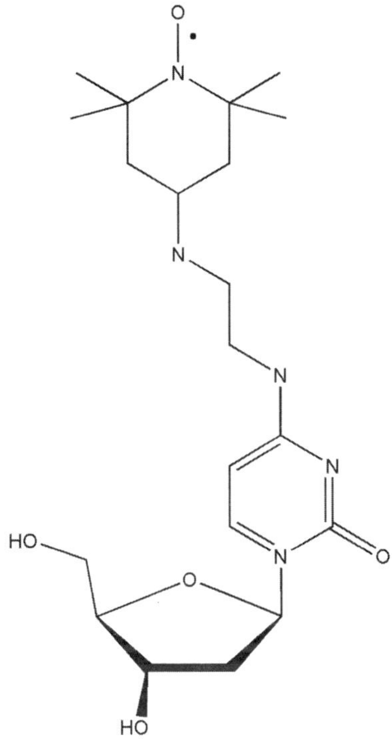

5DSA: (2-[3-carboxyl propyl]-4,4-dimethyl-2-tridecyl-3-oxazolidinyloxyl

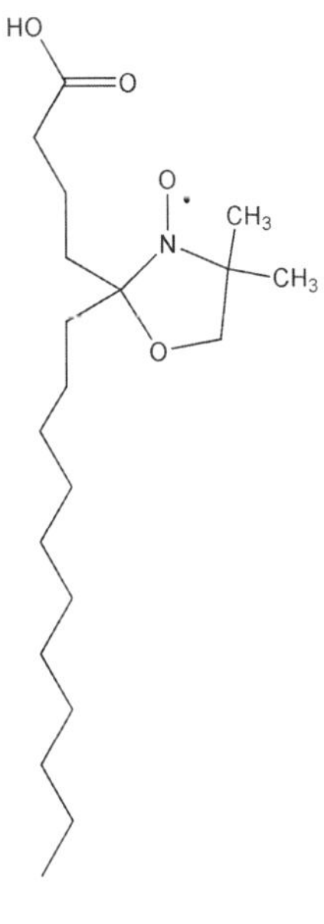

CSL: 3β-doxyl-5α-cholestane

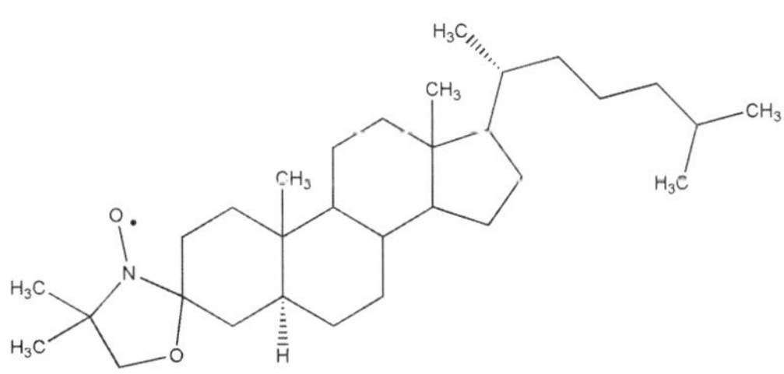

16DSA: (2-[14-carboxytetradecyl]-2-ethyl-4,4-dimethyl-3-oxazolidinyloxyl

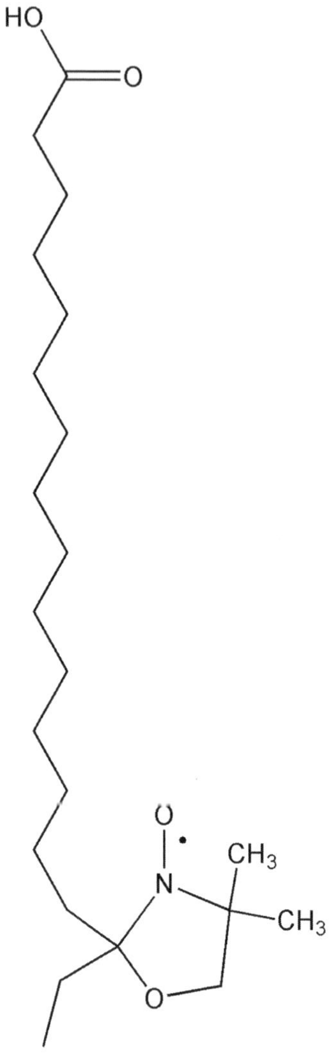

5PC: 1-palmitoyl-2-stearoyl-(5-doxyl)-sn-glycero-3-phosphocholine

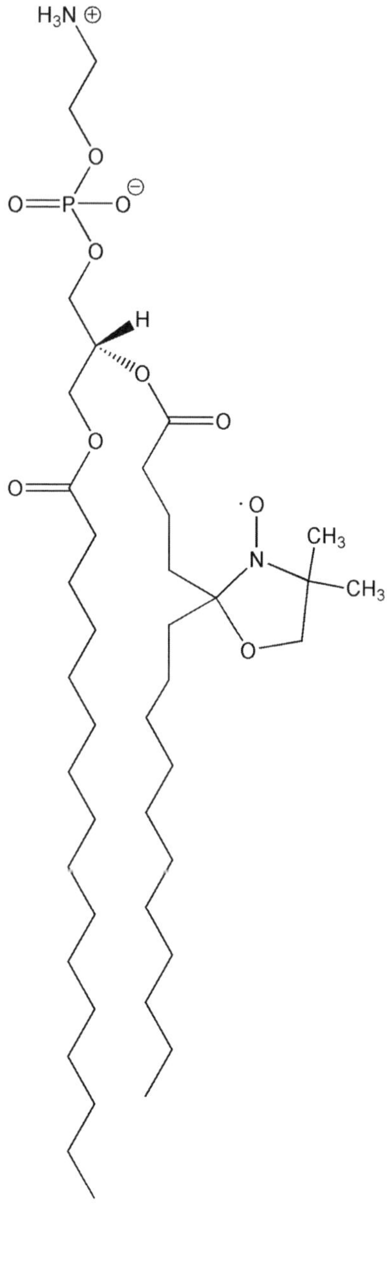

DHPC: 1,2-dihexanoyl-sn-glycero-3-phosphocholine

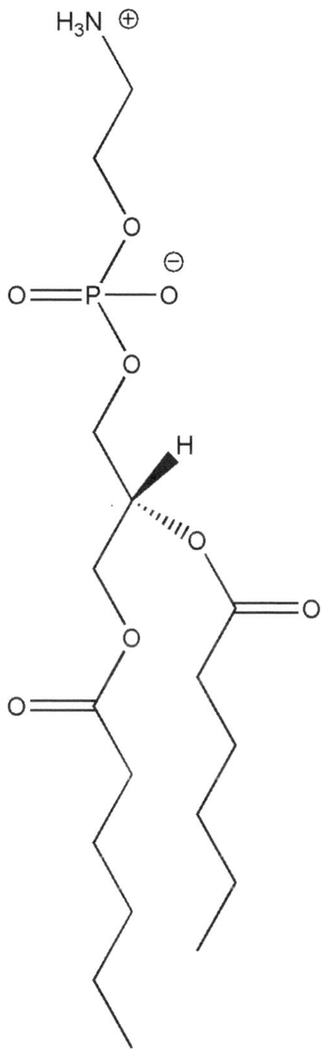

DMPC: 1,2-dimyristoyl-sn-glycero-3-phosphocholine

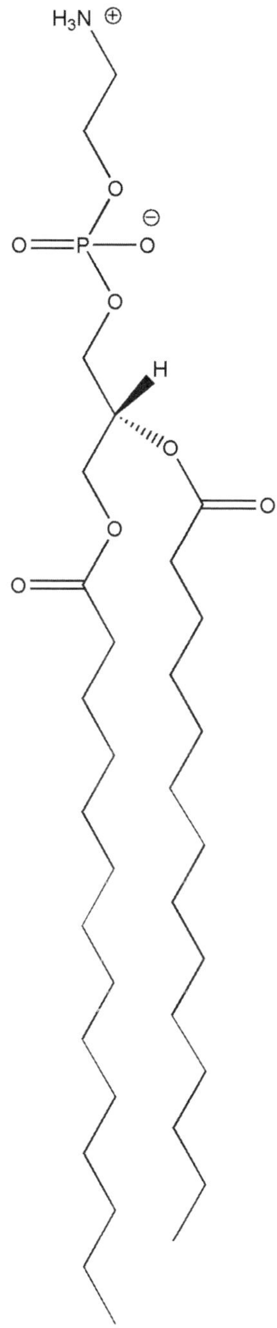

DPPC: 1,2-dipalmitoyl-sn-glycero-3-phosphatidylcholine

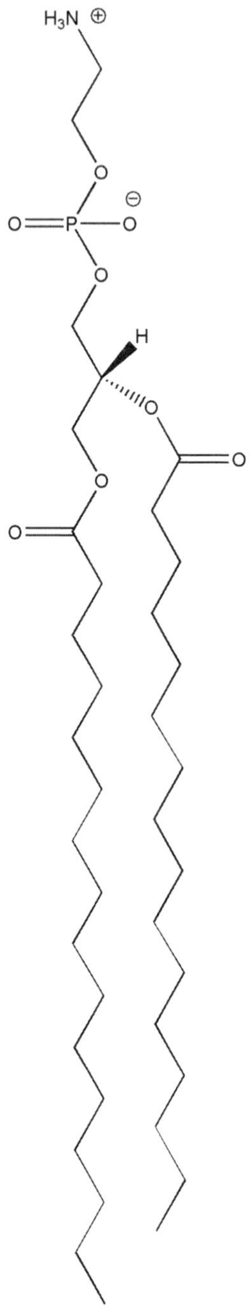

DMPS: 1,2-dimyristoyl-sn-glycero-3-phosphatidylserine

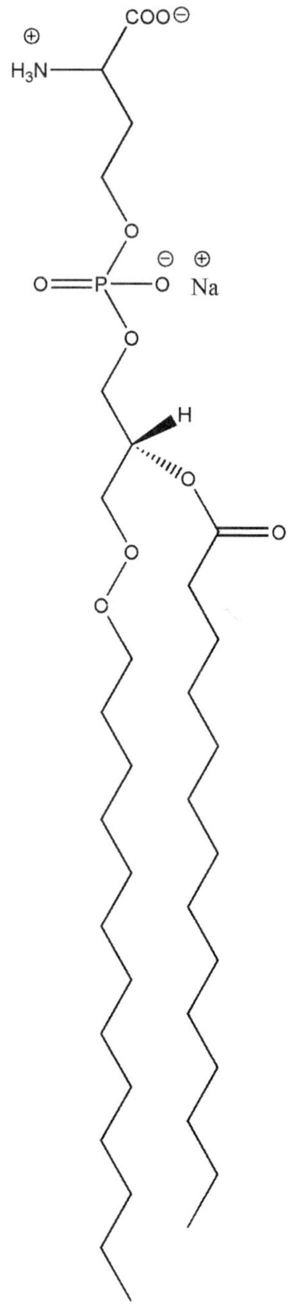

DMPG: 1,2-dimyristoyl-sn-glycero-3-phospho-rac-(1-glycerol)

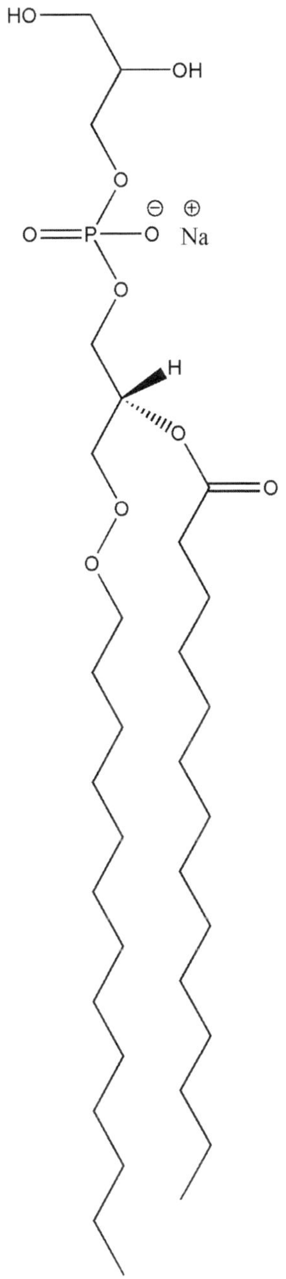

PBN: phenyl-tert-butylnitrone

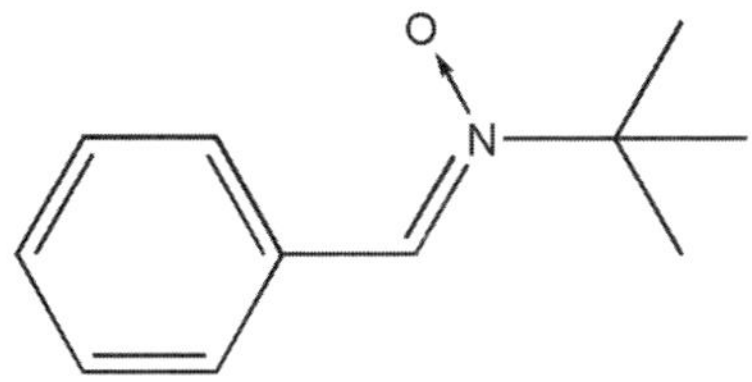

Phenyl-PBN

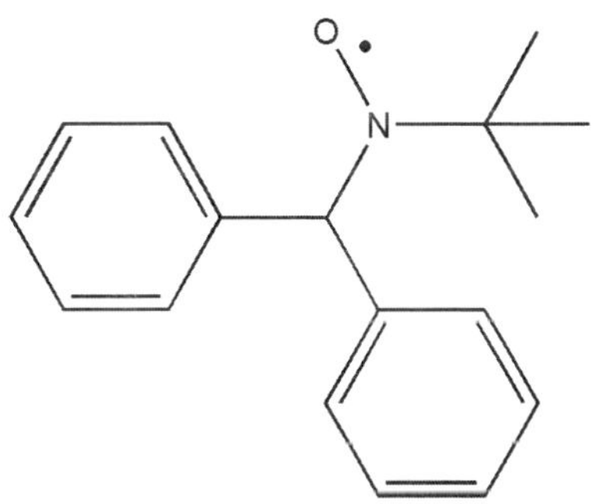

Trichloromethyl-PBN

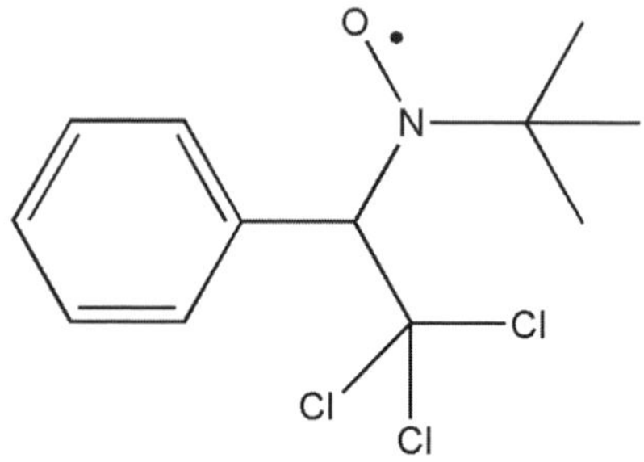

Methyl-PBN

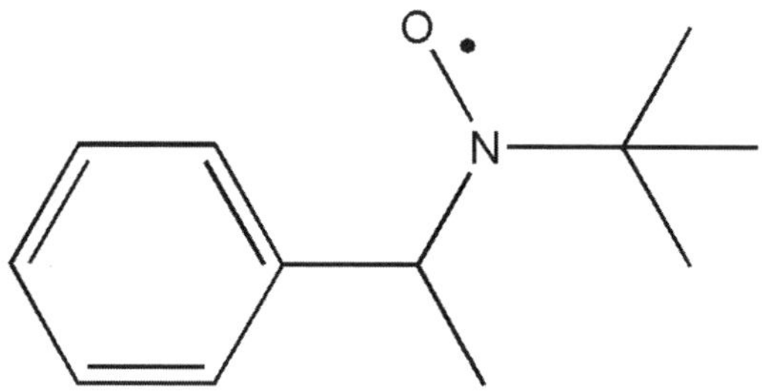

Ethyl-PBN

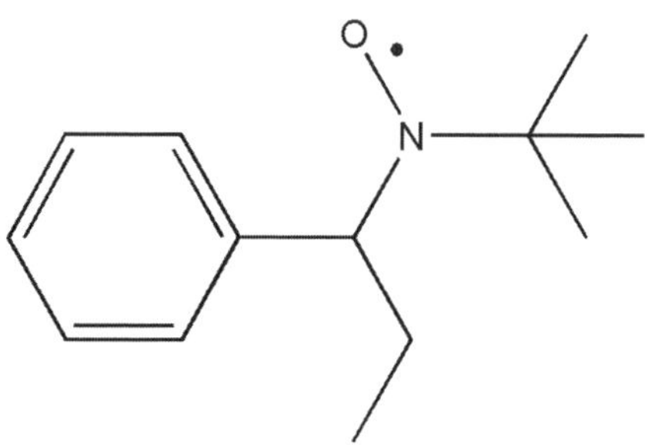

Tetradecyl-PBN

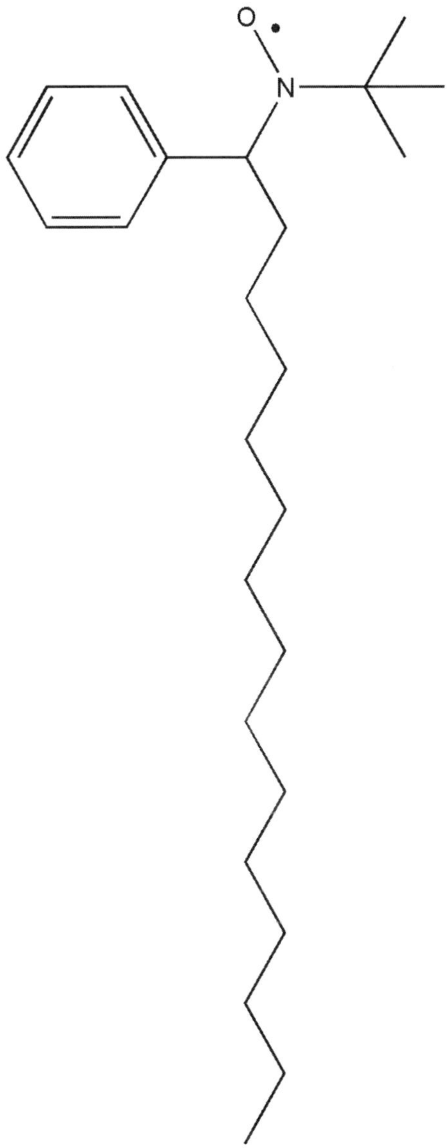

APPENDIX 1

MOLECULAR MODELING OF SPIN LABELS

Mikolai I. Fajer[1], Kenneth L. Sale,[1,2] and Piotr G. Fajer[1]
[1]Florida State University and National High Magnetic Field Laboratory, Tallahassee, Florida, USA; [2]Sandia National Laboratories, Livermore, California USA

Site-directed spin labeling has become an increasingly popular tool for structural and functional studies of macromolecules. However, the use of spin labels complicates spectral interpretations because the ESR signal originates from the unpaired electron in the pπ orbital of the nitroxide NO bond rather than from the protein backbone, which may be 5–10 Å away. A spin label is essentially a modified amino acid side chain, and as such spin label behavior can be simulated using standard molecular modeling tools provided the necessary molecular topology and force field parameters are available. This appendix reviews the efforts of our laboratory to use conformational search and molecular dynamics strategies to predict the conformation and dynamics of spin-label modified side chains. These methods can be used in a variety of applications: (a) "rational spin labeling," where we can predict whether the site/probe combination will result in either a rigidly attached or floppy spin label as required by a given experiment (Sale et al. 2002); (b) for rigidly attached labels, determine the orientation of the label within the protein (Sale et al. 2002) — a necessary prerequisite to interpreting ESR spectra in terms of absolute molecular orientation (Baumann et al. 2004); (c) predicting the librational motion of spin labels with respect to the protein, which is needed for deconvolution of the protein or domain mobility from the ESR signal (Budil et al. 2005); (d) accounting for the tether length and orientation of spin labels in interpretation of spin–spin distances from dipolar CW ESR and from pulsed ESR (Sale et al. 2005).

Since there are a number of excellent textbooks and websites dedicated to molecular modeling, we do not review the theoretical principles behind these strategies. Instead we present a guide to spin-label modeling organized in four sections:

(i) preparation of molecular models, including spin labeling in silico;
(ii) simulation of spin labels using Monte Carlo rotamer searches and Molecular Dynamics;

(iii) analysis of the molecular trajectories to extract: spin label coordinates, solvent-accessible areas, side chain dihedral angles and spin–spin distances;
(iv) discussion of spin-label-specific modifications to popular modeling programs to include label-cysteine adducts in topology files and force field parameters.

An expanded version of the appendix including source code and point-by-point instructions appears at http://fajerpc.magnet.fsu.edu and on the accompanying CD.

1. SYSTEM REQUIREMENTS

An Intel/AMD-based PC or an OS X Macintosh is sufficient to perform all the computations. Monte Carlo rotamer searches are run in CHARMM and take approximately 6 hours on a 3-GHz PC. Molecular dynamics simulations, which can be run using either CHARMM or NAMD, take roughly the same time for 1-ns trajectories using implicit solvent. Molecular dynamics simulations with explicit solvent take significantly longer, and parallel codes are recommended. Nanosecond simulations typically require several days on an eight-processor cluster.

The software used for simulation and analysis includes the commercial software InsightII and CHARMM (v. 27). An academic license for CHARMM can be obtained from http://www.charmm.org/. Academic freeware software includes NAMD, PSFgen, and VMD, available at http://www.ks.uiuc.edu/. Many of our analysis tools were developed in Matlab; however, we strived to give equivalent functionality using VMD or CHARMM.

2. PREPARATION OF STRUCTURES

Several steps are required to prepare the protein structure, usually downloaded for the Protein Data Bank, for molecular modeling:

(a) If the crystal structure of the isoform used in ESR experiments is not available, the correct isoform has to be built by homology to an available isoform(s). Homology modeling is available on a number of web-accessible servers listed on the accompanying CD or at the website.
(b) The desired residue needs to be mutated to spin-labeled cysteine. We use PSFgen or InsightII to introduce the cysteine spin label adduct. Regardless of how the spin label is introduced, it is important to use the atom names provided in the cysteine label topology files.
(c) If the protein structure has missing atoms, they should be built or mutated to the correct residue, using PSFgen.

(d) The protein structure is split into contiguous fragments using VMD or CHARMM. This is necessary to prevent modeling programs from replacing missing loops with a single covalent bond.
(e) The CHARMM-compatible protein structure file and coordinate file need to be created using CHARMM.
(f) The structure is energy minimized with either the protein backbone or Cα positions constrained in order to relieve bad interatomic contacts without any undue changes in backbone structure.

3. SIMULATIONS

3.1. Monte Carlo Rotamer Searches

The goal of the rotamer searches is to rapidly search the spin-label conformational space. Rotamer searches are performed over all rotatable bonds between the Cα and the nitroxide ring of the spin label using a Monte Carlo conformation search procedure. The search performed in CHARMM consists of iterations in which a randomly chosen torsion angle is set to a random value. Energy minimization follows the randomization and uses the adopted basis Newton Raphson (ABNR) algorithm. The Metropolis criterion then determines whether the new structure is rejected or used as the seed structure for the next round of randomization. Between 1000 and 2000 distinct (within a user-determined RMSD) energy-minimized rotamers are collected to sample exhaustively the rotamer space. This is essentially a "rigid cage" search in which the atoms of the protein structure are constrained in space and only the label is allowed to move. The "rigid cage" restriction is released in subsequent molecular dynamics simulations.

All the rotamers are compiled into a single CHARMM-formatted structure history file (DCD file) that can be read by most visualization programs (e.g., VMD). The DCD file contains the coordinates of all atoms of the molecule for each rotamer. The values of torsion angles and coordinates of defined atoms will be extracted from DCD files for further analysis using CHARMM or VMD. The lowest-energy rotamers are classed into conformational families of similar torsional angles. The 2–3 lowest energy conformational families are used as starting structures for MD.

3.2. Molecular Dynamics

MD simulations have two functions: (a) to relieve the "rigid cage" constraint to allow the protein to adapt to the specific conformation of a spin label; (b) to map out the local space in which the label is free to move. Either of two methods of accounting for solvent can be used: (i) a distance-dependent dielectric as an implicit solvent — a simple and fast simulation that has proven adequate for our calculations of preferred spin-label orientation within the protein; (ii) the TIP3 explicit solvent for more detailed simulations.

Details of the simulations include a radius for non-bonding interactions of 15 Å and a smooth switching function over a cutoff range of 13 to 15 Å. For large proteins the residues further than 15 Å from the spin label are fixed, thus focusing on local interactions of the spin label. The Verlet leapfrog method with a 2-fs time-step is used for integration. The SHAKE algorithm is used to fix bonds involving hydrogen atoms. The system is heated, equilibrated, and followed by a 1–2 ns production run. Heating to 300 or 500 K is accomplished over 6000 steps (12 ps), and production runs are carried out without velocity rescaling. The coordinates are corrected for global translation and rotation every 200 fs so that the MD trajectory reflect only the spin label motion relative to the protein.

Implicit solvent MD uses CHARMM19 forcefields and is performed using CHARMM27c or later versions. Explicit solvent MD uses CHARMM27 force fields and is performed using the NAMD program (Phillips et al. 2005). The protein is solvated in a rectangular water box of at least 10 Å thickness. We use periodic boundary conditions and a Particle Mesh Ewald grid in our calculations.

4. ANALYSIS OF NITROXIDE TRAJECTORIES

The output of the Monte Carlo and MD trajectories is stored in the DCD-formatted coordinate/trajectory file, which is used for all subsequent analysis. Analysis examples and the corresponding scripts are on the accompanying CD and at the website, and include:

(a) extracting atomic distances, including interspin distances, used in modeling of distance constraints from CW dipolar ESR and DEER (Sale et al. 2005)
(b) analyzing nitroxide atomic coordinates in applications such as determination of the nitroxide tensors within the protein frame of reference (Sale et al. 2002; Baumann et al. 2004)
(c) calculating the solvent-accessible surface area of the spin label for use in modeling of the trends in NiEDDA relaxation of nitroxides along the polypeptide chain (Brown et al. 2002)
(d) analyzing dihedral angles that describe the rotamer structures and fluctuations about each spin-label bond

The provided scripts are for the VMD program; however, all of these scripts are also available for use in the CHARMM and Matlab environments. Extraction of the desired atom coordinates from a trajectory file into an ASCII file allows user analysis in any desired program.

5. FORCE FIELDS

Incorporation of spin labels into molecular mechanics calculations requires a description of the molecular topology and a set of force field parameters. The spin labels provided are parameterized for use with the CHARMM19 (Chemistry at

HARvard Molecular Mechanics) force field developed by the Karplus group (Brooks et al. 1983). CHARMM19 is an extended atom, empirical force field. Extended atom means that non-polar hydrogen atoms are incorporated into the parameters of the heavy atom to which they are attached. The force field was developed and parameterized to calculate the potential energy of nucleic acids and proteins and has been shown to perform well with a distance-dependent dielectric constant (MacKerell 2001).

The amount of effort expended on parameterization should be proportional to the depth and accuracy expected from the scientific problem. Accurate calculation of thermodynamic properties requires the best set of parameters obtainable, while qualitative descriptions of reasonable geometries may require less work (Kollman et al. 2001). Thus, the general principle used to develop CHARMM19 forcefield parameters for nitroxide spin labels was to "clone" as many parameters as possible by analogy to existing molecular structures: reference values such as bond lengths and bond angles were derived from tabulations of experimental values in the literature, and force constants were derived by analogy to similar parameters in the force field. In cases where reference values were not available in the literature, both the reference value and the force constant were estimated from existing parameters.

6. SPIN LABEL TOPOLOGIES

Topology files describe the spatial relationship of all atoms in the molecule in terms of bond lengths, angles, and dihedral angles. A topology file provides the molecular mechanics program with the atom names and types. Atom types are a device used by molecular mechanics programs to define force field parameters among unique sets of atoms with similar chemical properties rather than for arbitrary sets of atom names. The topology file also contains the partial charge of each atom and a set of internal coordinates. The internal coordinates allow an atom with undefined coordinates to be added given the coordinates of three atoms to which it is attached across three bonds.

6.1. Atom Names

Atom names can be chosen arbitrarily but the naming convention we used was to name nitroxide ring atoms according to their positions in the ring starting at the nitroxide nitrogen (N1), and all other atoms were named sequentially as the *n*th appearance of a given element. Atoms native to the labeled cysteine residue retained their names as per the CHARMM19 cysteine topology file.

6.2. Atom Types

With the exception of the nitroxide nitrogen and the nitroxide oxygen atoms, the atom types of the spin labels used in the studies presented here are in large part typical, in terms of hybridization, mass, van der Waals radii and geometry, of those seen in amino acids. Most of the spin label atom types were parameterized by analogy to existing atom types.

There are two characteristics of the NO moiety needing special attention. First, the unpaired electron of the nitroxide is localized on the diatomic functional NO moiety, making the nitrogen and oxygen atoms and their topologies atypical of atom types in macromolecular force fields. Second, both the preference for a planar or pyramidal environment and the spin density between the N and O atoms are dependent on the nature of its substituents (Barone et al. 1997, 1998).

We added three new atom types to the CHARMM19 force field (ON, NN5, and NN6) to account for the peculiarities of these atoms and their unique interactions. The ON atom type corresponds to the nitroxide's oxygen. The NN5 and NN6 atom types correspond to the nitroxide's nitrogen in a five- and six-membered ring, respectively.

6.3. Partial Charges

Partial charges for the five- and six-member nitroxide rings were assigned based on published values (Barone et al. 1997, 1998), while the partial charges on the atoms of chemical groups attached to the nitroxide rings were assigned by analogy to existing amino acid residue topologies. Based on this approach a set of rules for assigning partial charges to spin label structures was developed and is presented in Table 2-2 on the accompanying CD or at the website.

6.4. Internal Coordinates

Initial coordinates of spin-label-modified cysteine residues were generated using the Insight II (Accelrys Inc., San Diego, CA) Builder module, which allows building of molecules by piecing together molecular fragments. These structures were then energy minimized and added to the amino acid fragment library from where they can be easily accessed for in silico mutation of any amino acid to a spin-labeled cysteine using the "residue replace" pull-down in the Insight II Biopolymer module.

An alternative method avoids the use of commercial software. The initial spin label coordinates are generated using PSFgen with topology files that contain internal coordinates. Further details are included on the accompanying CD or at the website.

7. REFERENCES

Barone V, Capecchi G, Brunel Y, Dheu ML, Subra R. 1997. Development and validation of force-field parameters for molecular simulations of peptides and proteins containing open-shell residues. *J Comput Chem* **18**(14):1720–1728.

Barone V, Bencini A, Cossi M, Di Matteo A, Mattesini M, Totti F. 1998. Assessment of a combined QM/MM approach for the study of large nitroxide systems in vacuo and in condensed phases. *J Am Chem Soc* **120**:7069–7078.

Baumann BA, Liang H, Sale K, Hambly BD, Fajer PG. 2004. Myosin regulatory domain orientation in skeletal muscle fibers: application of novel electron paramagnetic resonance spectral decomposition and molecular modeling methods. *Biophys J* **86**(5):3030–3041.

Brooks BR, Bruccoleri R, Olafson B, States D, Swaminathan S, Karplus M. 1983. CHARMM: a program for macromolecular energy, minimization and dynamics calculations. *J Comput Chem* **4**(2):187–217.

Brown LJ, Sale KL, Hills R, Rouviere C, Song L, Zhang X, Fajer PG. Structure of the inhibitory region of troponin by site directed spin labeling electron paramagnetic resonance. *Proc Natl Acad Sci USA* **99**(20):12765–12770.

Budil DE, Sale KL, Khairy KA, Fajer PG. 2006. Calculating slow-motional electron paramagnetic resonance spectra from molecular dynamics using a diffusion operator approach. *Phys Chem A* **110**(10):3703–3713.

Kollman P. 2001. *Parameter development.* ???please provide coherent citation with full information???

MacKerell A. 2001. Atomic models and force fields. In *Computational biochemistry and biophysics*, chap. 2. Ed O Becker, A MacKerell, B Roux, M Watanabe. New York: Marcel Dekker.

Ponder JW, Case DA. 2003. Force fields for protein simulations. *Adv Prot Chem* **66**:27–85.

Phillips JC, Braun RB, Wang W, Gumbart J, Tajkhorshid E, Villa E, Chipot C, Skeel RD, Kale L, Schulten K. 2005. Scalable molecular dynamics with NAMD. *J Comput Chem* **26**:1781–1802.

Sale K, Sar C, Sharp KA, Hideg K, Fajer PG. 2002. Structural determination of spin label immobilization and orientation: a Monte Carlo minimization approach. *J Magn Reson* **156**(1):104–112.

Sale K, Song L, Liu YS, Perozo E, Fajer P. 2005. Explicit treatment of spin labels in modeling of distance constraints from dipolar EPR and DEER. *J Am Chem Soc* **127**(26):9334–9335.

APPENDIX 2

SIMPOW6: A SOFTWARE PACKAGE FOR THE SIMULATION OF ESR POWDER-TYPE SPECTRA

Mark J. Nilges, Karen Mattson, and R. Linn Belford
University of Illinois at Champaign-Urbana, USA

1. INTRODUCTION

ESR spectra encountered in membrane biophysics can be very complex or fairly straightforward to analyze. The ESR-active centers may include metal ions and radicals as well as multiple, sometimes interacting, species. Over more than three decades, our group has devised programs for simulation and spectral fitting for a number of cases. The QPOW family of ESR simulation has been rather widely disseminated. As originally conceived, it dealt with metal ion centers and interacting ligands, and used full diagonalization of a core spin Hamiltonian matrix over a grid of orientations designed for fairly uniform coverage of the Euler angles on the sphere. Accumulation of the spectra generated for each orientation via a numerical integration scheme provided an efficient simulation for an ESR powder-type spectrum. Later, fitting of experimental spectra by simulated ones was facilitated by automatic adjustment of the electronic Zeeman, hyperfine, and nuclear quadrupole coupling parameters. Modifications were developed to handle high-spin cases with zero-field splittings and strain (distribution of inhomogeneities). Here we describe a software package called "SIMPOW6," which has been devised to handle an ESR-active center that can be adequately modeled as a central ion of effective spin ½ and anisotropic electronic Zeeman (g) factor coupled to several (up to six) nuclear spins characterized by anisotropic hyperfine factors. We find that some of the features of the QPOW family are useful for this case. But it does not necessarily require other features — such as full diagonalization of spin Hamiltonian matrices — that consume much computational time. In the interest of computational efficiency, SIMPOW6 was written for those special but commonly occurring situations.

In this appendix we describe SIMPOW6, explain the input parameters, show how to use the program, and illustrate with several examples. Examples include a radical in frozen or glassy solution, a metal-ion-based center with g-strain and hyperfine strain, separation of overlapping multiple species (anionic flavin semiquinone free radical and a mixed valence iron-sulfur center), resolution of hyperfine structures with the aid of second-derivative spectral presentation, and a system that has multiple metal isotopes.

2. SPIN HAMILTONIAN FOR SIMPOW6

SIMPOW6 is set up to calculate a spectrum for a system with an electronic spin, *S*, of ½ and up to six nuclear spins. The spin Hamiltonian for such a system can be written as:

$$\mathbf{H} = \beta\mathbf{B}_{res}\cdot\mathbf{g}\cdot\mathbf{S} + h\mathbf{S}\cdot\mathbf{A}\cdot\mathbf{I}_a + h\mathbf{S}\cdot\mathbf{B}\cdot\mathbf{I}_b + h\mathbf{S}\cdot\mathbf{C}\cdot\mathbf{I}_c + h\mathbf{S}\cdot\mathbf{D}\cdot\mathbf{I}_d + h\mathbf{S}\cdot\mathbf{E}\cdot\mathbf{I}_e + h\mathbf{S}\cdot\mathbf{F}\cdot\mathbf{I}_f,$$

where a, b, c, d, e, and f refer to the six nuclear spins, which are referenced in the program as SPINA, SPINB, SPINC, SPIND, SPINE, and SPINF. **A, B, C, D, E,** and **F** are the hyperfine coupling matrices for SPINA, SPINB, SPINC, SPIND, SPINE, and SPINF, respectively. B_{res} is the magnetic field vector for external magnetic field, and **g** is the electronic Zeeman *g* matrix. This spin Hamiltonian is applicable to systems with any symmetry down to C1 or triclinic. The electronic Zeeman and hyperfine coupling are not true tensors but in general should be described as asymmetric matrices. If the symmetry is orthorhombic or higher, the asymmetry in the matrices vanishes. In this program, the external coordinate system is set equal to that which diagonalizes the $\mathbf{g}^{\dagger}\cdot\mathbf{g}$ matrix so that the *g*-matrix is fully described by only its three principal values (parameter g(1), g(2), g(3)). The input data for the hyperfine matrices are the three principal values obtained from diagonalizing the square of the hyperfine matrix (parameters A(1), A(2), A(3) for SPINA, B(1), B(2), B(3) for SPINB, and so on) and the three Euler angles (Parameters ANGA(1), ANGSA(2), ANGSA(3) for SPINA and ANGB(1), ANGB(2), ANGB(3), and so on) that form the similarity transformation, **R,** which rotates the square of the diagonal hyperfine matrix, $\mathbf{A}_D$, into the coordinate system for which the $\mathbf{g}^{\dagger}\cdot\mathbf{g}$ matrix is diagonal:

$$\mathbf{A}^{\dagger}\cdot\mathbf{A} = \mathbf{R}\cdot\mathbf{A}_D\cdot\mathbf{A}_D\cdot\mathbf{R}^{\dagger}.$$

The three Euler angles are defined using the convention of Rose (1957), and represent three consecutive rotations. The first angle, usually referred to as α, represents a rotation around z to give the new axes x′,y′,z′; the second angle, β, represents a rotation around y′ to give the new axes x″,y″,z″, and the third angle, γ, represents a rotation around z″, to give the new axes X, Y, Z, where xyz is the coordinate system for which the g-matrix is diagonal and X,Y,Z refer to the coordinate system where the hyperfine matrix is diagonal. Note that Euler angles as used here refer to the rotation of the axes of the g-matrix into those of the hyperfine matrix. The Euler angles corresponding to the reverse rotation, the rotation of axes of the hyperfine matrix into those of the g-matrix, will be simply –γ, –β, and –α. Since the asymmetry of the g and hyperfine matrices cannot be measured in a simple ESR experiment, the possible asymmetry of the g and hyperfine matrices is ignored and $\mathbf{A}\cdot\mathbf{A}$ is assumed equal to $\mathbf{A}^{\dagger}\cdot\mathbf{A}$.

The spin Hamiltonian is calculated for up to nine different spin systems. The individual spectra are weighted (parameter TINT(j)) and summed. This allows simulation of spectra with multiple components or spectra in which features arising from different nuclear isotopes are resolved. As there will be a complete set of input parameters for the g and hyperfine matrices for each species, g(1), g(2), and g(3) will become g(1,j), g(2,j), and g(3,j).

3. CALCULATION OF RESONANCE FIELDS

The resonance fields for the above spin Hamiltonian are solved using second-order perturbation theory for terms involving SPINA and SPINB, while first-order perturbation theory is used for terms involving SPINC, SPIND, SPINE, and SPINF. The first-order contribution to the resonance fields from SPINA and SPINB are given as (Weil 1975):

$$B^{(1)}(m_a,m_b) = h\nu/g_{eff}\beta - (K_a m_a + K_b m_b)/\, g_{eff}\beta,$$

where

$$K_a = [\mathbf{l{\cdot}g{\cdot}A{\cdot}A{\cdot}g{\cdot}l}]^{1/2}/g_{eff},$$

$$K_b = [\mathbf{l{\cdot}g{\cdot}B{\cdot}B{\cdot}g{\cdot}l}]^{1/2}/g_{eff},$$

$$g_{eff} = [g_x^2\, l_x^2 + g_y^2\, l_y^2 + g_z^2\, l_z^2]^{1/2},$$

and $\boldsymbol{l} = (l_x,l_y,l_z) = |B|/\mathbf{B_{res}}$, where l_x,l_y,l_z are known as the direction cosines. The second-order contributions to the resonance field are given by

$$B(m_a,m_b) = B^{(1)} - [(Tr(A{\cdot}A) - k_a)(I_a(I_a+1)-m_a^2)+(k_a^2-K_a^2)m_a^2$$
$$+ (Tr(B{\cdot}B) - k_b)(I_b(I_b+1)-m_b^2)+(k_b^2-K_b^2)m_b^2]/2g_{eff}\beta\, B^{(1)}$$
$$- (L_{ab} - K_a K_b)m_a m_b/2g_{eff}\beta\, B^{(1)}$$

where

$$k_a = [\mathbf{l{\cdot}g{\cdot}A{\cdot}A{\cdot}A{\cdot}A{\cdot}g{\cdot}l}]^{1/2}/K_a/g_{eff},$$

$$k_b = [\mathbf{l{\cdot}g{\cdot}B{\cdot}B{\cdot}B{\cdot}B{\cdot}g{\cdot}l}]^{1/2}/K_b/g_{eff},$$

$$L_{ab} = [l{\cdot}g{\cdot}(A{\cdot}A{\cdot}B{\cdot}B + B{\cdot}B{\cdot}A{\cdot}A){\cdot}g{\cdot}\, l]^{1/2}/K_a/K_b\,/2g_{eff}.$$

Finally, the first order-contributions from SPINC to SPINF are added in:

$$B_{res} = B(m_a,m_b) - (K_c m_c + K_d m_d + K_e m_e + K_f m_f)/\, g_{eff}\beta.$$

SPINC, SPIND, SPINE, and SPINF are calculated to first order only, and this readily allows for calculation of hyperfine splitting for equivalent nuclei, as for example, the three protons on a freely rotating methyl group. The numbers of equivalent nuclei for the *i*th species are given by input parameters NEC(i), NED(i), NEE(i), and NEF(i) for SPINC, SPIND, SPINE, and SPINF, respectively.

4. LINESHAPES

SIMPOW6 can use either a Lorentzian or Gaussian line shape. When line broadening is due to short relaxation times (homogeneous broadening), a Lorentzian line shape is expected. For ESR the first-derivative of the absorption is recorded and the first derivative Lorentzian line shape will be:

$$F'(\Delta B) = 2\,\Delta B/w_l^3 \ln2\, e[-\ln2(\Delta B/w_l)^2]\,(\ln2/\pi)^{½}.$$

When line broadening is due to unresolved hyperfine splitting or random distortions of the paramagnetic species (inhomogeneous broadening), a Gaussian lin shape is expected. The first-derivative Gaussian line shape will be

$$F'(\Delta B) = 2\,\Delta B/w_g^3/((\Delta B/w_g)^2 + 1)^2/\pi.$$

For either choice of a Lorentzian or a Gaussian line shape (parameter LS: L or G, respectively) an anisotropic orientation-dependent line shape can be used. For a Lorentzian line shape the anisotropic line width w_l is treated in a way similar to that of an anisotropic hyperfine and is given by

$$w_l = [g_x^2\, w_x^2\, l_x^2 + g_y^2\, w_y^2\, l_y^2 + g_z^2\, w_z^2\, l_z^2]^{1/2}/g_{eff}/g_{eff}\beta,$$

where w_l is the half-width at half-height (HWHH) of the absorption line shape in units of MHz. For a Gaussian line shape the anisotropic line width w_g is calculated as:

$$w_g = [w_x^2\, l_x^2 + w_y^2\, l_y^2 + w_z^2\, l_z^2]^{1/2}/2,$$

where w_g is the peak-to-peak line width in units of Gauss. The three anisotropic line widths w_x, w_y, and w_z for the *i*th species are read in as parameters W(1,i), W(2,i), and W(3,i). (Note that indices 1, 2, and 3 are equated to x, y, and z throughout this description.) In addition to the first-derivative spectrum, the absorption spectrum and the second-derivative spectrum can be calculated. Simulation of the absorption spectrum is useful when the spectrum contains components with broad line widths, while simulation of the second derivative spectrum is useful when fitting small partially resolved superhyperfine structure. The parameter IDERIV determines which harmonic of the line shape is calculated. Values of IDERIV equal to 0, 1, and 2 correspond to the absorption, first-derivative, and second-derivative spectrum, respectively.

In many cases the experimental line shape is not purely Gaussian or Lorentzian and the experimental line shape may have both Gaussian and Lorenztian character, giving what is known as a Voigtian line shape. The parameter WL allows one to either convolute an isotropic Lorentzian when an anisotropic Gaussian is used or an isotropic Gaussian when a anisotropic Lorentzian is used. As for the anisotropic line width, the units of WL for an isotropic Lorentzian are in MHz (HWHH) and for addition of an isotropic Gaussian in Gauss (peak-to-peak).

In many systems, particularly those involving transition metal ion complexes in a frozen solution or glass, random molecular distortions of the complex can result in line broadening, in particular a m_I-dependent line width. This broadening is often referred to as g- or A-strain. A random set of molecular distortions is expected to give a Gaussian distribution for each spin Hamiltonian parameter. The resulting distribution in resonance fields can be expressed as a Taylor series expansion of the spin Hamiltonian parameter x:

$$B(\Delta x) = B_0 + B_x'\Delta x + 1/2\, B_x''\Delta x^2 + \ldots.$$

If Δx is small, terms involving Δx^2 and higher-order terms can be neglected, and the resulting distribution in fields will be that of a Gaussian with line width ($B_x'\delta x$), where δx is the variance in x. The value of B_x', where x is equal to one of the principal values of the g or hyperfine matrices, can be evaluated by differentiating the above expressions for the resonance fields:

$$B_x' = \partial B_{res}/\partial x$$

For g-strain there are three input parameters: DELG(1,i), DELG(2,i), and DELG(3,i). Instead of reading in the variance in the three principal values of g directly, DELG(1,i), DELG(2,i), and DELG(3,i) are read in as $\delta(g_x - g_e)/g_x$, $\delta(g_y - g_e)/g_y$, and $\delta(g_z - g_e)/g_z$, respectively. This is done because it is the shift in g from that of the free electron that is related to the crystal field splittings. Hyperfine strain for SPINA and SPINB can also be included. The input parameters DELA(1,i), DELA(2,i), and DELA(3,i) represent the fractional variance in A_x, A_y, and A_z, respectively, while DELB(1,i), DELB(2,i), and DELB(3,i) represent the fractional variance in B_x, B_y, and B_z.

The variances between g_x, g_y, g_z, between A_x, A_y, and A_z, and between B_x, B_y, and B_z, are set to be completely correlated. The input parameters for g- and A-strain can be set either negative or positive. If, for example, DELA(1,i) and DELA(2,i) have opposite signs, then the variances will be anti-correlated. The correlation between the g- and A-matrices, ε_{gA}, between the g- and B- matrices, ε_{gB}, and the A- and B-matrices, ε_{AB}, is set with parameters ECGA, ECGB, ECAB, respectively. The values for these parameters will range between −1 and +1. The total Gaussian line width can be written as:

$$w_g^T = [w_g^2 + ws_g^2 + ws_A^2 + ws_B^2 + 2\varepsilon_{gA}\, ws_g\, ws_A + 2\varepsilon_{gB}\, ws_g\, ws_B + 2\varepsilon_{AB}\, ws_A\, ws_B\,]^{1/2},$$

where w_g is the anisotropic Gaussian line width from above, ws_g, ws_A, and ws_B are the strain contributions for the g-, A-, and B-matrices, respectively. (If an anisotropic Lorentzian is chosen, parameter LS = L, the contributions from g- and hyperfine strains is ignored.) While the values of the anisotropic Gaussian or Lorentzian line widths as well as values of the g- and hyperfine strains can be set individually for each species, the value of line shape type parameter, LS, the line shape harmonic, IDERIV, and the correlation parameters ECGA, ECGB, ECAB are set globally. In particular, only the values listed for the last species is used and values set for species listed earlier are ignored.

5. INTENSITY (TRANSITION MOMENT)

In the perturbation treatment for the resonance fields, it is assumed that the electronic Zeeman interaction is much larger than the hyperfine couplings. Under this assumption the observed transitions will be those for which $\Delta m_S = \pm 1$ and $\Delta m_I = 0$. We have not included nuclear Zeeman or nuclear quadrupole coupling terms in the calculation of resonance fields, and any forbidden $\Delta m_I \neq 0$ transitions are ignored. When forbidden $\Delta m_I \neq 0$ transitions are observed either because the hyperfine coupling or the nuclear quadrupole coupling is large, the spectrum should be calculated using diagonalization of the spin Hamiltonian matrix. The intensity is proportional to the square of the transition moment averaged over all possible orientations of the oscillating microwave magnetic field. Under the condition that the electronic Zeeman is the dominant term in the spin Hamiltonian, the intensity will be

$$<TM^2> \propto \partial B/\partial E\ [g_x^2 g_y^2(1\text{-}l_z^2) + g_y^2 g_z^2(1\text{-}l_x^2) + g_z^2 g_x^2(1\text{-}l_y^2)]/g_{eff}^2,$$

where the term $\partial B/\partial E$ arises in converting a frequency-swept spectrum to a field-field spectrum (Aasa and Vangard 1975) and is equal to $1/g_{eff}\beta$ for the perturbation level used here.

6. GENERATION (INTEGRATION) OF A POWDER SPECTRUM

A powder spectrum results from a distribution of randomly oriented molecules. Experimentally this results either from a microcrystalline solid or from a frozen solution or glass. The powder spectrum can be represented by an integral over all possible orientations

$$\text{Intensity}(B) = \int\int \sum_{m_a}\sum_{m_b}\sum_{m_c}\sum_{m_d}\sum_{m_e}\sum_{m_f} F^n\left\{B_{res}(m_a, m_b, m_c, m_d, m_e, m_f) - B\right\} < TM^2 > \sin\theta\, d\theta\, d\phi,$$

where m_a runs from $-I_a$ to I_a and so on for m_b, etc. and F^n is either the *n*th harmonic Gaussian or Lorentzian line shape function. The powder spectrum can be calculated by replacing the integrals over theta and phi by a double sum over a grid of theta and phi values, providing the grid ($\Delta\theta$, $\Delta\phi$) is fine enough that $\Delta\theta\partial B_{res}/\partial\theta$ and $\Delta\phi\partial B_{res}/\partial\phi$ are smaller than the line width. For optimal integration efficiency, one would like the elements of the theta and phi grid to be of constant size, but for a simple theta and phi grid, the element of surface area, which is $\sin\theta\ \Delta\theta\ \Delta\phi$, is not constant. To get around this problem, we use what is often referred to as the Igloo grid (Nilges 1979) in which the number of elements per tier is proportional to sin θ. For orthorhombic or greater symmetry, the g and hyperfine matrices are collinear, and integration over theta and phi need only be made over a range from 0 to π/2. For triclinic symmetry the integration ranges must increased to 0 to π.

To better improve integration accuracy, we have used four-point Gauss-Point integration, in which the theta and phi integration blocks are further subdivided into unequal intervals. The weighting factors and subinterval size are determined by the zeros of a fourth-degree Legendre polynomial (Margenau and Murphy 1959). Gauss-Point integration can use any number of points. A four-point Gauss-Point scheme was chosen to give the best compromise between spectral accuracy and calculation time (Nilges 1979).

The final integrated powder spectrum is then convoluted with a Lorenztian line shape if an anisotropic Gaussian line shape is chosen and a Gaussian line shape if an anisotropic Lorentzian line shape is chosen. Note again that g-strain and hyperfine strain are included only if an anisotropic Gaussian line shape is chosen (LS = G).

7. SPECTRAL OPTIMIZATION

Program SIMPOW6 performs a real-time graphic fitting of an experimental spectrum. Minimization of the sum-of-square differences is performed using the Simplex method, which can use any number of parameters, n, to optimize the response. The method was originally developed by G. B. Dantzig and first presented by Spendley, Hext, and Himsworth in 1962. The name of the Simplex method is based on the simplex polyhedron, which has n + 1 vertices, where n is the dimension of the fitting space. For a single dimension (variable) the shape is a line, for two dimensions it is a triangle, for three dimensions a tetrahedron, and so on. The Simplex method starts with a simplex whose position is chosen by the starting values of the variables and is contained with the boundary values of the variables. The response function (RFn) is calculated at each of the vertices of the simplex. For spectral fitting here, the RFn is chosen as the sum of squared differences. The basic Simplex method uses a reflection (R) of the simplex opposite to the vertex with the poorest RFn. Implementation of the Simplex method here (Mattson, 1990) uses the modified Simplex method, which allows for contraction (C) and expansion (E) of the reflected simplex. In addition to contracted reflection away from the worst vertex (C_R), one can also have a contracted reflection toward the worst vertex (C_w). The ability to adjust the size/shape of the simplex allows for faster convergence. Unlike other methods, partial derivatives or gradients of the RFn with respect to the individual variables need not be calculated either analytically or by finite differences. The Simplex method still effectively minimizes the gradient. The method is rather efficient, especially when many variables are used.

The most important problem encountered with the Simplex method is that of finding a local minimum instead of the global minimum of the RFn, and to a certain extent this is a problem with almost every minimization routine. There are three important factors that determine the ability to find the correct fit of an experimental spectrum. The first factor is the choice of the correct spin Hamiltonian. In many cases, partially resolved structure can be ascribed to more than one source. Observed splittings in the spectrum may be due to hyperfine structure, or they could be due to a mixture of somewhat different paramagnetic species. In many cases, partially resolved hyperfine splitting patterns can be quite similar yet arise from quite different sets of nuclei. For example, a partially resolved pattern from one ^{14}N nucleus can be quite similar to that of two nearly equivalent protons. Generally, the best method to reduce uncertainties about the correct spin system needed to describe an experimental spectrum is the collection of multifrequency ESR data or ENDOR data. The second factor is to use the closest and the most reasonable initial values for the variables to be fitted. This will not only minimize the chance of getting stuck in a local minimum, but will also speed convergence. Deliberate use of starting values that give rise to a spectral pattern that is quite different from the observed spectrum can easily lead to being stuck in a local minimum. The third factor is the choice of boundary values. If too small a range or set of boundary values is used, this can result in the global minimum being put out of range. Too large a range and the convergence can be slow.

8. SUMMARY OF INPUT PARAMETERS

The input data file for SIMPOW6 can be broken into two parts. The first part inputs the spin Hamiltonian values. A detailed listing of parameters for the first part of the input file is given in Appendix A herein. The second part of the input data file is used to control and the list the parameters to be optimized. In addition to setting the parameters that are to be varied and the starting values and range of those parameters, there is the option of inputting equations that can be used to restrict the number of independent variables. This can simplify the fitting process for spin systems where the number of possible input parameters can be quite large. Depending upon the symmetry of the system, these restrictions can be either strict or approximate. In the case that they are approximate, the equations can be removed at later stages of fitting to test the validity of the approximations. A detailed listing of parameters used in the optimization input part of the input file is given in Appendix B. Several example input files are supplied with the program. These can be used as templates for making up new input files using any text editor, for example, NOTEPAD.

9. RUNNING THE PROGRAM

SIMPOW6 runs under 32-bit Windows. It is compiled using the QuickWin library of Visual Fortran, which allows implementation of basic windows functions including graphics. Two issues are known when running under windows. First, the color quality in the display settings of the control panel should be set no higher than 16-bit, or else blank windows will appear. Second, if the print or save function is to work (which is not really necessary for ordinary running of the program), the color need be set at only 256. Executing the program brings up the "Input Window," asking for the input experimental file. This can be one of three formats, the default of which is a "Dataeg" or SIMPOW6 file. The format of this ASCII file is given in Appendix C. The second type of file is an "EW" file (Scientific Software Services), which must be entered with an .fls extension. The third type of file is a simple list of data points, where the first line is the number of points, n, and the next n lines list the n intensity values. This file must be entered with an .smp extension. If a return is entered instead of a file name, an open file window will appear from which a file may be selected. Next the name of the input data file is asked for, the format of which is detailed in Appendices A and B. Next, the user is asked to give a name for the log file. This file gives detailed fitting information for each iteration; it will give the RFn and the type of simplex (Reflected(R), Expanded(E), Contracted toward the worst vertex (CW), Contracted away from the worst vertex (CR), and Shrunken (S)). Also given are the values of all the varied parameters and their relative position within the bounded range. Finally, a name for the output simulation file is requested. The format of the simulation file is the same as that used in the DOS data manipulation program Dataeg.

Once the file names have been entered, iterative fitting will begin and two new windows will open. The first window will list the RFn's for successive iterations, while the other window will present a graphic presentation of the fitting. The latter will present the experimental file (red), the simulated spectrum (green), and the residual (yellow). Also displayed are the values of the varied parameters (up to 12) at each iteration. The iterations will continue until either a minimal RFn is reached, or the maximum number of iterations is reached, or until the end of the next iteration after soft abort is checked on the "run" menu. In addition to the log and simulated output files, two other files are generated. One is the ARC.DAT file. This is the input data file updated with the values of the parameters from the end of the last iteration. The other file is RESTART.DAT, which can be used to restart the simulation using the vertices from the last run. Generally, when first starting to fit a spectrum it is better to use the ARC.DAT file as a new starting point rather then adding the RESTART command in the data file; the latter can easily lead one into getting trapped in a local minimum.

10. EXAMPLES

10.1. Example 1: Simple Frozen Solution Spectrum of a Nitroxide

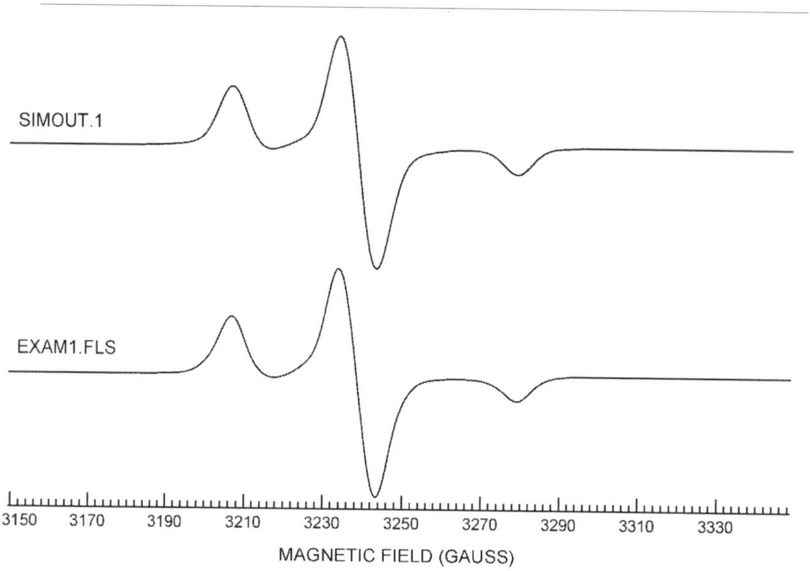

This basic example is that of a proxyl spin label (3-carbamoyl-proxyl) in a frozen water/alcohol glass. The g- and ^{14}N hyperfine matrices are assumed to be coincident, i.e., the symmetry is assumed to be orthorhombic. The line broadening in such spin labels is dominated by unresolved proton hyperfine from the methyl groups and as such the anisotropic line width is set to Gaussian. Using the equation W(2,1)=W(1,1), the line width is set to be axial ($W_x = W_y$). A small isotropic Lorenztian line broadening is also added (3.3 MHz — HWHH). Note that Gaussian line widths are measured in the field domain as peak-to-peak line width in units of Gauss, while Lorentzian line widths are measured in the frequency domain as half-width-at-half-height in units of MHz. The data file for this simulation, EXAM1.DAT, is included with the program and can be used as a template for simulating similar systems with one hyperfine nucleus.

10.2. Example 2: Using Equations for Generating a Solution Spectrum

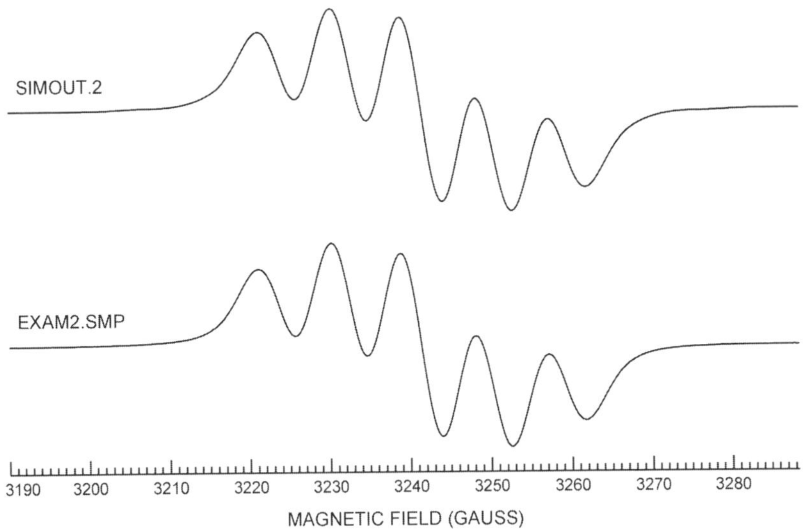

By setting equations in the data file, the number of parameters that need to be varied can be reduced if the symmetry is higher than orthorhombic. In the case of axial symmetry, equations like A(1,1) = A(2,1) can be used. In the case of cubic or higher symmetry, the g and hyperfine matrices are isotropic. The example here is the solution spectrum of DPPH, which shows hyperfine structure from two nonequivalent ^{14}N nuclei. The g, the two hyperfine matrices, and the anisotropic Gaussian line widths are all varied, yet set to be isotropic using the following equations in EXAM2.DAT:

```
EQUATIONS=08
G(2,1)=G(1,1)
G(3,1)=G(1,1)
A(2,1)=A(1,1)
A(3,1)=A(1,1)
B(2,1)=B(1,1)
B(3,1)=B(1,1)
W(2,1)=W(1,1)
W(3,1)=W(1,1)
```

10.3. Example 3: Separating Multiple Species

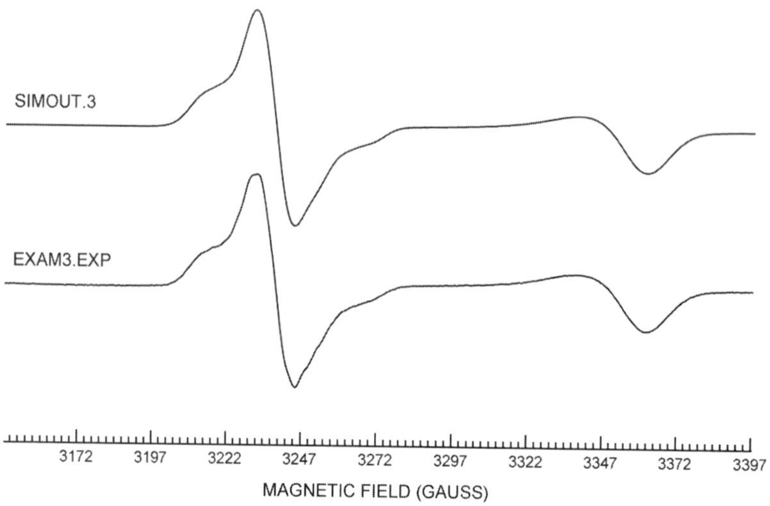

This example is that of two overlapping species, one being a free radical (anionic flavin semiquinone radical) and the other being a mixed valence iron–sulfur center. Often it is desirable to determine the concentration of one or both of the two species, but it is difficult to separate the contributions of the two species to the double integral of the spectrum. The simulation of the spectrum of a mixture can be simplified if a pure spectrum of either species can be simulated first. In this case a W-band spectrum that shows only the free-radical signal was simulated first. The anisotropy of the g-matrix as determined from the W-band simulation as well as the assumption that the ^{14}N hyperfine matrices are axial were set with the following equations in EXAM3.DAT:

```
EQUATIONS=04
A(2,1)=A(1,1)
B(2,1)=B(1,1)
G(1,1)=G(3,1)+.208000E-02
G(2,1)=G(3,1)+.174000E-02
```

Further details on the multifrequency simulation of these flavin radicals can be found in (Barquera et al. 2003). With the ratio determined by simulation and the double integral of the spectrum, the absolute number of spins for either species can be determined.

10.4. Example 4: g- and Hyperfine Strain in Glassy Copper Spectra

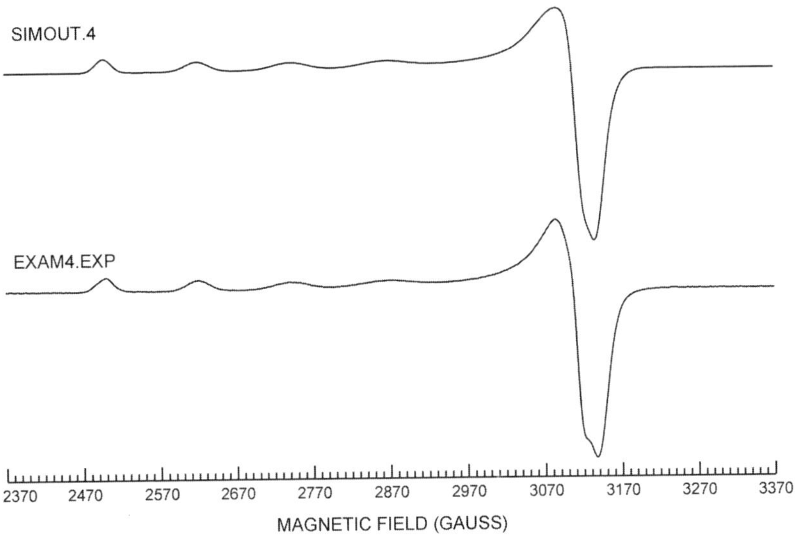

SIMPOW6 also allows input of g-strain and hyperfine strain broadening when an anisotropic Gaussian line width is chosen. This sample simulation is that of a frozen solution of CuSO4·5H2O in water/glycerol. The four parallel features show a marked m_I-dependent line width. This is accounted for in the simulation with the parameters X(3,1) = 0.042, Y(3,1)= –0.068. The fractional strain in the g-shift of 0.042 corresponds to a variance in g_z of 0.0172; the value of the fractional strain in A_z of –0.068 corresponds to a variance in A_z of 27.9 MHz. X(3,1) and Y(3,1) have opposite signs, indicating that g- and A-strains are anti-correlated. Alternatively, both X(3,1) and Y(3,1) could have been set positive and correlation parameter ECGA set equal to –1. The g-strain and A-strain cannot be readily determined for the perpendicular components in g and A as the copper hyperfine is not resolved. Once more the effects of nuclear quadrupole coupling that can alter the shape of the perpendicular feature are not included in the program. To simulate the effects of nuclear quadrupole coupling usually requires the use of matrix diagonalization of the Hamiltonian matrix, which is done in the analogous SIMPIPM program.

10.5. Example 5: Multiple Isotopes

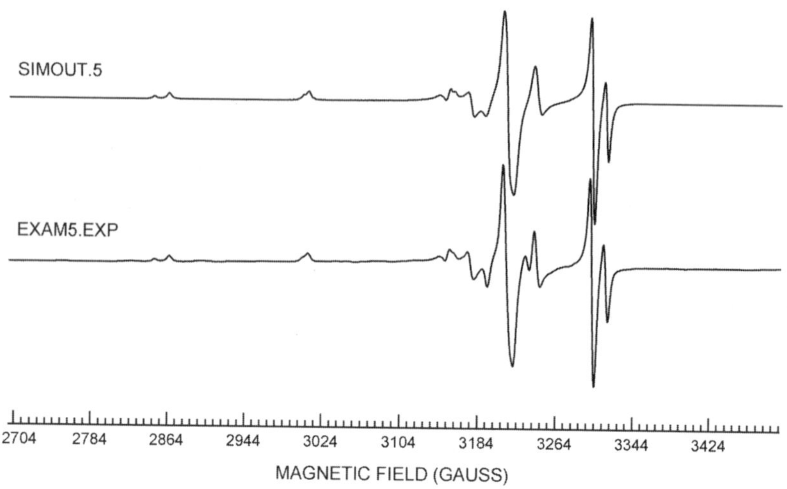

In a number of instances, nuclei may have more than one naturally abundant isotope. Examples of such nuclei include copper (^{63}Cu and ^{65}Cu) and molybdenum (^{95}Mo and ^{97}Mo). Example 5 is that of a powder of copper doped into crystals of $Zn(dtc)_2$. Here the line width is narrow enough that the splitting between ^{63}Cu and ^{65}Cu parallel peaks is readily resolved. In particular, the lowest and highest field peaks are split by 14 Gauss. In order to simulate the experimental spectrum of both isotopes simultaneously, we use the following equations in EXAM5.DAT:

```
EQUATIONS=09
G(1,2)=G(1,1)
G(2,2)=G(2,1)
G(3,2)=G(3,1)
W(1,2)=W(1,1)
W(2,2)=W(2,1)
W(3,2)=W(3,1)
A(1,2)=A(1,1)* 1.07130
A(2,2)=A(2,1)* 1.07130
A(3,2)=A(3,1)* 1.07130
```

Here, 1.07130 is the ratio of magnetic moments between the two isotopes.

10.6. Example 6: Use of 2nd Derivative to Resolve Hyperfine Structure

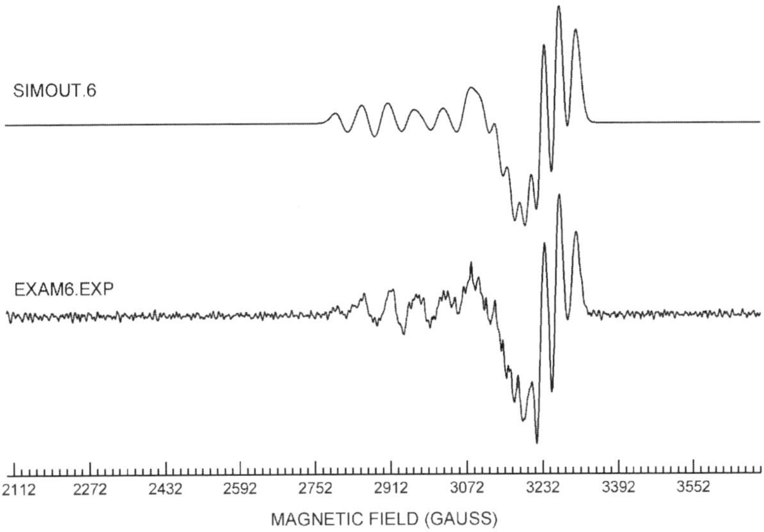

Coupling constants for partially resolved hyperfine structure can often best be determined by simulating the second-derivative spectrum in addition to the first-derivative spectrum. Example 6 is that of a mixed valence copper-copper dimer (CuA; see Hwang et al. 2005). The two coppers, which are both calculated to second order, give rise to a seven-line copper hyperfine pattern. Unlike the first-derivative spectrum, the copper hyperfine splitting along the perpendicular directions appears as well resolved. In this example, the spectral simulation using the following equations allowed the two coppers to be inequivalent and the z-axes of the g and hyperfine matrices to be non-coincident:

S(3,1)=S(1,1)*–1.00
T(3,1)=T(1,1)*–1.00

This sets the first and third Euler angles such that the rotation is limited to only a rotation of the z axis about the axis determined by angle S(1,1) for the first copper hyperfine interaction and angle T(1,1) for the second copper hyperfine interaction (see EXAM6.DAT). The second-derivative spectrum can be obtained experimentally by double modulation or better by numerically modulating the first-derivative spectrum. The latter requires that the signal-to-noise ratio of the corresponding first-derivative spectrum be quite high.

11. APPENDIX A: FORMAT OF THE INPUT FILES: SPIN HAMILTONIAN PARAMETERS

The first part of the input file lists the simulation parameters:

NU					
FACTR	FACT	CUTOFF			
NTH	NPH	NSI			
HLW	HLH	NTOT			
SPINA(1)	SPINB(1)	SPINC(1)	SPIND(1)	SPINE(1)	SPINF(1)
NEC(1)	NED(1)	NEE (1)	NEF(1)		
G(1,1)	G(2,1)	G(3,1)	TINT(1)		
A(1,1)	A(2,1)	A(3,1)	ANGA(1,1)	ANGA(2,1)	ANGA(3,1)
B(1,1)	B(2,1)	B(3,1)	ANGB(1,1)	ANGB(2,1)	ANGB(3,1)
C(1,1)	C(2,1)	C(3,1)	ANGC(1,1)	ANGC(2,1)	ANGC(3,1)
D(1,1)	D(2,1)	D(3,1)	ANGD(1,1)	ANGD(2,1)	ANGD(3,1)
E(1,1)	E(2,1)	E(3,1)	ANGE(1,1)	ANGE(2,1)	ANGE(3,1)
F(1,1)	F(2,1)	F(3,1)	ANGF(1,1)	ANGF(2,1)	ANGF(3,1)
W(1,1)	W(2,1)	W(3,1)	LS	IDERIV	WL
X(1,1)	X(2,1)	X(3,1)	ECGA	ECGB	ECAB
Y(1,1)	Y(2,1)	Y(3,1)	Z(1,1)	Z(2,1)	Z(3,1)

Lines 4 through 15 are repeated for each additional species. This part of the data file is formatted in six columns having a width of 8 characters, and it is important that the input data is properly aligned. The parameters are defined:

NU	microwave frequency in GHz.
FACTR	vertical scaling; full scale = 100.0.
FACT	not used; set to zero.
CUTOFF	number of line widths out from the center that the line shape is calculated.
NTH	number of integration units along the theta coordinate, each unit is four unequally weighted points.
NPH	number of integration units along the phi coordinate. If nth and nph are too small, extra peaks, known as "graining", will occur.
NSI	number of different paramagnetic species or spin systems to be calculated; maximum is 9.
HLW	starting field in gauss.
HLH	ending field in gauss.
NTOT	number of points in spectrum, maximum is 4096.
SPINA(J)	nuclear spin for nuclei A for the j'th species, which is calculated to second order.
SPINB(J)	nuclear spin for nuclei B for the j'th species, which is calculated to second order.
SPINC(J)	nuclear spin for nuclei C for the j'th species, which is calculated to first order.

SPIND(J)	nuclear spin for nuclei D for the j'th species, which is calculated to first order
SPINE(J)	nuclear spin for nuclei E for the j'th species, which is calculated to first order.
SPINF(J)	nuclear spin for nuclei F ,for the j'th species, which is calculated to first order.
NEC(J)	number of equivalent nuclei for SPINC(J).
NED(J)	number of equivalent nuclei for SPIND(J).
NEE(J)	number of equivalent nuclei for SPINE(J).
NEF(J)	number of equivalent nuclei for SPINF(J).
G(1,J)	g_x for the j'th species.
G(2,J)	g_y for the j'th species.
G(3,J)	g_z for the j'th species.
TINT(J)	relative weighing factor for the j'th species.
A(1,J)	first principal value (A_x) of the hyperfine matrix (j'th species) in MHz.
A(2,J)	second principal value (A_y) of the hyperfine matrix (j'th species) in MHz.
A(3,J)	third principal value (A_y) of the hyperfine matrix (j'th species) in MHz.
ANGA(1,J)	Euler angle alpha relating the principal axes of the hyperfine matrix to g matrix.
ANGA(2,J)	Euler angle beta.
ANGA(3,J)	Euler angle gamma.
B(1,J)	first principal value (B_x) of the hyperfine matrix (j'th species) in MHz.
B(2,J)	second principal value (B_y) of the hyperfine matrix (j'th species) in MHz.
B(3,J)	third principal value (B_y) of the hyperfine matrix (j'th species) in MHz.
ANGB(1,J)	Euler angle alpha for rotation of the B hyperfine matrix.
ANGB(2,J)	Euler angle beta.
ANGB(3,J)	Euler angle gamma.
C(1,J)	first principal value (C_x) of the hyperfine matrix (j'th species) in MHz.
C(2,J)	second principal value (C_y) of the hyperfine matrix (j'th species) in MHz.
C(3,J)	third principal value (C_y) of the hyperfine matrix (j'th species) in MHz.
ANGC(1,J)	Euler angle alpha relating the principal axes of the hyperfine matrix to g matrix.
ANGC(2,J)	Euler angle beta.
ANGC(3,J)	Euler angle gamma.
D(I,J)	three principal values of the D hyperfine matrix in MHz

ANGD(I,J)	three Euler angles relating the principal axes of the D hyperfine matrix to g matrix.
E(I,J)	three principal values of the E hyperfine matrix in MHz
ANGE(I,J)	three Euler angles relating the principal axes of the E hyperfine matrix to g matrix.
F(I,J)	three principal values of the F hyperfine matrix in MHz
ANGF (I,J)	three Euler angles relating the principal axes of the F hyperfine matrix to g matrix.
W(1,J)	line width along g_x (j'th species).
W(2,J)	line width along g_y.
W(3,J)	line width along g_z.
LS	line shape: L=Lorentzian with half-width-at-half-height in MHz. G=Gaussian with the peak-to-peak line width in Gauss.
WL	if main line shape is Gaussian adds an additional Lorentzian component of line width (MHz) (i.e. Voight line shape). If main line shape is Lorentzian adds an additional Gaussian component of line width (Gauss).
IDERIV	0 for absorption, 1 for first derivative, 2 for second derivative.
DELG(1,J)	fractional strain of shift of g_x from g_e.
DELG(2,J)	fractional strain of shift of g_y from g_e.
DELG(3,J)	fractional strain of shift of g_z from g_e.
ECGA	is the correlation in strain between the g matrix and the A hyperfine matrix(-1 to +1).
ECGB	is the correlation in strain between the g matrix and the A hyperfine matrix(-1 to +1).
ECAB	is the correlation in strain between the A and the B hyperfine matrix (-1 to +1).
DELA(1,J)	fractional strain of the first principal value of the A hyperfine matrix.
DELA(2,J)	fractional strain of the second principal value of the A hyperfine matrix.
DELA(3,J)	fractional strain of the third principal value of the A hyperfine matrix.
DELB(1,J)	fractional strain of the first principal value of the B hyperfine matrix .
DELB(2,J)	fractional strain of the second principal value of the B hyperfine matrix.
DELB(3,J)	fractional strain of the third principal value of the B hyperfine matrix.

12. APPENDIX B: FORMAT OF THE INPUT FILES: OPTIMIZATION PARAMETERS AND CONTROL

The first part of the input file lists the optimization control parameters and the initial starting values and search ranges for the parameters to be varied. Note that the starting value of a parameter listed in this section overrides the value of given

VARIABLES=n ITER=m GRAPHOFF(optional) RESTART(optional)
M(i,j)=x RANGE= y

.

.

N(k,l)=u RANGE=v
EQUATIONS=ne
M(i,j)=M(k,j)+OFFSET

.

.

N(i,j)=N(k,j)*MULTIPLIER

Where

VARIABLES	number of parameters to be optimized. Can range from 0 to 30.
ITER	number of iterations to be performed. Ranges from 0 to 999. If set to 0 then only one spectrum is calculated using the initial input parameters. If it is set between 1 and the number of variables, n, n spectra will be calculated representing the vertices of the simplex. First new simplex is calculated on the n+1 iteration.
GRAPHOFF	if listed the graphic display of the experimental and simulated spectra is turned off.
RESTART	if listed the optimization is restarted using the vectors from the previous run. Requires the file RESTART.DAT as well as the input data and experimental file from the previous run.
EQUATIONS	number of equations to be defined. Used to reduced the number of independent variables. Can ranger from 0 to 30. Two parameters can be directly equated or can be equated using an offset or multiplier.
RANGE	range or window over which the variable can be changed.
M(i,j)	any parameter to be varied or equated, where i can range from 1 to 3 and typically represents the first, second, or third principal value of a spin Hamiltonian matrix and j represents the species number. For use a variable must be followed by RANGE. For use as an equation, an example would be A(1,1)=A(2,1) which sets A_x equal to A_y for the first species. Valid values of M are listed below.
G(i,j)	principal values of the g-matrix.
A(i,j)	principal values of the A hyperfine matrix.

B(i,j)	principal values of the B hyperfine matrix.
C(i,j)	principal values of the C hyperfine matrix.
D(i,j)	principal values of the D hyperfine matrix.
E(i,j)	principal values of the E hyperfine matrix.
F(i,j)	principal values of the F hyperfine matrix.
W(i,j)	three values of the anisotropic line width
S(i,j)	three Euler angles for the A hyperfine matrix. (Equated to ANGA(I,J).)
T(i,j)	three Euler angles for the B hyperfine matrix. (Equated to ANGB(I,J).)
X(i,j)	fractional strains for the g-matrix (Equated to DELG(I,J).)
Y(i,j)	fractional strains for the A hyperfine matrix (Equated to DELA(I,J).)
Z(i,j)	fractional strains for the B hyperfine matrix (Equated to DELB(I,J).)

13. APPENDIX C: FORMAT OF THE OUTPUT SIMULATION FILES

FIRST LINE:
---- NTOT - number of points, FACT - scaling factor , PL – (set to 0)
FORMAT(I8,F8.4,F8.2)

SECOND LINE:
---- HLW - starting field (gauss), HLH - ending field, NU - frequency
FORMAT(2F8.1,F8.4)
NEXT NTOT/6 LINES:
---- spectral intensity data points
FORMAT(10I6)

LAST 3+NST(2+NSP) LINES:
---- simulation parameters
FORMAT – HOLLERITH

14. REFERENCES

Aasa RA, Vanngard ME. 1975. EPR signal intensity and powder shapes: a reexamination. *J Magn Res* **19**:308.

Barquera B, Morgan JE, Lukoyanov D, Scholes CP, Gennis RB, Nilges MJ. 2003. X- and W-band EPR and Q-band ENDOR studies of the flavin radical in the Na^+-translocating NADH:quinone oxidoreductase from *Vibrio Cholerae*. *J Am Chem Soc* **125**:265.

Hwang HJ, Berry SM, Nilges MJ, Lu Y. 2005. Axial methionine has much less influence on reduction potentials in a Cu_a center than in a blue copper center. *J Am Chem Soc* **127**:7274.

Mattson K. 1990. Doctoral dissertation. University of Illinois, Urbana-Champaign.

Nilges MJ. 1979. Doctoral dissertation. University of Illinois, Urbana-Champaign.

Rose ME. 1957. Elementary theory of angular momentum. New York: Wiley.

Spendley W, Hext GR, Himsworth FR. 1962. Sequential application of simplex designs in optimisation and evolutionary operation. *Technometrics* **4**:441.

Weil JA. 1975. Comments on second-order spin-Hamiltonian energies. *J Magn Reson* **18**:113.

APPENDIX 3

ACERT SOFTWARE: SIMULATION AND ANALYSIS OF ESR SPECTRA

Jack H. Freed

Cornell University, Ithaca, New York, USA

This appendix contains a description of the ACERT software for simulation and analysis of ESR spectra. It includes:

1. Original versions of well-known programs described in the papers *Calculating slow motional magnetic resonance spectra: a user's guide* by Schneider and Freed [Schneider DJ, Freed JH. 1989. *Biol Magn Reson* **8:**1–76], *Nonlinear least-squares analysis of slow-motional epr spectra in one and two dimensions using a modified Levenberg-Marquardt algorithm* [Budil DE, Lee S, Saxena S, Freed JH. 1996. *J Magn Reson* **A120**:155–189], and their most recent updates: **EPRLL, PC, NLSL, PC.NEW, NL2DR, NL2DC**.
2. Basic programs for CW spectral fitting using the SRLS model as described in *An assessment of the applicability of multifrequency ESR to study the complex dynamics of biomolecules* [Liang Z, Freed JH. 1999. *J Phys Chem B* **103**:6384–6396]. **NLSL.SLRS.**
3. Software for Tikhonov regularization (TIKR) and maximum entropy (MEM) methods. The software was developed to extract pair distributions from pulsed ESR experiments. See *The determination of pair distance distributions by pulsed ESR using tikhonov regularization* [Chiang Y-W, Borbat PP, Freed JH. 2005. *J Magn Reson* **172:**279–295] and *Maximum entropy: a complement to Tikhonov regularization for determination of pair distances distributions by pulsed ESR* [Chiang Y-W, Borbat PP, Freed JH. 2005. *J Magn Reson* 177:184–196]. **PD_Tikhonov, DPD_pkg.**

The programs are contained on the bundled CD-ROM in the following subdirectories:

EPRLL — the basic simulation programs for CW spectrum calculation including the EPRBL program used to determine truncated basis sets. A detailed description of the method and Version 1.0 of the program is given in the "User's guide" by Schneider and Freed (1989). See file UPDATE.DOC for cumulative changes through Version 1.6b that have been made in EPRLL and related programs since publication of Version 1.0.

PC — the original PC version of the CW spectrum simulation programs (11/3/95).

NLSL — the least-squares version of the above CW program described by Budil et al. (1996). See the subdirectory EXAMPLES for fitting examples to test.

PC.NEW — a version of the NLSL programs suitable for running on a PC with WINDOWS 98/2000/NT/XP. Besides executables and calculation examples, the folder contains files necessary for Fortran Powerstation 4.0 to compile and link the source files from the ***NLSL*** folder. See README.TXT in this directory.

NL2DC — contains the two-dimensional Fourier Transform (2D-FT) version of the least squares program using the conjugate gradients matrix solution method. For 2D basis set pruning see the subdirectory BASIS in NL2DR.

NL2DR — contains the 2D-FT version of the least-squares program using the Rutishauser matrix solution method. For 2D basis set pruning see the subdirectory BASIS.

The program source codes for both **NL2DC** and **NL2DR** are designed for running under **LINUX**. They are stored in uncompressed format and may be obtained by ASCII ftp transfer. Tar files are provided for compressed data transfer and contain all files in the specific directories and subdirectories except the executables to save space.

READ.ME files are provided in each of the above directories giving further information on their contents and usage.

NLSL.SRLS contains the basic programs for CW spectral fitting using the SRLS model as described by Liang and Freed (1999). **NLSL.SRLS** is the fitting program, and **ESRBS.SRLS** is used to prune the basis set. **LBSR.SRLS** is for inputting all the parameters used in an **ESRBS.SRLS** calculation.

PD_Tikhonov package: includes all necessary functions required to extract the pair distributions from pulsed ESR experiments using the Tikhonov regularization method. MATLAB version 6.5 or higher is required to perform the task. PD_Tikhonov requires the use of the Regularization Toolbox package obtainable from http://www2.imm.dtu.dk/~pch/Regutools/index.html. The toolbox is well documented with a detailed manual and illustrated with numerous examples. We strongly suggest users first get acquainted with the Regularization Toolbox. After unzipping the package, please read **README.pdf** file.

DPD_pkg package: includes all necessary functions required to perform Tikhonov regularization (TIKR) method and maximum entropy method (MEM) to extract the pair distributions from pulsed ESR experiments. DPD_pkg can be considered an upgraded version of the PD_Tikhonov package since both TIKR and MEM are included in the package. MATLAB version 6.5 or higher is required to perform the task. The DPD_pkg requires the use of the Regularization Toolbox package obtainable from http://www2.imm.dtu.dk/~pch/Regutools/index.html, and which is very well documented with a manual and illustrated with numerous examples. We strongly suggest users first get acquainted with the Regularization Toolbox. After unzipping the package, please read the **PD_demo.m** file.

Every effort has been made to ensure that these programs are correct and thoroughly tested. However, the programs are distributed "AS IS," and all warranties, whether expressed or implied, as to correctness or fitness for any specific purpose are specifically disclaimed. In no event shall the authors be liable for any direct, consequential, or incidental damage arising from the use of these programs. Free use and distribution of these programs is permitted with suitable reference to the original publication (see above) in any published work resulting from the use of these programs or programs derived from them. The programs may be copied and distributed, so long as (1) due credit is given by retaining the comment lines at the beginning of each source file in all copies, and (2) all copies must be distributed free of charge.

Questions and comments may be directed to ACERT at acert@cornell.edu, which will forward the correspondence to the most appropriate person.

ACKNOWLEDGMENT

This work is supported by a grant from NCCRR/NIH.

APPENDIX 4

DEERANALYSIS 2006: DISTANCE MEASUREMENTS ON NANOSCOPIC LENGTH SCALES BY PULSE ESR

Gunnar Jeschke
Max Planck Institute for Polymer Research,
Mainz, Germany

A few methods for structure analysis are capable of characterizing distance distributions in the range between 1 and 10 nm in disordered systems, such as polymers or non-crystalline proteins. As a ruler on such length scales, we can use the dipole–dipole interaction between two electron spins that depends on the inverse cube of the distance. At 1 nm distance this interaction is approximately 50 MHz and is usually resolved in continuous-wave ESR spectra of nitroxide spin probes. For distances larger than 2 nm, the spectral broadening by the hyperfine and anisotropic electron Zeeman interaction has to be eliminated, and for distances larger than 3 nm, relaxational broadening must be avoided. In disordered systems, it may often be important to observe the dipole–dipole time evolution without any dead time to obtain reliable distance distributions.

Based on a pulse electron–electron double resonance (ELDOR) experiment introduced by Milov, Salikhov, and Shirov in the early 1980s and developed mainly by the Tsvetkov group in Novosibirsk, we have recently introduced the four-pulse DEER experiment that fulfils all these requirements.

On the bundled CD-ROM we supply the program DeerAnalysis 2006. This program can extract distance distributions from dead-time free-pulse ELDOR data (constant-time and variable-time four-pulse DEER) (Pannier et al. 2000; Jeschke et al. 2004a). Furthermore, it can be used for direct comparison of primary data of similar samples (Jeschke et al. 2004b). For a series of related samples, an average distance distribution can be computed taking into account the signal-to-noise ratio of the individual data sets. With some caution (Jeschke et al. 2001) the program may also be applied to the analysis of dead-time-free double-quantum coherence ESR experiments (Borbat and Freed 1999). It should not be used for data from experiments that have a significant dead time, $t_d > 2r^3$ ns nm^{-3}, where r is the shortest distance in the distribution.

The underlying mathematical problem is (moderately) ill posed, i.e., the quality of the analyzed data is very crucial. Pre-processing tools are implemented to correct for experimental imperfections (phase errors, displacements of the time origin of the modulation) and to separate the intramolecular distances of interest from the intermolecular background contribution. Furthermore, the program provides several independent approaches for extracting the distance distribution, which helps to get a feeling for the reliability of the distribution. Characterization of the distance distribution in terms of its mean value $\langle r \rangle$ and width (standard deviation σ_r) is usually reliable (Jeschke et al. 2004) and is therefore a standard output. The performance of the different approaches for data analysis depends on the type of distance distribution (narrow or broad peaks or both) and was discussed in some detail in Jeschke et al. (2004a,b).

The most recent version of the software can be found at the following website: http://www.mpip-mainz.mpg.de/~jeschke/distance.html.

REFERENCES

Borbat PP, Freed JH. 1999. Multiple-quantum ESR and distance measurements. *Chem Phys Lett* **313**:145–154.

Jeschke G, Koch A, Jonas U, Godt A. 2001. Direct conversion of EPR dipolar time evolution data to distance distributions. *J Magn Reson* **155**:72–82.

Jeschke G, Bender A, Paulsen H, Zimmermann H, Godt A. 2004a. Sensitivity enhancement in pulse EPR distance measurements. *J Magn Reson* **169**:1–12.

Jeschke G, Panek G, Godt A, Bender A, Paulsen H. 2004b. Data analysis procedures for pulse ELDOR measurements of broad distance distributions. *Appl Magn Reson* **26**:223–244.

Pannier M, Veit S, Godt A, Jeschke G, Spiess HW. 2000. Dead-time free measurement of dipole–dipole interactions between electron spins. *J Magn Reson* **142**:331–340.

EWVOIGT AND EWVOIGTN: INHOMOGENEOUS LINE SHAPE SIMULATION AND FITTING PROGRAMS

Alex I. Smirnov
Department of Chemistry, North Carolina State University, Raleigh, NC 27606 USA

1. INTRODUCTION

The EWVoigt package contains two programs, EWVOIGT and EWVOIGTN. EWVOIGTN is an extended version of EWVOIGT and is suitable for least squares simulations of two overlapping spin label nitroxide spectra. Some of the extended features of EWVOIGTN include:

- Second-order perturbation approximation for calculating nitroxide spectra
- Simulation of two nitroxides with different inhomogeneous line width parameters, for example, different width of the Gaussian envelope
- Explicit accounting for ^{13}C satellite lines in the simulation
- Extended equality constraints for fitting parameters

2. CONVOLUTION-BASED FITTING OF CONTINUOUS WAVE EPR SPECTRA

EWVOIGT is designed for simulation/optimization of inhomogeneously broadened EPR spectra, which are very common in EPR spectroscopy (e.g., anisotropy broadening, hyperfine interaction, inhomogeneities in the magnetic field). With the help of this program one can accurately and quickly extract spectral parameters from one class of continuous-wave EPR spectra $I(B)$ — those that are uniformly broadened along the magnetic field coordinate and can be described by a convolution integral:

$$I(B) = \int p(B-B')f(B')dB' = p(B) \otimes f(B)\,, \qquad (1)$$

where ⊗ is the convolution symbol, $p(B)$ is the envelope function (or an unbroadened spectrum), and $f(B)$ is the line shape of an individual spin packet (or broadening function).

The model given by Eq. (1) is more common than it first appears. One type of data that is well modeled by a convolution integral is an isotropic inhomogeneously broadened EPR spectrum. Normally, electron spin packets are distributed under an envelope whose shape is determined by the details of the inhomogeneous broadening mechanisms. The model describing the inhomogeneous line shape requires that the broadening must come from interactions outside the spin system and must be adiabatic during the time of the spin transition (Portis 1953). Examples of inhomogeneous broadening include, among others, anisotropy broadening, hyperfine interaction, and inhomogeneities in the magnetic field. Ultimately, all EPR spectra are inhomogeneously broadened at least to some extent. However, Eq. (1) works well only for spectra that are broadened uniformly such as, for example, individual nitrogen hyperfine components of spin-label EPR spectra in the fast-motional limit. Then envelope $p(B)$ can be assigned to the proton hyperfine structure. This envelope can be depicted by a stick diagram and directly used in the simulations or approximated by a Gaussian envelope (Bales 1989). The resonance line profile of an individual spin packet is given by a Lorentzian function with a center at B_0 and width $\Delta B_{1/2}$ at half line height:

$$f(B) = \frac{2A}{\pi} \frac{\Delta B_{1/2}^2}{\Delta B_{1/2}^2 + 4(B_o - B)^2}, \tag{2}$$

where A is the value of the double integral, which is proportional to the number of spins, and $\Delta B_{1/2}$ is the line width, which is determined by the rotational relaxation and spin–spin interactions. For spin labels in the fast-motional limit the individual line widths, $\Delta B_{1/2}(m_I)$ should be assigned to each m_I nitrogen hyperfine transition.

Slow-motion EPR spectra require more sophisticated simulation analysis, which generally involves a larger number of parameters than fast-limit spectra (Budil et al. 1996). However, even these spectra may contain isotropically broadened components. For example, if some hyperfine interaction (e.g., with protons of methyl groups) can be considered as "pre-averaged" by a fast (on the EPR time scale) intramolecular rotation, then those interactions will contribute to anisotropic slow-motional EPR spectra as uniform broadening. This kind of broadening can be modeled by Eq. (1), in which the slow-motional spectrum is treated as envelope $p(B)$ and $f(B)$ is assigned to the additional uniform broadening.

The second, and perhaps one of the most practically important examples, is uniform broadening of EPR spin-label spectra by magnetic interactions with similar or dissimilar electron spins. These include Heisenberg spin exchange (this is the dominant mechanism of broadening of spin-label EPR spectra by molecular oxygen) and isotropic dynamic dipolar interaction. The latter dominates the magnetic interactions of spin labels with some paramagnetic metal ions such as Gd^{3+} in non-viscous liquids (Hyde et al. 1979). Spin exchange between the nitroxides in solutions at concentrations corresponding to slow exchange also leads to line broadening, but the shape remains Lorentzian only to a first approximation (Molin et al. 1980). Another case of uniform spectral broadening is caused by magnetic interactions between two nitroxide labels introduced into a protein by site-directed spin-labeling methods. Under certain conditions, e.g., if the spins in such a pair are separated by ca. >10–12 Å and the rotational correlation time of the interspin vector is short enough, the static dipolar interaction is averaged out and the broadening function is again a Lorentzian and can be considered to be uniform across the nitroxide spectrum (e.g., Mchaourab et al. 1997). In all these cases, the "envelope" spectrum can be assigned to that taken in absence of magnetic interactions. For example, in EPR oximetry experiments, the envelope is measured in the absence of oxygen or at a lower oxygen concentration while in spin-label pair distance measurements the envelope is the sum of double-integral-normalized EPR spectra from single-labeled protein mutants.

In 1995 Rabenstein and Shin suggested that, even when the static part of a dipolar interaction is not averaged out, under certain conditions the dipolar broadening effects on the EPR spectra can be approximated by a convolution integral (although more rigorous treatment shows that, generally, these effects are not uniform across the nitroxide spectra). In their experiments, Rabenstein and Shin (1995) measured EPR spectra of single- and double-labeled mutants at low temperatures (ca. 77 K) to eliminate dynamic averaging of dipolar interactions. Then these authors followed the original paper by Pake (1948) and treated polarization of dipoles as static, so only shifts in the resonance frequency are produced. The EPR spectrum in the presence of the dipolar interaction was approximated by a convolution of an unbroadened EPR spectrum (which we can consider as an envelope in Eq. (1)) with the Pake's dipolar pattern (Pake 1948). A somewhat similar approach to that of Rabenstein and Shin was independently developed by Steinhoff (Steinhoff et al. 1991, 1997). In their studies of frozen spin-label samples, Steinhoff and coworkers also treated the dipolar interaction as static and approximated the EPR spectrum from spin-label pairs as a convolution integral of an unbroadened spectrum and the dipolar broadening function; however, the broadening function was constructed digitally under the assumption of a Gaussian distribution for interspin distances.

The third example is inhomogeneous broadening due to local heterogeneity of biological samples. In some cases this heterogeneity can manifest in strain broadening — a distribution in hyperfine coupling constants and *g*-factors due to local variations in the electrostatic and hydrogen bonding environment of spin labels. Heterogeneity of biological membranes can also result in a distribution of line widths due to different mobilities of spin probes located in different membrane domains. Superposition of EPR spectra with different widths could be treated as an additional inhomogeneous broadening (Sankaraman and Marsch 1999). Characteristic parameters of such a spectrum such as, for example, an intensity ratio (ibid.), can be used to deduce the statistical distribution of spin labels.

Overall, the above examples demonstrate that in many cases EPR spin-label spectra can either be fully described by a convolution integral or have a spectral contribution given by Eq. (1). The purpose of this appendix is (i) to provide a practical guide to least-squares simulations of such spin-label spectra and (ii) to discuss the range of problems to which this convolution-based simulation approach can be applied. Smirnov and Belford (1995) have shown that this approach is flexible enough to accommodate a variety of experiments and is very accurate (e.g., Gaussian-Lorentzian convolution often being calculated with an accuracy better than 10^{-6}). The main advantage of the method is that it allows one to extract the amount of uniform broadening from the spectra directly and with fewer parameters than simulations of the full spectra (Smirnov et al. 1996; Smirnov and Smirnova 2004). The algorithm is combined with an efficient Levenberg-Marquart optimization algorithm, which converges rapidly and allows for estimates of parameter uncertainties (ibid.). A more detailed description is provided on the accompanying CD and illustrates applications of this technique, with selected examples ranging from studies of rotational diffusion to EPR oximetry in membranes and structural protein studies.

3. BRIEF DESCRIPTION OF EWVOIGT CAPABILITIES

The main application of EWVOIGT/EWVOIGTN is least-squares simulation of solution spectra in fast-motion regimes; however, it can also be successfully used for extracting oxygen and other homogeneous broadening from EPR spectra of nitroxides undergoing slow and restricted motion. The program is very advantageous in studies of microviscosity using the nitroxide spin label/probe method; it drastically improves the accuracy of EPR oximetry measurements (including low-frequency EPR oximetry with a significant phase distortion of the spectra) and spin-label bioreduction studies.

In brief, EWVOIGT and EWVOIGTN offer

- The most accurate way of computing Voigt functions using a precision Fast Convolution algorithm
- Speed: optimization of spectral shapes is completed in a matter of a few seconds
- Optimization of simulated parameters to fit experimental data is carried out by a modified Levenberg-Marquart algorithm that usually converges in 4–7 iterations
- Estimates of parameter uncertainties
- Sequential mode of processing of similar data: automatic fit to a sequence of up to 1000 data files

The current version of EWVOIGT is capable of

- Reading any sequential ASCII or EW (EPR data acquisition program supplied by Scientific Software Services) data files of up to 2048 data points
- Accounting for up to 3 different types of paramagnetic centers with a common envelope function

- Accounting for 2 types of nonequivalent nuclei with 1/2 spins (such as protons) and optimization of corresponding hyperfine structure
- Optimizing spectra in "first integral" or "first derivative" mode
- Optimizing a dispersion contribution to the experimental spectra
- Accounting for a "fixed" hyperfine pattern that can be entered as any function digitized in a sequential file
- Saving all fitting parameters, status of fitting parameters, borders of variable fitting intervals, and other program parameters under a separate file available for easy retrieval by the user
- Saving experimental spectrum, best-fit spectrum, and residual in a comma-separated table format for easy import into other graphical programs
- Saving fitting parameters after sequential run as a separate file in a comma-separated table format for easy import into other graphical programs

4. CONVOLUTION ALGORITHM WITH LEVENBERG-MARQUARDT OPTIMIZATION FOR FITTING INHOMOGENEOUS EPR SPECTRA

Briefly, the general function for fitting inhomogeneously broadened EPR lines is an extension of a convolution integral given by Eq. (1) with addition of a polynomial baseline term of order N (typically, $N = 1$ or 2 is sufficient):

$$I(B) = f(B) \otimes p_1(B) \otimes p_2(B) \otimes \ldots \otimes p_k(B) + \sum_{i=0}^{N} l_i B^i \; . \qquad (3)$$

EWVOIGT employs a polynomial with $N = 1$ (i.e., a linear baseline correction), while EWVOIGTN employs a polynomial with $N = 2$. The Lorentzian line shape is given by $f(B)$, and the $p_j(B)$'s are optional envelope functions. Typically, one of the envelope functions is a Gaussian, which is a very common model of inhomogeneous broadening. Other envelopes are assigned to isotropic hyperfine interactions with magnetic nuclei and may also include a non-adjustable envelope function, which is read as a file. The latter could be, for example, an experimental spin-label EPR spectrum in the absence of oxygen and/or some other broadening function due to isotropic dipolar interactions.

Continuous-wave EPR spectra are usually detected in the form of a first derivative, $I'(B)=\partial I(B)/\partial B$. Since

$$I'(B) = p'(B) \otimes f(B) = p(B) \otimes f'(B) \, , \qquad (4)$$

only one of the two functions in the right-hand side of Eq. (4) is differentiated in order to describe the experimental first-derivative spectrum. This property of the convolution integral is also very useful for efficient calculations of partial derivatives required for Levenberg-Marquardt optimization.

The model of inhomogeneous line shape we describe here is general and allows for different line shapes *f*(*B*). One practically important example is a dispersion shape or a mixture of dispersion and absorption. Dispersion and absorption signals are related by Kramers-Kronig equations. If the contribution from dispersion is measured as a phase shift $\Delta\varphi$ ($\Delta\varphi = 0$ corresponds to an absorption line), then the first-derivative Lorentzian EPR line is given by

$$f(B) = \frac{2A}{\pi} \frac{8\Delta B^L_{1/2}(B_0 - B)\cos\Delta\varphi + \left[4(B_0 - B)^2 - (B^L_{1/2})^2\right]\sin\Delta\varphi}{\left[(\Delta B^L_{1/2})^2 + 4(B_0 - B)^2\right]^2}, \tag{5}$$

where $\Delta B^L{}_{1/2}$ is the Lorentzian width measured at half height of an absorption signal, A is the area under the resonance absorption curve or (double-integral) intensity of the signal, B_0 is the resonance field. Then the inhomogeneous EPR spectrum for an arbitrary $\Delta\varphi$ is given by a convolution of $f(B,B_0,\Delta B^L{}_{1/2},\Delta\varphi)$ with an inhomogeneous envelope. Thus, only one function in the convolution integral has to be corrected for the dispersive component. Although the dispersion line shape can be computed for both Gaussian and Lorentzian functions, expression for the dispersive Lorentzian shape is much simpler while calculations of the dispersive Gaussian shape require evaluations of the error function integral.

Accounting for the dispersive contribution in least-squares fitting of spin-label EPR spectra is especially important because those spectra are often collected from aqueous samples, which have high dielectric constants. For these samples the EPR signal always has some dispersive component because the microwave phase varies along the sample. However, even in this case the observed EPR signal *I*(*B*) appears as a mixture of absorption, *A*(*B*), and dispersion, *D*(*B*), signals with an effective phase shift $\Delta\varphi_e$. Indeed, if $\Delta\varphi(\vec{r})$ is the microwave phase expressed as a function of spatial vector $\vec{r}$, then

$$\begin{aligned} I(B) &= \int_V \left[A(B)\cos(\Delta\varphi(\vec{r})) + D(B)\sin(\Delta\varphi(\vec{r}))\right] d\vec{r} = \\ &= A(B)\langle\cos(\Delta\varphi(\vec{r}))\rangle + D(B)\langle\sin(\Delta\varphi(\vec{r}))\rangle = \\ &= R\left[A(B)\cos(\Delta\varphi_e) + D(B)\sin(\Delta\varphi_e)\right], \end{aligned} \tag{6}$$

where

$$R^2 = \langle\cos(\Delta\varphi(\vec{r}))\rangle^2 + \langle\sin(\Delta\varphi(\vec{r}))\rangle^2, \tag{7}$$

$$\Delta\varphi_e = \tan^{-1}\frac{\langle\sin(\Delta\varphi(\vec{r}))\rangle}{\langle\cos(\Delta\varphi(\vec{r}))\rangle}.$$

Thus, the shift observed in such an experiment cannot be always experimentally compensated by adjusting the spectrometer reference arm. The phase shift problem becomes much more severe with high-frequency (95–250 GHz) EPR spectrometers and also can be a problem in certain low-frequency (0.5–2.0 GHz) EPR experiments when a single channel is used for signal detection. These phase shifts can significantly distort line shapes and lead to data misinterpretation if they are not accounted for. This can be overcome by using Eq. (5) in the simulations or by correcting experimental spectra by mixing in out-of-phase components calculated using Kramers-Kronig equations.

For simulations initial values of the parameters in the fitting function given by Eq. (3) are determined by a search for crude local extrema in the experimental spectrum $E(B)$. The initial microwave phase, $\Delta\varphi$, is usually set to 0 or π. Both values correspond to pure absorption spectra at different modulation phases. Baseline coefficients are set to zero. A multiparameter line shape function, $f(B,\alpha_i)$, is derived analytically and then, if desired, digitally convoluted with optional envelope functions $p_i(B)$ using the FFT convolution algorithm. Envelope functions are derived analytically or read as a file. Some of the envelopes, such as Gaussian and modulation and/or time-constant broadening, should be centered over the spectral window and are easily simulated in the frequency domain. The use of FFT requires that the simulated spectrum $I(B_i)$ has n data points, where n is a power of 2. If the experimental spectrum was collected with a different number of data points, the EWVOIGT program automatically generates additional data points via linear extrapolation to make the data file of appropriate length.

Simulated function $I(B_i, \vec{a})$ is compared with experimentally measured spectrum $E(B_i)$ by a square norm χ^2 test:

$$\chi^2(\vec{a}) = \sum_{i=1}^{N} \left[E(B_i) - I(B_i,\vec{a})\right]^2 , \tag{8}$$

where $\vec{a}$ is a parameter vector, and the standard deviations at each experimental data points are assumed to be the same. Parameter vector $\vec{a}$ is then adjusted to minimize the value of $\chi^2(\vec{a})$.

For minimization we utilized the Levenberg-Marquardt algorithm (e.g., Press et al. 1986; Ball et al. 1973). Many modifications of this algorithm require a user-supplied subroutine, which, for input value B_i, returns the value of the model function $I_i = I(B_i,a_j)$ and vector of derivatives $\partial I(B_i,a_j)/\partial a_j$ (e.g., see Press et al. 1986). In order to take the advantage of the FFT convolution algorithm, we modified this procedure as follows. Instead of calculating one value of I_i and vector $\partial I(B_i,a_j)/\partial a_j$ at each value of B_i at a time, we calculate these functions for the whole set of B_i; $I = 1, \ldots, n$. Then the calculation procedure is organized as follows:

(1) Functions $p_k(B_i)$, $f(B_i,a_j)$, and $\partial f(B_i,a_j)/\partial a_j$ are generated from analytically derived expressions for all values of B_i, $I = 1, \ldots, n$. Alternatively, a non-adjustable envelope function is read as a file.

(2) Convolution $p(B)$ of functions $p_1(B)$, ... $p_k(B)$ ($p(B) = p_1(B) \otimes p_2(B) \otimes \ldots \otimes p_k(B)$) is computed digitally by the FFT convolution algorithm, and all intermediate Fourier images of envelopes and their products are stored in computer memory.

(3) All functions generated in step (1) are convoluted with $p(B)$; $f(\omega,a)$ is stored. Since $p(\omega)$ was previously stored during step (2), each convolution requires only two Fourier transforms instead of three. Convolution of $p(B)$ with $\partial f(B_i,a_j)/\partial a$ gives a matrix of partial derivatives $\partial I(B_i,a_j)/\partial a_j$ for all values of B_i, since

$$\frac{\partial I(B)}{\partial a_j} = p_1(B) \otimes p_2(B) \otimes \ldots \otimes p_k(B) \otimes \frac{\partial f(B,a_j)}{\partial a_j}. \tag{9}$$

The vector of simulated function $I(B_i)$ is computed by adding a baseline term (see Eq. (3)).

(4) If envelopes contain adjustable parameters σ_j, vectors $\partial p_k(B_i, \sigma_j)/\partial\sigma_j$ are computed from analytical expressions and then are convoluted digitally with other envelopes $p_m(B)$ and $f(B, \vec{a})$. The number of required Fourier transforms can be decreased by using vectors $p_m(\omega_i)$ and $f(\omega_i, \vec{a})$ stored during steps (2) and (3).

(5) Partial derivatives with respect to the baseline parameters and with respect to the amplitude parameter are computed without additional convolutions. A complete matrix of partial derivatives $\partial I(B_i,a_j)/\partial a_j$ is formed and then supplied for Levenberg-Marquardt optimization. The iterations are stopped after the value of $\chi^2(\vec{a})$ is improved insignificantly (i.e., by 10^{-3} of the previous value) for a second time. The estimates of parameter uncertainties are calculated in a standard manner (Press et al. 1986).

The advantages of the described algorithm are such that it does not require digital differentiation of simulated functions with respect to the simulation parameters. Most functions required for calculations are derived in analytical forms. The final step involves digital convolution of the generated functions by FFT. Saving the intermediate results can be used to reduce the number of Fourier transforms. For convolutions with Gaussian and/or modulation and time-constant broadening functions, simulating these functions directly in the frequency domain from the analytical expressions, and then digitally transforming the product back to the field domain, can further decrease computational time. In EWVOIGTN fitting function (3) is extended by summation over nonequivalent line shape functions $f_j(B,a_i)$ and/or nonequivalent paramagnetic centers.

The basis of the computation algorithms has been initially described by Smirnov and Belford (1995). A comprehensive review of the use of convolution-based fitting in EPR has been also published (Smirnov and Smirnova 2004).

5. REFERENCES

Bales BL. 1989. Inhomogeneously broadened spin-label spectra. *Biol Magn Reson* **8**:77–130.

Ball WE, Kuester JL, Mize JH, eds. 1973. *Optimization techniques with fortran.* New York: McGraw-Hill.

Budil DE, Lee S, Saxena S, Freed JH. 1996. Nonlinear-least-squares analysis of slow-motion EPR spectra in one and two dimensions using a modified Levenberg-Marquardt algorithm. *J Magn Reson A* **120**:155–189.

Hyde JS, Swartz HM, Antholine WE. 1979. The spin-probe-spin-label method. In *Spin Labeling II: Theory and Applications*, pp. 71–114. Ed LJ Berliner. New York: Academic Press.

Mchaourab HS, Oh KJ, Fang CJ, Hubbell WL. 1997. Conformation of T4 lysozyme in solution: hinge-bending motion and the substrate-induced conformational transition studied by site-directed spin labeling. *Biochemistry* **36**:307–316.

Molin YuN, Salikhov KM, Zamaraev KI. 1980. *Spin exchange.* New York: Springer.

Pake GE. 1948. Nuclear resonance absorption in hydrated crystals: fine structure of the proton line. *J Chem Phys* **16**:327–336.

Portis AM. 1953. Electronic structure of F centers: saturation of the electron spin resonance. *Phys Rev* **91**:1071–1078.

Press WH, Teukolsky SA, Vetterling WT, Flannery BP. 1986. *Numerical recipes in fortran*, 2nd ed. Cambridge: Cambridge UP.

Rabenstein MD, Shin Y-K. 1995. Determination of the distance between two spin labels attached to a macromolecule. *Proc Natl Acad Sci USA* **92**:8239–8243.

Sankaraman MB, Marsch D. 1999. Analysis of spin label line shapes with novel inhomogeneous broadening from different component widths: application to spatially disconnected domains in membranes. *Biol Magn Reson* **14**:5–22.

Smirnov AI, Belford RL. 1995. Rapid quantitation from inhomogeneously broadened EPR spectra by a fast convolution algorithm. *J Magn Reson A* **98**:65–73.

Smirnov AI, Clarkson RB, Belford RL. 1996. EPR linewidth (T_2) method to measure oxygen permeability of phospholipid bilayers and its use to study the effects of low ethanol concentrations. *J Magn Reson B*. **111**:149–157

Smirnov AI, Smirnova TI. 2004. Convolution-based algorithm: from analysis of rotational dynamics to EPR oximetry and protein distance measurements. *Biol Magn Reson* **21**:277–348.

Steinhoff H-J, Dombrowsky O, Karim C, Schneiderhahn C. 1991. Two-dimensional diffusion of small molecules on protein surfaces: an EPR study of the restricted translational diffusion of protein-bound spin labels. *Eur Biophys J* **20**:293–303.

Steinhoff H-J, Radzwill N, Thevis W, Lenz V, Brandenburg D, Antson A, Dodson G, Wollmer A. 1997. Determination of interspin distances between spin labels attached to insulin: comparison of electron paramagnetic resonance data with the X-ray structure. *Biophys J* **73**:3287–3298.

APPENDIX 6

EasySpin: Simulating cw ESR Spectra

Stefan Stoll and Arthur Schweiger
Physical Chemistry Laboratory, ETH, Zürich, Switzerland

1. INTRODUCTION

Only the simplest of cw ESR spectra, those with few, sharp, well-resolved and sufficiently intense lines, can be analyzed in a straightforward manner. The magnetic parameters of the paramagnetic species can directly be obtained from line positions, amplitudes, and widths using simple analytical formulae. However, such formulae do not exist for all cases. Often, physical information is implicit and hidden in the details of the line shape. It is not always possible to assign lines, especially in spectra from molecules with many magnetic nuclei. When several interactions are of similar strength, line positions and intensities behave in a nontrivial manner. A straightforward analysis by read-out may also be impeded by high spectral noise levels and by very broad or strongly overlapping lines.

Thus, for many cw ESR spectra, a full-scale quantum-theoretical numerical simulation, possibly combined with least-squares fitting, is necessary to reproduce the spectrum and to recover the underlying structural and dynamic parameters. Spectral simulation is therefore an important ingredient of the successful application of ESR spectroscopy.

In this overview, we describe EasySpin, a software package that is able to simulate cw ESR spectra of most paramagnetic systems under a wide range of dynamic conditions. EasySpin is a collection of functions and programs intended to run on Matlab, a commercial software platform for technical and scientific computation and visualization, available for several operating systems (Windows, Linux, and others). EasySpin 2.5 is included on the CD accompanying the book, and that version is also the one referred to and used in this overview. The latest version is available from www.easyspin.ethz.ch.

This overview summarizes the core concepts of the ESR simulation algorithms implemented in EasySpin. Full details can be found in Stoll and Schweiger (2006) and the references cited therein, as well as in the EasySpin documentation included on the CD. Many of the examples in this overview are related to simple nitroxide radicals, as often applied in membrane biophysics, but the programs in EasySpin can handle complex paramagnetic molecules with several unpaired electrons and several magnetic nuclei.

The Matlab code for all simulated spectra shown in this overview is included in the text. The code snippets should be almost self-explanatory to readers who are familiar with the basics of Matlab.

2. FOUR DYNAMIC REGIMES IN CW ESR

All the static ESR spectroscopic properties of a paramagnetic molecule are determined by the strengths and geometries of the interactions between the various magnetic moments present. For a molecule with one unpaired electron, these include the g tensor of the electron, its hyperfine couplings to nearby magnetic nuclei, their magnetogyric ratios, and their nuclear quadrupole tensors. For systems with more than one unpaired electron, couplings between the electrons are present as well.

All static interactions are to some extent anisotropic. They consist of an isotropic, orientation-independent part and a part that depends on the orientation of the paramagnetic molecule with respect to the spectrometer with its fixed magnetic field direction. Thus, a cw ESR spectrum is in general orientation-dependent: the resonance lines shift if the molecule is rotated.

The appearance of the ESR spectrum does not only depend on static interactions in the paramagnetic molecule, but on dynamic processes on the timescale of the ESR experiment as well. This ESR timescale is determined by the spectral anisotropy, $\Delta\omega$, i.e., the maximum difference between resonance line positions when the orientation of the molecule is varied. Any dynamic process that is not extremely slow ("frozen") on this timescale will affect the shape of the ESR spectrum.

The most important and most visible dynamic process in solutions or membranes observable by ESR is the tumbling of molecules. This rotational motion modulates the resonance frequencies and gives rise to fluctuating local magnetic fields that induce transitions. The rotational diffusional motion is a random process, and its timescale is characterized by rotational correlation time τ_c, denoting the characteristic time after which molecules with initially identical orientations lose their alignment.

By comparing the speed of the rotational diffusion (τ_c^{-1}) to the spectral anisotropy ($\Delta\omega$), we can distinguish four dynamic regimes with distinct types of ESR spectra: (1) isotropic limit, (2) fast motion, (3) slow motion, and (4) rigid limit. They are illustrated in Figure 1, together with sample spectra of a nitroxide radical.

In the *isotropic limit*, paramagnetic molecules are tumbling extremely fast, so that any anisotropic property is completely averaged out. Only the isotropic parts of all interactions are observable, that is, the isotropic g factor and the isotropic hyperfine coupling constants. The spectrum consists of a series of symmetric lines with equal widths. This is the case, for example, for many organic radicals in low-viscosity solutions at room temperature.

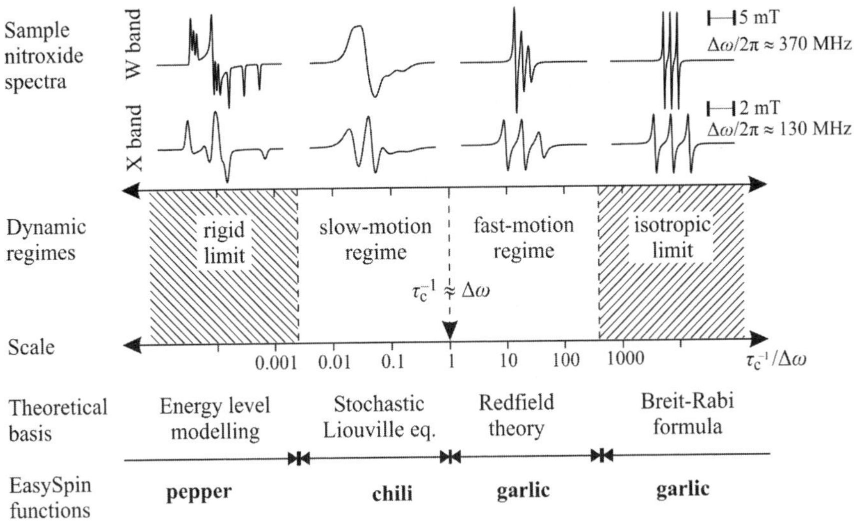

Figure 1. Schematic representation of the timescale of rotational dynamics accessible by cw ESR. τ_c: isotropic rotational correlation time, $\Delta\omega$: spectral anisotropy. The borders between the four regimes are only approximate. For each regime, representative X- and W-band cw ESR spectra of a nitroxide radical are shown.

At the other end of the range, in the *rigid limit*, the paramagnetic molecules are immobilized. There is neither translational nor rotational motion. The ESR spectra reveal the full anisotropy of all interactions. This is generally the case in powder materials, glasses, and frozen solutions at low temperatures.

Between these two extremes, there are dynamic regimes where the rotational motion is visibly affecting the spectrum but is too slow to completely average the anisotropic interactions. In the *fast-motion regime*, the rotational motion is fast enough so that the spectrum still appears similar to an isotropic-limit spectrum, although with line widths varying substantially from line to line.

In the *slow-motion regime*, the rotational diffusion is slower, so that the spectrum loses its similarity to the isotropic limiting case. The lines broaden and distort in an intuitively not easily understandable way. As the rotational motion slows down further, the spectrum successively approaches its rigid-limit shape.

The terms "slow" and "fast" do not refer to the absolute speed of reorientation, but are understood relative to the spectral broadening. Hence, a "fast" motion at the X band may become a "slow" motion at high spectrometer frequencies. Note that the borders between the four regimes in Figure 1 are not sharp. They also depend on the type of experiment: saturation-transfer ESR can detect movements in systems that appear completely "frozen" in a cw ESR experiment.

3. SIMULATION OF CW ESR SPECTRA

For the four dynamic regimes, different theoretical approaches and, consequently, different algorithms are used to simulate the resulting cw ESR spectra. All of them compute line positions, intensities, and widths before the spectrum is constructed. However, depending on the regime, various simplifications can be made. In the rigid and isotropic limits, rotational dynamics can be entirely ignored. In the fast-motion regime, the rotational diffusion can be treated as a small perturbation affecting only line widths. Only in the slow-motion regime is a full quantum-dynamical treatment necessary. In the following, we present each of the four regimes and their associated algorithms, including some illustrative simulations using EasySpin.

3.1. Isotropic limit

The simulation of cw ESR spectra of paramagnetic centers that tumble fast enough to completely average out all anisotropic interactions is simple. Spectral line intensities and widths are identical for all transitions. There are two methods for computing the line positions. Usually the hyperfine couplings are much smaller than the electron Zeeman splitting. Therefore, for a long time, the method of choice was based on explicit formulae for the energy levels obtained from perturbation theory:

$$E(m_S, m_I) \approx g_{iso}\mu_B m_S B_0 + E^{(1)} + E^{(2)} + E^{(3)} + \cdots,$$

$$E^{(1)} = a_{iso} m_S m_I, \qquad E^{(2)} = a_{iso}^2 \frac{I(I+1) - m_I^2}{2 g_{iso}\mu_B B_0},$$

where the first term in the sum is the electron Zeeman splitting, followed by perturbational correction terms of increasing order for the hyperfine splitting. To obtain the resonance line positions, the above expressions are inserted into resonance condition

$$E(+\tfrac{1}{2}, m_I) - E(-\tfrac{1}{2}, m_I) = h\nu_{mw},$$

which is then solved for B_0. For order n, this involves solution of a polynomial of degree n. For small hyperfine couplings, an evaluation to second order is sufficient to obtain accurate resonance fields. The convergence, though, can be very slow for large hyperfine couplings. To achieve convergence, high orders have to be incorporated.

There is a second, more general and more accurate way to obtain the resonance fields (Stoll et al. 2006). The isotropic spin Hamiltonian of a system with an unpaired electron and one nucleus with arbitrary spin can be diagonalized analytically, even for a large hyperfine coupling. The resulting energy levels as a function of the magnetic field are given by the Breit-Rabi formulae (Weil 1971). It is not possible to rearrange the resonance condition with these expressions to obtain an explicit solution for the resonance field, but numerically, a few fixed-point iterations based on

$$B_{k+1}(m_\mathrm{I}) = \frac{ha_\mathrm{iso}}{\gamma\left(1-\xi_k^2\right)}\left\{-m_\mathrm{I} \pm \sqrt{m_\mathrm{I}^2 + \left(1-\xi_k^2\right)\left[(2\xi_k)^{-2} - \left(I+\tfrac{1}{2}\right)^2\right]}\right\},$$

$$\xi_k = \frac{a_\mathrm{iso}/2}{\nu_\mathrm{mw} + \mu_\mathrm{n} g_\mathrm{n} B_k / h}, \qquad \gamma = \mu_\mathrm{B} g_\mathrm{iso} + \mu_\mathrm{n} g_\mathrm{n}$$

can solve the equation quickly. An excellent starting B_0 value is obtained by setting g_n to zero. Numerically, this method is very efficient: after two to four iterations, the value of B_k converges to the exact resonance field within numerical accuracy (i.e., 10^{-16} relative), even for hyperfine couplings almost as large as the electron Zeeman interaction.

In EasySpin, isotropic-limit spectra are computed by the function **garlic**, which implements this second approach. The following Matlab code simulates the isotropic spectrum of a nitroxide radical as shown in Figure 2a:

```
Nitroxide = struct('g', 2.0059, 'Nucs', '14N', 'A', 40, 'lw', [0.1, 0.15]);
XBand = struct('mwFreq', 9.5, 'CenterSweep', [338.4, 8]);
[B,spc] = garlic(Nitroxide,XBand);
```

The first line defines the nitroxide radical parameters. The *g* value, the nuclear isotope, the *A* value (in MHz), and the line width (in mT) are given. Here, the line width is specified to be a Lorentzian broadening with an FWHM (full width at half maximum) of 0.1 mT, convoluted with a Gaussian one with an FWHM of 0.15 mT. This way, the effect on the line shapes of unresolved hyperfine splittings, e.g., from protons, is easily accounted for. Experimental conditions are listed in the second line: the microwave frequency (in GHz) and the magnetic field range in terms of the center field and the sweep width (both in mT). The third line calls **garlic**, the EasySpin function that simulates the spectrum. The magnetic field axis and the spectrum are returned in B and spc, respectively.

garlic also works for more complicated radicals with several magnetic nuclei. The following code computes the solution spectrum of the anthracene radical anion. The resulting spectrum is shown in Figure 2b:

```
A = mt2mhz([-0.5337, -0.2740, -0.1509], gfree);
Anthracene = struct('g', gfree, 'Nucs', '1H,1H,1H', 'n', [2, 4, 4], 'A', A, 'lw', 0.02);
Params = struct('mwFreq', 9.6, 'nPoints', 16384);
[B,spc] = garlic(Anthracene,Params);
```

Often, hyperfine couplings of radicals are given in magnetic field units. In the first line, they are converted from mT to MHz using **mt2mhz**, a utility function of EasySpin. The second line defines the spin system. There are three types of protons. For each proton type, the hyperfine coupling and the equivalence number are given. The remaining lines are similar to those of the nitroxide example. Note that the field range is not specified: in this case, **garlic** determines it automatically.

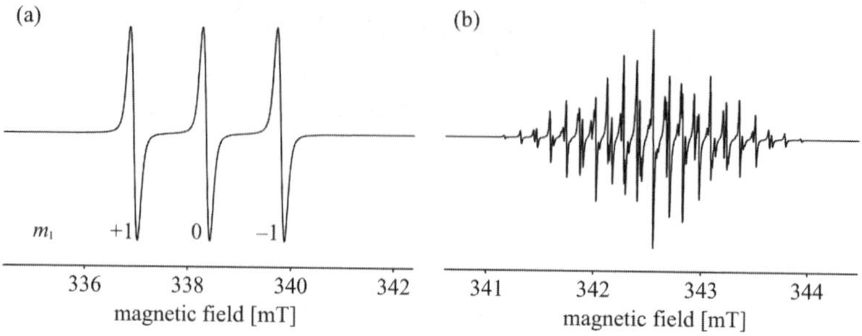

Figure 2. Isotropic cw ESR spectra simulated with EasySpin: (**a**) a generic nitroxide radical, (**b**) the anthracene radical anion. All simulation parameters are given in Matlab code in the text.

In such many-nuclei systems, **garlic** treats each group of equivalent nuclei separately and then combines the results in a first-order manner, neglecting any effect of one group of nuclei on the others. However, if one nucleus has a sufficiently large hyperfine coupling that it strongly influences the unpaired electron, it will also affect other nuclei via the electron. These indirect nuclear–nuclear cross-effects are neglected by **garlic**. In such cases, the rigid-limit algorithm described in §3.4 should be used. It is accurate for any number of nuclei and any magnitude of the hyperfine couplings, though much slower.

3.2. Fast-Motion Regime

If, starting from the very fast tumbling in the isotropic limit, the rotation of the paramagnetic molecules is slowed down either by cooling or by increasing the viscosity of the solvent, we enter the fast-motion regime. The spectral lines do not shift, and their shapes are still Lorentzian (possibly with a Gaussian contribution due to unresolved hyperfine couplings), but their widths increase relative to the isotropic limiting case. This broadening depends on the nuclear magnetic quantum numbers m_I and is thus different for each line.

In this fast-motion regime, the tumbling can be treated as a small perturbation, as long as the spectral anisotropy is small compared to the microwave frequency. The theory (Redfield theory) leads to practical expressions for the line widths. For the case of one nucleus with small hyperfine coupling, they are given by the quadratic formula (see, e.g., Atherton 1993)

$$\frac{1}{T_2} = A_0 + A + Bm_I + Cm_I^2 ,$$

where $1/(\pi T_2)$ is the FWHM of the Lorentzian line in the frequency domain. The terms with coefficients A, B, and C describe the broadening contribution from the rotational motion, whereas the first term, A_0, collects all other broadening effects.

With accurately known g and A tensors, the isotropic rotational correlation time τ_c can be obtained from A, B, and C according to

$$A = \frac{\mu_e^2}{h^2} B_0^2 (\Delta \mathbf{g}{:}\Delta \mathbf{g})\left[\tfrac{2}{15} j_0 + \tfrac{1}{10} j_1\right] + I(I+1)(\Delta \mathbf{A}{:}\Delta \mathbf{A})\left[\tfrac{2}{15} j_0 + \tfrac{7}{60} j_1\right],$$

$$B = \frac{16\pi^2}{15}\frac{\mu_e}{h} B_0 (\Delta \mathbf{g}{:}\Delta \mathbf{A})\left[\tfrac{5}{15} j_0 + \tfrac{1}{5} j_1\right],$$

$$C = (\Delta \mathbf{A}{:}\Delta \mathbf{A})\left[\tfrac{1}{12} j_0 - \tfrac{1}{60} j_1\right],$$

with spectral densities $j_0 = \tau_c$ and $j_1 = \tau_c/(1+\omega_0^2\tau_c^2)$. $\Delta \mathbf{g}$ and $\Delta \mathbf{A}$ are the anisotropic parts of the g and A tensors, i.e., $\Delta \mathbf{g} = \mathbf{g} - g_{iso}\mathbf{1}$ and $\Delta \mathbf{A} = \mathbf{A} - a_{iso}\mathbf{1}$. ($\Delta \mathbf{A}$ is assumed to be given in frequency units). The symbol $\mathbf{X}{:}\mathbf{Y}$ indicates element-by-element multiplication of the matrices followed by summation, giving a number characterizing the corresponding anisotropy.

From the formulae we see that the line widths are affected by both anisotropic hyperfine coupling $\Delta \mathbf{A}$ and g shift $\Delta \mathbf{g}$. Coefficients A and B are a function of magnetic field B_0, and, as a consequence, a larger magnetic field leads to larger values for A and B. The broadenings for a given molecule are thus different at different spectrometer frequencies. This is illustrated in Figure 3.

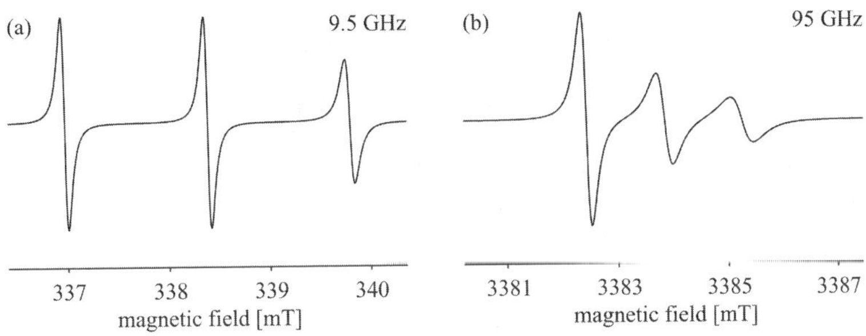

Figure 3. Fast-motion examples: (**a**) a generic nitroxide radical at 9.5 GH, (**b**) the same nitroxide radical at 95 GHz. All simulation parameters are given in Matlab code in the text.

EasySpin's simulation algorithm for the fast-motion regime is identical to the one used for isotropic spectra, with the only difference that each spectral line is constructed with a different line width according to the expressions given above. Only the isotropic rotational correlation time needs to be supplied to the function **garlic**, which then computes the line widths and builds the spectrum. In the following is the code for a nitroxide radical spectrum, with a rotational correlation time of 1 ns:

```
A = mt2mhz([5.8, 5.8, 30.8]/10);
Sys = struct('g', [2.0088, 2.0061, 2.0027], 'Nucs', '14N', 'A', A, 'lw', 0.1);
Sys.tcorr = 1.63-10;
Exp = struct('mwFreq', 9.5);
[B, spc] = garlic(Sys, Exp);
```

The resulting spectrum is shown in Figure 3a.

Note that here, in contrast to the isotropic limit, **garlic** has to be given the full anisotropic g and A tensors in terms of their principal values, since they are needed to compute the line width parameters.

If we go from the X to W band by changing the spectrometer frequency in the above code to 95 GHz, but leaving the correlation time untouched, the line widths change significantly (see Figure 3b), as is expected due to their dependence on the magnetic field: The spectral anisotropy increases, the ESR timescale shrinks, and the same rotational motion is now "slower" relative to it.

Naturally, this method is not limited to nitroxide radicals. e.g., the fast-motion spectra of vanadyl or copper complexes can be computed by setting the corresponding isotope and changing the numerical values of g and A in the above code.

For paramagnetic molecules in which the unpaired electron interacts with more than one nucleus, the broadenings are not a simple sum over all nuclei of the quadratic line width formula. Cross-terms between the hyperfine anisotropies of different nuclei have to be added (Atherton 1993). **garlic** can handle this many-nuclei case. A simulation example, for the radical anion of p-fluoronitrobenzene, is shown in Figure 4. The code used contains full tensor information for all nuclei:

```
A_N = 40.40 + [24, -12, -12]; A_F = 22.51 + [34.9, -19.8, -15];
A_oH = [1, 1, 1]*9.69; A_mH = [1, 1, 1]*3.16;
Sys = struct('g', [2.0032, 2.0012, 2.0097], 'lw', 0.01);
Sys.Nucs = '14N,19F,1H,1H,1H,1H';
Sys.A = [A_N; A_F; A_oH; A_oH; A_mH; A_mH];
Sys.tcorr = 8e-11;
Exp = struct('mwFreq', 9.5, 'nPoints', 1e4);
[B, spec] = garlic(Sys, Exp);
```

3.3. Slow-Motion Regime

When the rotational motion of the paramagnetic molecules slows down further, we leave the fast-motion regime and enter the slow-motion regime. The symmetric Lorentzian lines broaden substantially, become asymmetric, distort, and gradually converge to rigid-limit powder line shapes, revealing the full anisotropy of the underlying magnetic interactions (see the sample spectra in Figure 1).

The theory necessary to compute cw ESR spectra in this slow-motion regime is rather complicated. A full exposition of all aspects can be found in (Gamliel and Levanon 1995). The theory was developed and implemented by Freed and coworkers in a well-known program (Schneider and Freed 1989). We will shortly summarize the main aspects of the theory, as EasySpin's slow-motion simulation algorithm is based on it.

The central equation is the stochastic Liouville equation (SLE)

$$\frac{\partial}{\partial t}\rho(\boldsymbol{\Omega},t) = -\mathrm{i}[H(\boldsymbol{\Omega}(t)),\rho(\boldsymbol{\Omega},t)] - \Gamma_{\Omega}\left[\rho(\boldsymbol{\Omega},t) - \rho_0(\boldsymbol{\Omega})\right],$$

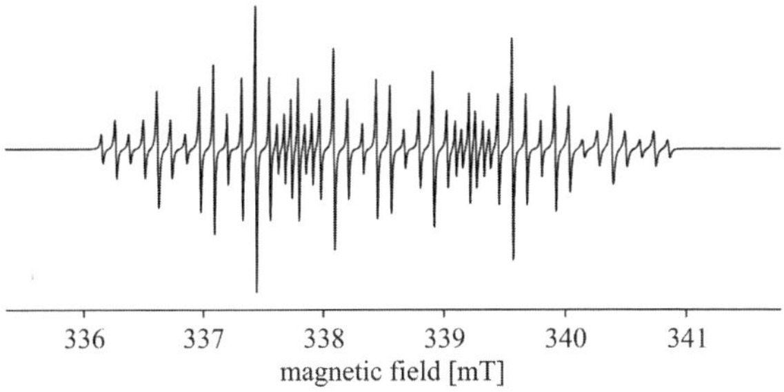

Figure 4. Fast-motion spectrum of the para-fluoronitrobenzene radical anion. All simulation parameters are given in Matlab code in the text.

which describes the time dependence of density matrix ρ in terms of spin Hamiltonian H and diffusion superoperator Γ_Ω. ρ_0 is the density matrix at thermal equilibrium, and all quantities depend on orientation $\boldsymbol{\Omega}$.

The SLE can be recast into a simpler form:

$$\frac{\partial}{\partial t}\chi(\boldsymbol{\Omega},t) = -\left[\mathrm{i}H^\times(\boldsymbol{\Omega}(t)) + \Gamma_\Omega\right]\chi(\boldsymbol{\Omega},t) = -L(\boldsymbol{\Omega},t)\chi(\boldsymbol{\Omega},t)\,,$$

with out-of-equilibrium density $\chi = \rho - \rho_0$, Hamiltonian commutation superoperator $H^\times$, and stochastic Liouville superoperator L, a shorthand notation for the term in square brackets. The absorption signal measurable in a cw ESR experiment is given by

$$I(\Delta\omega) \propto \mathrm{Re}(\boldsymbol{v}^\dagger \boldsymbol{u}) \quad \text{with} \quad \left(L + \mathrm{i}\Delta\omega\right)\boldsymbol{u} = \boldsymbol{v}\,,$$

where $\boldsymbol{v}$ is the supervector describing the x magnetization, and $\Delta\omega$ is the frequency offset from the center of the spectrum. In numerical simulations, superoperators are represented by matrices, and supervectors by column vectors.

The diffusion matrix Γ_Ω depends on the exact model for the rotational motion. The most common is rotational Brownian diffusion, analogous to translational Brownian motion. Other models include jump diffusion (the molecule is at rest for a certain time and the jumps instantaneously to another orientation) and free diffusion (the molecule is at rest for a certain time and the rotates directly and unimpeded to another orientation).

In contrast to the isotropic limit and the fast-motion regime, we are dealing with orientation-dependent quantities $H^\times$, Γ_Ω, and $\boldsymbol{v}$. How can this orientation dependence be represented? The orientations are best described by three Euler angles, and the orientational distribution could be modeled using discrete orientations as basis functions. It is better, though, to use Wigner rotation functions $D^L{}_{KM}(\boldsymbol{\Omega})$, where $L \geq 0$ and $-L \leq K$, $M \leq L$. These functions are perfectly suited for the description, since they are the eigenfunctions of the rotational diffusion operator.

Each of these functions describes a continuous orientational distribution; by linear combinations any distribution can be modeled. Rotation functions for some values of L, K, and M are illustrated in Figure 5. (EasySpin's function **spherharm** was used to compute the plots.) The higher the parameters L, K, and M, the stronger the functions oscillate.

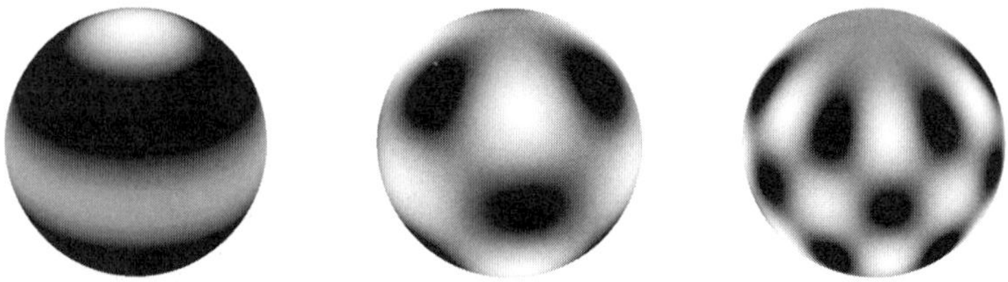

$D^4_{0,0}(\boldsymbol{\Omega})$ $D^5_{3,0}(\boldsymbol{\Omega})$ $D^8_{6,0}(\boldsymbol{\Omega})$

Figure 5. Illustration of rotational basis functions. Black (white) corresponds to negative (positive) values. The depicted functions have $M = 0$ und thus depend only on two of the three Euler angles describing the orientation of the paramagnetic molecule.

In general, a given orientational distribution is represented by a linear combination of an infinite number of such rotational basis functions. This is of course impractical, and the basis size is usually truncated to a finite number N of functions by specifying maximum values for L, K, and M. These values are chosen so that the error introduced is acceptable. The degree of truncation has to be determined by trial and error and depends on the speed of rotational diffusion and on the diffusion model used. If too few orientational basis functions are used, the spectrum will exhibit oscillating noise, just as in the case of the rigid-limit spectra described in §3.4.

For each possible rotation function, all ESR transitions involved in the dynamic process are included in the computation. For a nitroxide radical, this includes three allowed transitions as well as several forbidden ones, giving a total of, say, P. The total number of basis functions for representation of the Hamiltonian and the diffusion operators in matrix form is thus PN. The larger P and N, the larger the matrices.

The overall simulation procedure for slow-motional ESR spectra is schematically shown in Figure 6. It consists of three steps. First, matrices $H^{\times}$ and Γ and vector $\boldsymbol{v}$ are constructed in the basis described above. Then matrix $L = iH^{\times} + \Gamma$ is reduced to tridiagonal form, using appropriate transformation algorithms. In these, vector $\boldsymbol{v}$ is used as a "starting vector." In the last step, each point in the spectrum is computed by evaluating a continued fraction expression containing the nonzero elements of the tridiagonal matrix.

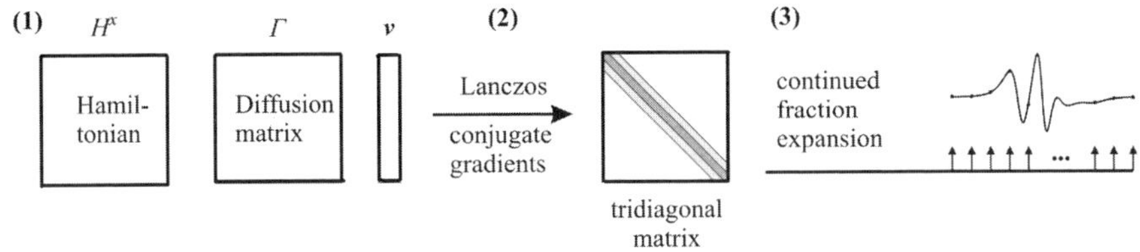

Figure 6. Illustration of the algorithm for solving the SLE: (1) H^x, Γ, and v are constructed in an appropriate basis, (2) matrix L is tri-diagonalised, (3) the tridiagonal matrix is used to compute the spectrum.

In EasySpin, the above algorithm for solving the SLE is implemented in the function **chili**. A simple simulation of a slow-motional ESR spectrum of a nitroxide radical, assuming Brownian rotational diffusion with an isotropic diffusion tensor, is shown in Figure 7a. The corresponding Matlab code is

```
Sys = struct('g', [2.008, 2.0061, 2.0027], 'Nucs', '14N', 'A', [16, 16, 86]);
Exp = struct('mwFreq',9.8);
Dynamics = struct('lw', 0.01, 'tcorr', 5.1e-9);
[x, spc] = chili(Sys, Exp, Dynamics);
```

The input is similar to that for the fast-motion regime. In the third line, all dynamic parameters are collected into a new structure, given as the third input argument to the function. In the above example, only the isotropic rotational correlation time and a residual line width are given. **chili** also supports an axial rotational diffusion tensor.

To rerun the simulation with another value for the rotational correlation time, two lines are sufficient:

```
Dynamics.tcorr = 15.9e-9;
[x, spc] = chili(Sys, Exp, Dynamics);
```

The spectrum depicted in Figure 7b is obtained.

If the hyperfine splittings from two or more nuclei are visible in the ESR spectrum, the simulation in the slow-motion regime gets more difficult, since the number of transitions, the dimensions of the matrices, and, consequently, the computation times increase tremendously — e.g., including another ^{14}N nucleus increases the matrix dimension by a factor of nine, giving 81 times more matrix elements.

Often, however, the spectral anisotropy is governed almost entirely by the anisotropy of the g tensor and of the hyperfine coupling of one nucleus only, and the hyperfine anisotropies of the other nuclei can be neglected. Then an approximative two-step technique can be used: the slow-motion spectrum is computed for the system with the first nucleus only. The hyperfine splitting pattern due to the other nuclei is accounted for by convoluting the slow-motion spectrum with a simple isotropic-limit stick spectrum due to all others. This post-convolution technique (Della Lunga et al. 1994) is implemented in **chili**. To simulate the slow-motional spectra of a planar Cu^{2+} complex, the following code is used. The main difference to the nitroxide example of Figure 7 is that now five nuclei (one ^{63}Cu and four ^{14}N) are included:

```
CuPc = struct('g', [2.0525, 2.0525, 2.1994], 'Nucs', '63Cu,14N', 'n', [1, 4]);
CuPc.A = [-54, -54, -608; 52.4, 41.2, 41.8];
Exp = struct('mwFreq', 9.878, 'Range', [260, 380]);
Dynamics.tcorr = 10^-8.15;
Dynamics.lw = 3;
Opt.LLKM = [14, 0, 0, 4];
[x, spc] = chili(CuPc, Exp, Dynamics, Opt);
```

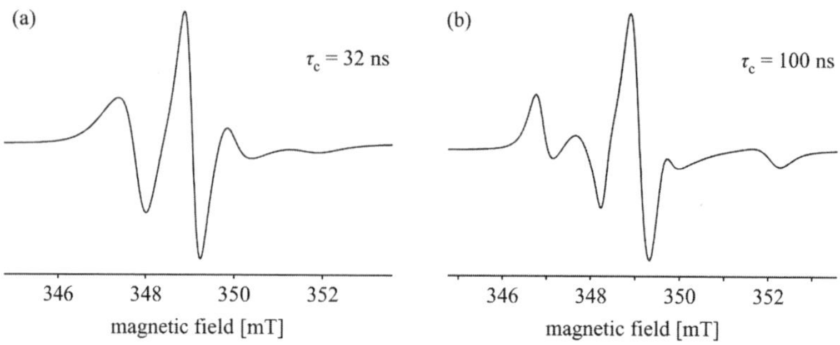

Figure 7. Two slow-motion examples of a generic nitroxide radical with isotropic Brownian rotational diffusion. All simulation parameters are given in Matlab code in the text.

The resulting spectrum is shown in Figure 8. The sixth line contains an explicit specification of the orientational basis size by giving maximum values for odd L, and even L, K, and M. Usually, however, **chili** chooses a reasonable basis size internally.

3.4. Rigid Limit

In the rigid limit, the paramagnetic molecules in the sample are fixed, and their orientations do not change with time. In crystals, only one or at most a few orientations are present, whereas in a powder, a glass, or a frozen solution all orientations occur with equal probability. As for the isotropic-limit and the fast-motion regime, the simulation of rigid-limit spectra entails two steps: the first positions, intensities, and widths of all resonance lines are computed for each orientation. In the second step, the total spectrum is constructed from this line information by summation or an equivalent procedure.

The following contains only an outline of the basics of EasySpin's computation of rigid-limit cw ESR line positions, intensities, and widths and the efficient procedure used to obtain the spectrum from these. More details can be found in Stoll et al. (2006) and the references cited therein.

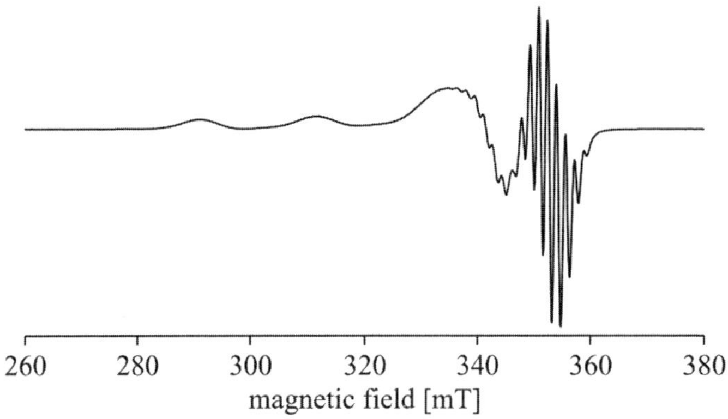

Figure 8. Slow-motion spectrum of Cu^{2+} phthalocyanine. All simulation parameters are given in Matlab code in the text.

To compute the resonance field positions for a given orientation of the paramagnetic molecule, various methods are commonly employed. For spin systems with hyperfine couplings and zero-field splittings small compared to the Zeeman splitting, explicit analytical formulae based on second-order perturbation theory can be used. EasySpin uses another much more general method (Stoll et al. 2003), which adaptively models the energy level diagram, from which the positions of resonance lines are then obtained.

The adaptive modeling of the energy levels as a function of the magnetic field works as follows (see Figure 9). First the energy levels at the left and right ends of the requested field range are computed by diagonalizing the associated spin Hamiltonians, giving the left and right ends of the diagram. Next, the levels at the center of the range are computed, yielding another set of energies. These are then compared to the approximate values obtained by cubic spline interpolation between the values at both ends of the range. If the interpolated and the exact energies are sufficiently close, the procedure terminates. If not, exact energies at the centers of the left and right half of the range are computed and then compared to values obtained by interpolation from the left and right ends of these segments. The algorithm works iteratively: any segment where the center modeling error is not small enough is subdivided further. The subdivisions continue until all segments are accurately modeled, resulting in a faithful cubic spline representation of the energy level diagram. Resonance positions for a given microwave frequency are readily obtained by searching though all the segments and solving the associated cubic equation if a resonance is possible.

This iterative segmentation method automatically adapts the number of field positions where energy levels are computed to the complexity of the energy level diagram. For simple systems only three diagonalizations are necessary, whereas in complex situations more than 50 can occur. Figure 9 illustrates this.

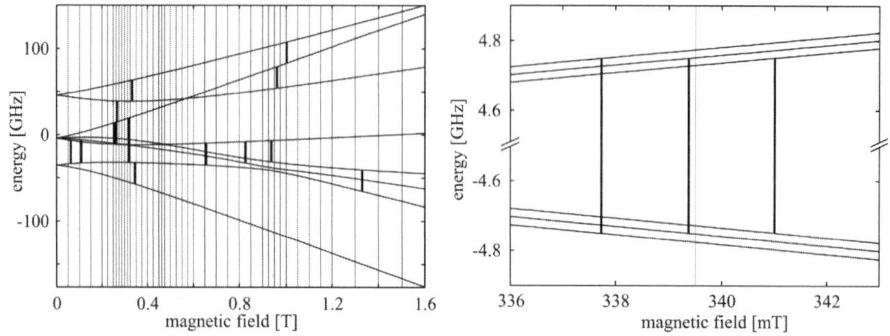

Figure 9. Energy level diagrams, resonances, and diagonalization positions of the iterative segmentation approach. (**a**) Tb^{4+} in ThO_2 (S = 7/2), 25 GHz (**b**) a generic nitroxide radical, 9.5 GHz

After the line positions, line intensities are computed. For a given transition between two energy levels, they are products of three factors: the ESR transition probability, the polarization term, and a frequency/field conversion factor.

The ESR transition probability is the modulus-square of the transition matrix element connecting the two energy levels. For nitroxide radicals, this probability is identical for all three allowed ESR transitions. However, for more complicated systems such as the one shown in Figure 9a, it can vary substantially from line to line.

The polarization term accounts for the fact that the line intensity is proportional to the population difference of the two levels involved in the transition. The populations can be in thermal (Boltzmann) equilibrium, or they can deviate from it due to a chemical or physical preparation process.

A frequency/field conversion factor is included since cw ESR spectra are not acquired as a function of the microwave frequency at constant magnetic field, but rather as a function of the magnetic field given constant microwave frequency. To obtain correct spectra, line intensities and line widths have to be multiplied by the difference of the inverse slopes of the energy levels involved. For simple systems, this factor is $1/g$.

Next come the line widths. Often, they are anisotropic due to the presence of many small and unresolved hyperfine couplings. Their orientation dependence is modeled by using a phenomenological anisotropic line width "tensor."

Another origin of anisotropic line widths are random static variations of the molecular structure around the unpaired electron, resulting in small distributions of the principal values of the interaction tensors (so-called g and A strain). In EasySpin simulations, these distributions are assumed to be Gaussian, and the dependence of the resonance line position on them is approximated as linear.

Both types of broadening give orientation-dependent line widths in frequency units. To convert them to magnetic field units, they have to be multiplied by the frequency/field conversion factor, as already discussed for line intensities.

In the next step, the ESR spectrum is constructed from the positions, intensities, and widths of all transitions. For single crystals, this is done by simply summing over all lines using Gaussian or Lorentzian line shapes for each of them. For disordered systems (powders, glasses, frozen solutions), these spectra have to be summed over all contributing orientations. The more orientations are included in this summation, the closer the total spectrum approximates the real spectrum, where more than 10^{11} molecules are usually contributing. If not enough orientations are included, the spectrum shows oscillating artefacts, so-called simulation noise.

First, a relatively small set of almost uniformly distributed orientations is set up by a very simple procedure. This is shown in Figure 10. Each orientation in this set is taken as a representative of all orientations that lie in its immediate neighborhood, which is called its Voronoi cell (see Figure 10). The area of the Voronoi cell (or a good estimate of it) is then used as a weight for the orientation's contribution to the powder spectrum.

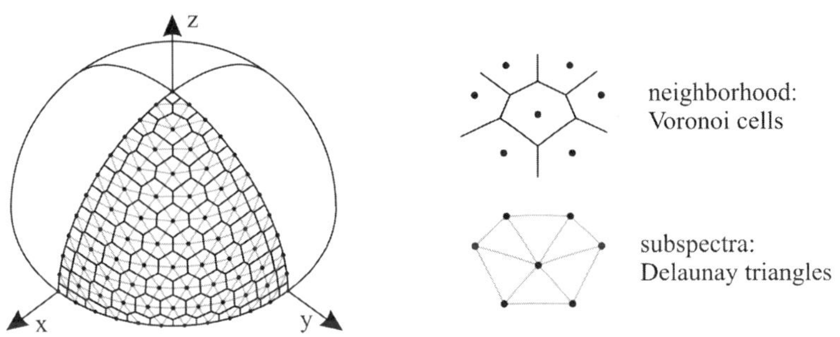

Figure 10. A simple spherical grid over one octant. Orientations are shown as dots; Voronoi cells and Delaunay triangles are indicated as well.

This weighted-sum-of-orientations approach is far from optimal: for small intrinsic line widths and large spectral anisotropies, thousands of orientations have to be evaluated and added. There exists a more efficient method, used in EasySpin, that automatically takes advantage of the anisotropy of the spectrum. Three neighboring orientations form the vertices of a small spherical triangular region, called a Delaunay triangle (see Figure 10). If the triangle is small enough, the total spectrum due to all orientations lying in it has a simple triangular shape, as shown in Figure 11. By constructing these subspectra for all Delaunay triangles and summing them up, the total powder spectrum is obtained (Stoll et al. 2006). This way, between 50 and 500 orientations per octant are sufficient for most simulations.

In EasySpin, the rigid-limit cw ESR spectra are computed by the function **pepper**. Its usage is similar to the one of **garlic** in the fast-motion regime. A rigid-limit powder ESR spectrum of a generic nitroxide radical is shown in Figure 12a. The necessary code is

```
Sys = struct('g', [2.008, 2.0061, 2.0027], 'Nucs', '14N', 'A', [16, 16, 86]);
Sys.lw = 0.1;
Exp = struct('mwFreq', 9.5, 'Range', [334 344]);
[B, spc] = pepper(Sys, Exp);
```

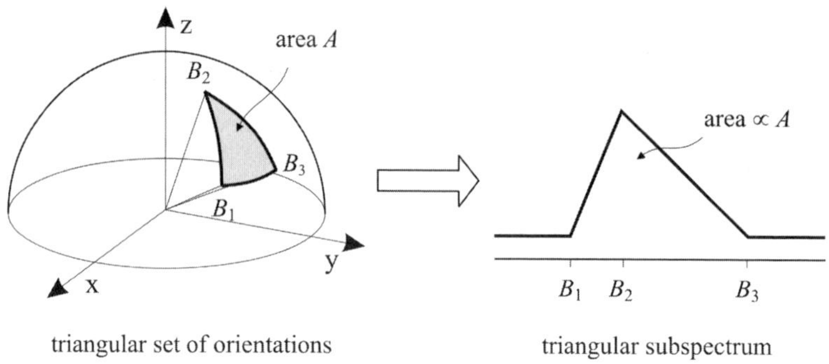

Figure 11. The set of orientations from a small spherical Delaunay triangle (left) give a powder subspectrum in the form of a triangle (right). B_1, B_2, and B_3 indicate resonance fields for the orientations corresponding to vertices of the spherical triangle.

It is not necessary to specify the number of orientations, which is a mandatory user-supplied parameter in many other powder cw ESR simulation programs. For most common systems, **pepper** is able to automatically choose this parameter to yield a smooth spectrum.

Instead of the entire powder spectrum, **pepper** can return one powder subspectrum for each ESR transition. This is demonstrated in Figure 12b. The following lines generate the group of spectra:

```
Sys = struct('g', [2.008, 2.0061, 2.0027], 'Nucs', '14N', 'A', [16, 16, 86]);
Sys.lw = 0.1;
Exp = struct('mwFreq', 9.5, 'Range', [334 344], 'Harmonic', 0);
Opt = struct('Output', 'separate');
[B, spcs] = pepper(Sys, Exp, Opt);
```

In the fourth line, it is specified that **pepper** should return each transition in a separate spectrum. For demonstration, the harmonic is set to zero in the third line.

An example of a transition metal complex is shown in Figure 13a. Instead of a single well-defined hyperfine coupling, the simulation includes a distribution of A values (A strain), as specified via AStr and AStrain in the second and third lines. In addition, an anisotropic line width to model unresolved hyperfine splittings is specified by res and HStrain in the first and third line:

```
g = [2.28, 2.19, 2.01]; res =[1 1 1]*60;
A = [70, 30, 309]; AStr = [45, 25, 40];
Sys = struct('S', 1/2, 'Nucs', '59Co', 'g', g, 'A', A, 'AStrain', AStr, 'HStrain', res);
Exp = struct('Range', [261, 411], 'mwFreq', 9.475);
Opt = struct('Output', 'separate');
[B, specs, Transitions] = pepper(Sys, Exp, Opt);
```

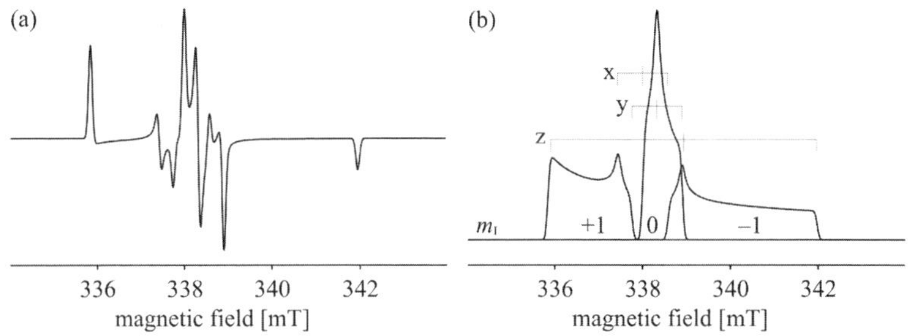

Figure 12. Powder cw ESR spectra of a generic nitroxide radical: (**a**) first-derivative spectrum, (**b**) absorption spectrum, separate transitions. All simulation parameters are given in Matlab code in the text.

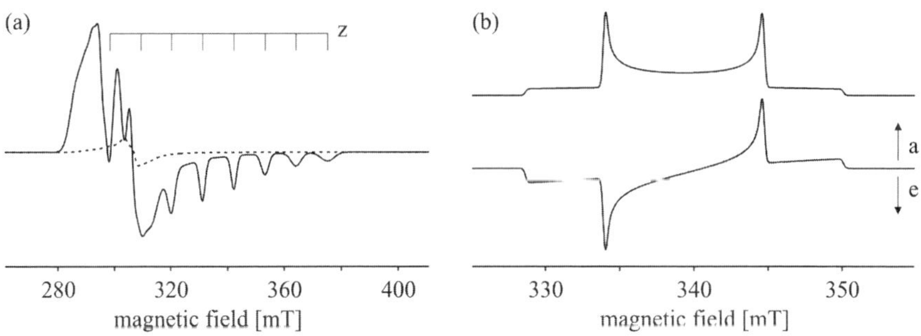

Figure 13. Powder cw ESR spectra simulated with **pepper**. (**a**) Co^{2+} complex including distributions of *A* values. The dashed spectrum represents the contribution of the forbidden ESR transitions. (**b**) A triplet state in thermal equilibrium (top) and with nonequilibrium populations of the three levels (bottom). All simulation parameters are given in Matlab code in the text.

Again, the simulation function was asked to return the simulated spectrum transition by transition. In the figure, the contributions from the forbidden transitions are shown separately. As a consequence of the *A* strain, the hyperfine lines in the high field region exhibit m_I-dependent broadenings. As mentioned above, in contrast to the fast-motion regime, these asymmetric broadenings are due to static distributions, and not to dynamic processes.

pepper can handle molecules with several unpaired electrons and several magnetic nuclei. High-order interaction terms for the electron spins are implemented. The simulation of spectra in oriented phases such as liquid crystals and membranes is supported by including appropriate weight factors during the summation of subspectra. Nonequilibrium populations are also possible: a sample simulation of the cw ESR spectrum of a molecule in an exited triplet state is

```
Sys = struct('S', 1, 'g', [2, 2, 2], 'lw', 0.3, 'D', [-1, -1, 2]*100);
Exp = struct('mwFreq', 9.5, 'Harmonic', 0, 'CenterSweep', [340, 30]);
Exp.Temperature = 298;
[B, spc1] = pepper(Sys, Exp);
Exp.Temperature = [0.85, 1, 0.95];
[B, spc2] = pepper(Sys, Exp);
```

In the third line, a temperature in Kelvin is given. In the fifth line, the same field is used to specify a nonequilibrium population vector for the zero-field states. The resulting two spectra are shown in Figure 13b.

4. OTHER EASYSPIN FUNCTIONS

Aside from the modules for simulating cw ESR spectra, EasySpin contains functions for a variety of ESR-related tasks. A list of all functions is included at the end of this overview. Here, we shortly focus on a few of them.

4.1. ENDOR Spectra

salt is an EasySpin function for simulating rigid-limit ENDOR spectra. Both disordered systems and single crystals are supported. The algorithm is very similar to that used for rigid-limit ESR spectra: The powder averaging is identical to the one in **pepper**, only the computation of line positions, intensities, and widths is different.

The ENDOR line positions of a nucleus in a paramagnetic molecule depend on its Larmor frequency and on its hyperfine coupling to the unpaired electron. Via the electron, it can feel the presence of other, strongly coupled nuclei, resulting in second-order line shifts. In addition, the nuclear quadrupole interaction can strongly affect the line positions. All these effects are accurately treated by **salt**, as it relies on a numerical computation of the energy levels by diagonalizing the spin Hamiltonian. Since the external magnetic field is kept constant in ENDOR experiments, the energy levels have to be computed only once for each orientation of the paramagnetic molecule. This makes the computation of powder ENDOR spectra much faster compared to the cw ESR case described above.

For each ENDOR transition, the line intensity is the product of three terms: (1) the ENDOR transition rate, which includes the hyperfine enhancement, (2) a factor giving the polarization of the transition, and (3) a selectivity factor taking into account that the only ENDOR transitions observed are those where one level is also involved in the ESR transition that is saturated. More details about the formulae implemented in **salt** can be found in Stoll et al. (2006).

4.2. Pulse ESR Spectra

Although EasySpin currently does not include a black-box function for the simulation of pulse ESR spectra, it provides various tools to support the necessary time-dependent simulations based on the density matrix formalism. The most important function is **evolve**, which evolves a given density matrix under a given Hamiltonian over a certain time. Both time-dependent and time-independent Hamiltonians are possible. **evolve** supports various pulse sequences.

4.3. Other Tools

levels and **levelsplot** are functions for computing and plotting energy level diagrams. The plot in Figure 9a was generated by **levelsplot**:

```
TbSys = struct('S', 7/2, 'g', [1, 1, 1]*2.0136);
b = -2527.53/60; c = -24.84/1260;
TbSys.B40 = b; TbSys.B44 = [5*b, 0];
TbSys.B60 = c; TbSys.B64 = [-21*c, 0];
levelsplot(TbSys, [1; 1]*pi/50, [0, 1600], 25);
```

Gaussian and Lorentzian line shapes, of almost daily necessity in ESR, are provided by **gaussian** and **lorentzian**. EasySpin also includes functions for Voigtian (convolution of Gaussian and Lorentzian, **voigtian**) and pseudo-Voigtian (linear combination of Gaussian and Lorentzian, **lshape**) line shapes.

Another convenient function, **pseumod**, emulates the effect of field modulation on a cw ESR spectrum (Hyde et al. 1990). The following Matlab lines generate the spectra shown in Figure 14, which demonstrates the influence of the modulation amplitude on the spectral line shape of a nitroxide radical:

```
Nit = struct('g', 2, 'Nucs', '14N', 'A', 40, 'lw', 0.1);
Exp = struct('mwFreq', 9.5, 'Range', [336, 343], 'Harmonic', 0);
[B, spc0] = garlic(Nit, Exp);
spc1 = pseumod(B, spc0, 0.05);
spc2 = pseumod(B, spc0, 0.4);
```

For data analysis, Savitzky-Golay smoothing filters (**smooth**) as well as polynomial and exponential fitting routines (**basecorr** and **exponfit**) are available. **rcfilt** is a digital equivalent to the common RC filter employed in cw ESR spectrometers. A set of functions (**erot**, **eulang**, **vec2ang**, **ang2vec**, **rotaxi2mat**, **rotmat2axi**) supports interconversion between rotation matrices, rotation axes and Euler angles. **sop** can be used to generate matrix representations of cartesian and shift spin operators for arbitrary spin systems.

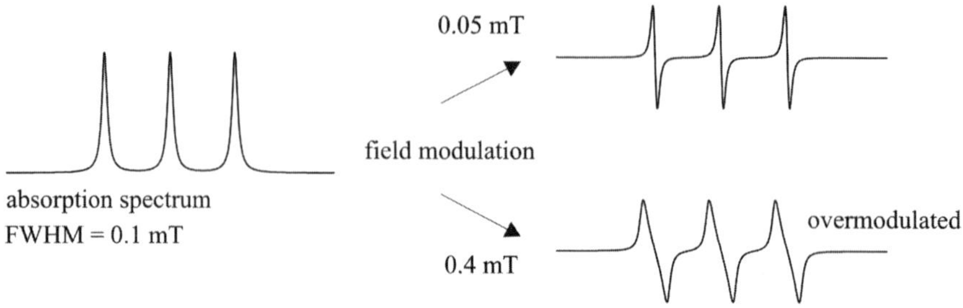

Figure 14. Effect of field modulation on a nitroxide spectrum. All simulation parameters are given in Matlab code in the text.

LIST OF EASYSPIN FUNCTIONS

General

easyspininfo	Information about EasySpin installation
eprload	Read spectrum files in common ESR data formats

Utilities

eprconvert	Graphical frequency/field/g value conversion utility
fastmotion	Fast-motion regime line widths
isotopes	A browsable list of isotopes
mhz2mt	Unit conversion from MHz to mT
mt2mhz	Unit conversion from mT to MHz

Spectral Simulations

chili	Slow-motion cw ESR spectra
garlic	Isotropic and fast-motion cw ESR spectra
pepper	Solid-state cw ESR spectra
salt	ENDOR spectra
nucfrq2d	Sketch of HYSCORE spectra

Spin Systems

nucspinadd	Adds a nucleus to a spin system
nucspinrmv	Removes a nucleus from a spin system

Data Analysis

basecorr	Polynomial baseline correction
ctafft	Cross-term averaged FFT

deriv	Numerical differentiation
exponfit	Exponential fitting
fdaxis	Frequency domain axis for FFT
pseumod	Pseudo-modulation of ESR absorption spectra
rcfilt	RC filtering of ESR spectra
smooth	Moving averages: smoothing and differentiation

Vectors, Orientations, and Angles

ang2vec	Converts polar angles to cartesian unit vector
erot	Computes rotation matrix from Euler angles
eulang	Euler angles from rotation matrix
rotaxi2mat	Convert rotation axis plus angle to rotation matrix
rotmat2axi	Convert rotation matrix to rotation axis plus angle
vec2ang	Polar angles from cartesian unit vector

Line Shapes and Windows

apowin	Apodization windows.
convspec	Convolute spectrum with line shape
gaussian	Gaussian line shape
lorentzian	Lorentzian line shape
lshape	General line shape function
makespec	Construct spectrum from peak positions and amplitudes
voigtian	Voigtian line shape function

Physical Constants

nucabund	Natural abundance of nuclear isotopes
nucdata	Nuclear spin data
nucgval	Nuclear g values
nucqmom	Nuclear electric quadrupole moments
nucspin	Nuclear spin quantum numbers
amu	Atomic unit of mass
avogadro	Avogadro constant
barn	Unit of nuclear quadrupole moments
bmagn	Bohr magneton
bohrrad	Bohr radius
boltzm	Boltzmann constant
clight	Vacuum speed of light
echarge	Elementary electric charge
emass	Mass of electron
faraday	Faraday constant
gfree	g value of the free electron
hartree	Atomic unit of energy
molgas	Molar gas constant

nmagn	Nuclear magneton
nmass	Mass of neutron
planck	Planck constant
pmass	Mass of proton
rydberg	Rydberg constant

Basic Spin Physics

commute	Commutator of two matrices
equivcouple	Coupling of equivalent spins
equivsplit	Splitting pattern for equivalent spins
hsdim	State space dimension of a spin system
larmorfrq	Larmor frequency of nuclear spins
sop	Spin operator matrices
spinvec	Spin quantum numbers in a spin system
stev	Extended Stevens operator matrices

Angular Momentum

clebschgordan	Clebsch-Gordan coefficients
plegendre	Legendre polynomials and Associated Legendre polynomials
spherharm	Spherical harmonics
wigner3j	Wigner 3-j symbols
wigner6j	Wigner 6-j symbols

Energy Levels, Resonances, and Propagations

eigfields	Resonance fields using eigenfield equation
endorfrq	ENDOR frequencies and amplitudes
evolve	Evolves density matrices in time-domain
levels	Energy levels computation
levelsplot	Energy levels and resonances plot
orisel	Orientation selection
propint	Compute pulse propagator
resfields	General resonance fields and amplitudes

Hamiltonians and Densities

eeint	Electron-electron interactions
hfine	Hyperfine interactions
internal	Internal interactions
nquad	Nuclear quadrupole interactions
sham	Full Spin Hamiltonian
sigeq	Thermal equilibrium density matrix
symm	Hamiltonian symmetry determination
zeeman	Zeeman interactions
zfield	Zero field interactions

Orientations

sphgrid	Triangular orientational grids
sphrand	Random orientational grids
sphtri	Triangulation of orientational grids

REFERENCES

Atherton NM. 1993. *Principles of electron spin resonance*. Chichester: Ellis Horwood.

Della Lunga G, Pogni R, Basosi R. 1994. Computer simulation of EPR spectra in the slow-motion regime for copper complexes with nitrogen ligands. *J Phys Chem* **98**:3937–3942.

Gamliel D, Levanon H. 1995. *Stochastic processes in magnetic resonance*. Singapore: World Scientific.

Hyde JS, Pasenkiewicz-Gierula M, Jesmanowicz A, Antholine WE. 1990. Pseudo-field modulation in EPR spectroscopy. *Appl Magn Reson* **1**:483–496.

Schneider D, Freed J. 1989. Calculating slow motional magnetic resonance spectra: a user's guide. *Biol Magn Reson* **10**:1–76.

Stoll S, A Schweiger. 2006. EasySpin, a comprehensive software package for spectral simulation and analysis in EPR. *J Magn Reson* **178**:42–55.

Stoll S, Schweiger A. 2003. An adaptive method for computing resonance fields for continuous-wave EPR spectra. *Chem Phys Lett* **380**:464–470.

Weil JA. 1971. The analysis of large hyperfine splittings in paramagnetic resonance spectroscopy. *J Magn Reson* **4**:394–399.

APPENDIX 7

EPRSIM-C: A SPECTRAL ANALYSIS PACKAGE

Janez Štrancar
EPR Center, Laboratory of Biophysics,
"Jožef Stefan" Institute, Ljubljana, Slovenia

1. INTRODUCTION

ESR spectroscopy in combination with nitroxide spin labeling has been shown to be one of the most important techniques in exploring biological systems. This is due to the sensitivity of the ESR technique to local molecular motions as well as the fact that the technique is nondestructive, enabling one to study complex systems without splitting them up into their building blocks.

Although there was a substantial development in ESR spectral simulations during the last decades, users often faced the problem of being overwhelmed by the large number of spectral parameters. This arises on one hand due to the precision of the spectral simulation models and on the other due to the complexity that should be taken into account when exploring biological systems. The ESR simulation software packages that were available in this research field paid insufficient attention to the problem of extracting such a large number of spectral parameters — the optimization routines were not chosen according to the complexity of the phase space and the inverse problem appeared to be ill posed many times due to strong numerical correlations in the parameter space.

Therefore, we developed the EPRSIM software package that recently evolved into the more general EPRSIM-C software package. This package is capable of performing an automatic nitroxide-based characterization of a biological system by implementing simpler and still accurate simulation models, powerful optimization routines as well as data condensation techniques that enable the user to parameterize the complex system in a more understandable way. Many ideas of the whole approach can be found in Chapter 3 in this volume: "*Advanced ESR Spectroscopy in Membrane Biophysics.*"

The EPRSIM-C software package provides **a tool for nitroxide-based characterization** of:

- Specifically and nonspecifically labeled biological membranes
- Site-directed spin labeled proteins

- Biopolymer networks explored by specific labeling or concentration imaging
- Nanomaterials
- Other interesting complex biological systems labeled with different labeling techniques

The EPRSIM-C software package simulates nitroxide ESR spectra as well as solves the inverse problem involved in the characterization procedure. For increased efficiency the different tasks are handled within different programs (see Figure 1).

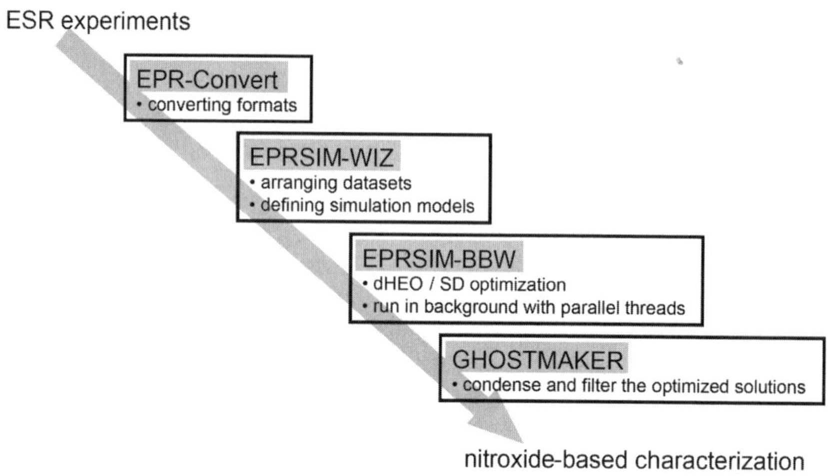

Figure 1. Illustration of the EPRSIM-C software package concept with the specific tasks of the individual program modules.

All the intensive numerical tasks – in fact, the optimizations – are done in background within **EPRSIM-BBW** in a way that one can submit the tasks from a network in a simple way. The simulation models are based on partial motional averaging, as described in Chapter 3 herein. The user should understand that the complex phase space can only be searched by a stochastic population-based optimization routine. This approach takes a great deal of computational time. However, much faster local search methods are not applicable in general, as the user can never be sure that there are not equally good solutions in other parts of the phase space.

The user interface of **EPRSIM-WIZ** helps to arrange and check the experimental and simulated data sets and is not meant to show the progress of the optimization routines. The main idea of having this kind of interface separated is to arrange clearly the experimental and simulation data. Various formats of experimental series can be read directly or translated into a proper format within **EPR-Convert**.

Inverse problem solving is separated from the data condensation routine implemented in **GHOSTMAKER** to increase the efficiency of the calculations. This module filters the data sets resulting from the optimization routine according to goodness of fit and solution density. In addition, it helps to condensate the solutions into groups of various complexity.

2. MAIN CHARACTERISTICS

The spectral simulation modules use various models applicable for different biological systems. An important aspect of the software is implementation of a fast-motion approximation with partial averaging. This makes it possible to use the software on personal computers. To get the idea of the computational demand of all steps in the characterization procedure, we provide here a list of time intervals typical for current desktop processors with a computing power of a few GFLOPS:

- 1–10 ms for the simulation model
- 1–10 s for one local search
- 1000 s for one stochastic population-based search
- 10 h for a complete characterization procedure

The user can easily scale the time demand for the characterization procedure if the time for the simulation model is increased dramatically. In addition, this time demand increases dramatically when the inverse problem becomes close to being ill-posed, i.e., when the number of spectral parameters is increased in such a way that strong correlations appear in the parameter space.

An autonomous optimization routine (implemented in **EPRSIM-BBW**) derived from hybridization of evolutionary optimization methods and a local search routine handle inverse problem solving with a large number of spectral parameters involved. In addition, many knowledge-based operators are introduced to improve the numerical consistency of the simulation models. Multiple optimizations are needed as the optimization routine belongs to the class of stochastic algorithms. Since the inverse problem solving routine requires a great deal of computing power, module **EPRSIM-BBW** can do multiple optimizations in parallel threads, making it more efficient on multiprocessor platforms. **EPRSIM-BBW** runs in the background and continuously scans the desired directory for tasks to be performed.

The user interface is carried out with the **EPRSIM-WIZ** module, which takes care of data handling and presentation of ESR spectral simulations. Multi-frequency experiments, multi-angle experiments, as well as other experimental series can be analyzed. Several formats can be read directly or translated with the **EPR-Convert** module.

As multiple optimization routines provide a huge number of data sets, one has to find which solutions are most significant and reliable. This is done by the

GHOST condensation algorithm implemented in **GHOSTMAKER**. The populations of solutions are compared to each other for goodness of fit, passing through only the solutions that are best. In addition, there is density filtering, which passes only those solutions that have many other solutions in their neighborhood (in the parameter space). Finally, the solutions are grouped, i.e., condensed to make parameterization more transparent.

3. EPRSIM-C LIBRARY

The general library of EPRSIM-C programs contains all the modules that are shared among the EPRSIM-C programs, like handling with spectral and MTP files, simulation models, and optimization routines.

3.1. MTP File Structure

The name "MTP" stands for multiple data sets. The MTP file is divided into two parts: system constants (block A) and spectral parameters (block B). As it is a text file, it can be easily modified manually without using EPRSIM-WIZ, if desired.

The system part (block A) includes:

- Optimization constants like different sizes, probabilities, switches, etc.
- Constants of the simulations models like limits, uncertainties, etc.
- Nitroxide tensor elements
- Artificial noise value
- Number of optimization runs (i.e., "copies") that should be made to achieve appropriate statistics

The spectral parameter part (block B) contains the data structure for each of the experimental data sets:

- The number of experiments performed on the same sample and should be simulated by the same group of spectral simulation parameters (as in multi-frequency experiments)
- Filenames of the experimental data sets
- Experimental constraints: center field, sweep width, frequency and angle
- Number of spectral components (up to 5)
- Spectral parameters, error, and switches (for optimization) for each of the spectral components, including the spectral simulation model that should be applied for each particular spectral component
- Type of measure of goodness of fit

For the precise structure see the implementation notes and manuals of the EPRSIM-C package on the accompanying CD.

When opening spectral data sets, the following implementations should be noted:

- Additional Gaussian noise is added to experimental spectral data with the amplitude as defined in MTP file
- The standard deviation of the noise is defined by fitting the outer wings ("zero-signal") of the spectrum by two quadratic functions
- A rough baseline correction is done according to the derivative of the third-degree polynomial fit to the outer wings of the first integral of the spectrum
- A rough correction of the center-field position is done by aligning the maximum of the first integral of the simulated spectrum to the maximum of the first integral of the experimental spectrum
- Fine correction of the amplitude (normalization), center-field, and baseline are done together with spectral simulations in the simulation module to adapt also to the long wings of the spectra and high noise amplitudes

3.2. The Simulation Models

The EPRSIM library contains the various algorithms for the **simulation models**. The physical background for the models is described in Chapter 3. In this software module, the following simulation models (with three-letter abbreviations) can be used:

- first model: isotropic tumbling (ISO)
- second model: anisotropic tumbling with full averaging over long axes (MEM)
- third model: isotropic spin-exchange label-label (LLE)
- fourth model: isotropic spin-exchange label-broadening agent (LBE)
- fifth model: anisotropic tumbling with partial averaging of all rotations (MES)
- ninth model: spin trap simulations (TRP)

Models 6-8 are not relevant, since they are meant for developmental purposes.

The following general principles are used in the implementation of all simulation models:

- Weighting of the spectral components is done according to double integrals (spectral intensities)
- Each spectral parameter has a "switch" that defines whether this particular parameter will be optimized within the optimization routine (see, e.g., Table 1); note that some models contain spectral parameters that will not be optimized, because they represent some variations of the models (these switches are false by default)

- Linear baseline correction is done together with amplitude normalization after weighting and composition of the spectral components
- The goodness of fit is calculated after normalization and baseline correction according to the χ^2 type (defined in block B of the MTP file for each experimental data set): $\beta = 0$ for standard χ^2 and $\beta = 1$ for first-order island-weighted χ^2

$$\chi^2_{(\beta)} = \frac{1}{N}\sum_{\text{island}}\left(\left(\lambda_{\text{island}}\right)^{\beta}\sum_{i=1}^{\lambda_{\text{island}}}\frac{\left(y_{\exp}-y_{\text{sim}}\right)^2}{\sigma^2}\right).$$

Table 1. Parameter Space of the First Model: Isotropic Tumbling (ISO)

Parameter name	Notation	Unit	Definition interval	Default switch
Proportion of Lorentzian/ Gaussian mixture	L_w		0 …1	T
Isotropic rotational correlation time	τ_c	ns	0.02 … 3	T
Additional broadening constant	W	G	0.1 … 4	T
Polarity correction p_A	p_A		0.8 … 1.2	T
Polarity correction p_g	p_g		0.9996 … 1.0002	T
Weight	d		0 … 1	T

3.2.1. First Model: Isotropic Tumbling (ISO)

The ISO simulation model can be used to simulate ESR spectral component of isotropic tumbling nitroxide spin labels in the fast motional regime.

Here are some implementation notes about the ISO model:

- The fast motion approximation determines the motional-sensitive part of the linewidth,
- An additional broadening constant W is added to the linewidth of all lines directly

$$\frac{1}{T_2} = A\left(\tau_c, \underline{A}, \underline{g}\right) + B\left(\tau_c, \underline{A}, \underline{g}\right)M + C\left(\tau_c, \underline{A}, \underline{g}\right)M^2 + W$$

- A mixture of a Lorentzian and Gaussian line shape is calculated at the resonant line positions; $L_w = 0$ for a pure Gaussian line

- The polarity correction factors are used to correct the scalar of the tensors (trace-scalar correction)
- A hyperfine structure is added for ^{13}C with the abundance determined in the MTP file (section "Tensors")

3.2.2. *Second Model: Anisotropic Tumbling with Full Averaging over Long Axes (MEM)*

The MEM simulation model can be used to simulate ESR spectra of nitroxide spin probes that are fast tumbling in an anisotropic environment like a membrane.

Note that the rotations around the long axes are assumed to be so fast as to allow full averaging of the magnetic tensor values over this direction. The anisotropy of the rotational motion is therefore described by one order parameter, S_{zz}.

Note that the model implements only one effective rotational correlation time, which describes the wobbling of the long axes.

Table 2. Parameter Space of the Second Model: Anisotropic Tumbling with Full Averaging over Long Axes (MEM)

Parameter name	Notation	Unit	Definition interval	Default switch
Order parameter S_{zz}	S		0 ... 1	T
Isotropic rotational correlation time	τ_c	ns	0.02 ... 3	T
Additional broadening constant	W	G	0.1 ... 4	T
Polarity correction p_A	p_A		0.8 ... 1.2	T
Polarity correction p_g	p_g		0.9996 ... 1.0002	T
Weight	d		0 ... 1	T

Here are some implementation notes about the MEM model:

- The fast motion approximation determines the motional-sensitive part of the linewidth
- An additional broadening constant, W, is added to the linewidth of all lines directly (see model 1)
- Polarity correction factors are used to correct all the tensor components by the same amount (trace-scalar correction)
- Order parameter S_{zz} is used to determine the partial averaging of the tensor component due to wobbling of the long molecular axes
- Complete averaging is used for motions around the long axes
- The stick spectrum is generated by numerical computation with the $\pi/2$ interval divided into 2000 steps by also taking into account transition probability correction factors up to second order

- A second-order correction to the stick spectrum is made near singularities by analytical expansions
- Renormalization is done to the stick spectrum after all corrections
- A convolution of a pure Lorentzian derivative and stick spectrum is done up to a maximum of ten linewidths from each point

3.3.3. Third Model: Isotropic Spin-Exchange Label-Label (LLE)

The LLE simulation model can be used to simulate ESR spectra of concentrated nitroxide spin probes in an isotropic environment. In this case, label–label spin exchange is present.

Table 3. Parameter Space of the Third Model: Isotropic Spin-Exchange Label-Label (LLE)

Parameter name	Notation	Unit	Definition interval	Default switch
Isotropic rotational correlation time	τ_c	ns	0.02 ... 3	*T*
Additional broadening constant (not exchange)	W	G	0.1 ... 4	*T*
Spin exchange rate	W_{ex}	G	0.1 ... 10	*T*
Polarity correction p_A	p_A		0.8 ... 1.2	*T*
Polarity correction p_g	p_g		0.9996 ... 1.0002	*T*
Weight	d		0 ... 1	*T*

Here are some implementation notes about the LLE model:

- The linewidth is determined as in model 1 using τ_c and W
- The spin exchange rate, W_{ex}, is used in the Bloch equations to determine the line shapes for a spin–spin interacting system with the use of complex magnetization G

$$\left(\frac{1}{T_2(M_1)}+i\left(B-B_{M_1}\right)\right)G_{M_1}-\sum_{M_1\neq M_2}W_{M_1M_2}\left(G_{M_2}-G_{M_1}\right)=-iB_1G_0\,,$$

- The polarity correction factors are used as in model 1.

3.2.4. Fourth Model: Isotropic Spin-Exchange Label-Broadening Agent (LBE)

The LBE simulation model can be used to simulate ESR spectra of nitroxide spin probes in an isotropic environment with a high concentration of broadening agents. Therefore label-broadening agent spin exchange is present.

Table 4. Parameter Space of the Fourth Model; Isotropic Spin-Exchange Label-Broadening Agent (LBE)

Parameter name	Notation	Unit	Definition interval	Default switch
Isotropic rotational correlation time	τ_c	ns	0.02 ... 3	*T*
Intrinsic linewidth of broadening agent	W_{Br}	G	100 ... 3000	*T*
Concentration of broadening agent	c_{Br}	mmol/l	100 ... 10000	*T*
Polarity correction p_A	p_A		0.8 ... 1.2	*T*
Polarity correction p_g	p_g		0.9996 ... 1.0002	*T*
Weight	d		0 ... 1	*T*

For the implementation notes see the LLE model.

3.2.5. Fifth Model: Anisotropic Tumbling with Partial Averaging of All Rotations (MES)

The MES simulation model can be used to simulate ESR spectra of nitroxide spin probes in an anisotropic environment like a membrane, or of nitroxide spin probes attached to a protein (site-directed spin labeling). It is a generalization of the MEM model.

Instead of describing the anisotropy of rotational motion by a single order parameter, this model has two cone angle parameters describing the wobble model (the main cone angle and asymmetry angle) to generalize restriction of the rotational motion. The reason for implementing a wobble model instead of a Gaussian distribution is that one can easily translate two cone angles into order parameters (assuming that the probability function is a nonzero constant up to two boundary angles and zero above their values). However, it is not possible to translate order parameters easily into cone angles, which would be a more natural choice in the case of site-directed spin labeling.

In order not to introduce a strong correlation in the parameter space, the implementation of a second rotational correlation time is not carried out.

Note that this model allows varying the geometry of the orientation distribution of spin probes by a geometrical parameter g_1 within the sample in the ESR spectrometer: it can be isotropic, planar, cylindrical, or pore/neck-like (see Chapter 3, Figure 16). The other geometrical parameter, g_2, is for developmental purposes only; it is not relevant here.

In addition, this model allows one to smoothly vary the orientation of the tensors relative to the coordinate system of fast-motional averaging by changing the two nitroxide rotation angles.

Table 5. Parameter Space of the Fifth model: Anisotropic Tumbling with Partial Averaging of All Rotations (MES)

Parameter name	Notation	Unit	Definition interval	Switch
Geometrical parameter 1	g_1		0 ...	*F*
Geometrical parameter 2	g_2			*F*
Nitroxide rotation angle 1	th	rad	0 ... π/2	*F*
Nitroxide rotation angle 2	fi	rad	0 ... π/2	*F*
Main cone angle	th_0	rad	0 ... π/2	*T*
Asymmetry angle	fi_0	rad	0 ... π/2	*T*
Effective rotational correlation time	τ_c	ns	0.02 ... 3	*T*
Additional broadening constant	*W*	G	0.1 ... 4	*T*
Polarity correction p_A	p_A		0.8 ... 1.2	*T*
Proticity prot	prot		–0.07 ... 0.07	*T*
Weight	*d*		0 ... 1	*T*

Here are some implementation notes about the MES model:

- Polarity correction factor p_A is used as a scalar to correct all the *A* tensor components

- Proticity correction factor prot is used in the following way together with g_{fe} (free electron *g*), p_A, and empirical constants:

  g_{xx}:=g_{xx}+(g_{xx}-g_{fe})*(-0.77*(p_A-1)-prot)

  g_{yy}:=g_{yy}+(g_{yy}-g_{fe})*(-0.06*(p_A-1))

  g_{zz}:=g_{zz}+(g_{zz}-g_{fe})*(-0.12*(p_A-1))

- Two nitroxide rotation angles (1 and 2) are used to get a proper orientation of the nitroxide coordinate system relative to the molecular axes (axes of fast motion)

- The spectral line shape is calculated using a fast-motion approximation by convolution of a Lorentzian line shape (linewidth defined in the motional narrowing approximation using a single effective rotational correlation time τ_c) with a distribution of resonant lines (stick spectrum) that is calculated using partial averaging of the tensor components and appropriate geometry of the orientation distribution

- Although the effective rotation correlation time τ_c is not allowed to exceed the spectroscopy time window (t_{win} = 3 ns), the linewidths *t[j]* (calculated from the motional narrowing approximation) are corrected empirically for larger values based on linewidths *tt[j]* at t_{win}:

t[j]:=tt[j]-(tt[j]-t[j])*Exp(-($\tau_c - t_{win}/3)/t_{win}$)

- The transformations between the coordinate systems used in the algorithm are based on the following angles:
  - Membrane-normal orientation dependence $\{\Theta, \Phi\}$
  - Orientation dependence of fast changing rotational conformations defined with Euler angles $\{\psi, \vartheta, \varphi\}$

Note that in the software a "text-like" notation is used:

- {psi, th, fi} for $\{\psi, \vartheta, \varphi\}$
- {TH, FI} for $\{\Theta, \Phi\}$

Maximal wobble angles are denoted by the bottom-right index "0".

- The applied Hamiltonian is:

$$\hat{H} = \mu_B B\ g^{\text{eff}}\left(\vec{\Omega}\right)\ \hat{S}_z + \hat{\vec{I}}\ A^{\text{eff}}\left(\vec{\Omega}\right)\ \hat{\vec{S}},$$

where partial averaging is taken into account with effective expressions of the *g* and *A* tensor components. However, the magnetic field orientation dependence remains explicit in vector $\vec{\Omega} = \{\Theta, \Phi\}$.

- The most important averages of the trigonometric functions of ϑ_0, φ_0, and ψ_0 (ψ_0 is assumed to be similar to φ_0) involve the following expressions:

$$\overline{\cos(\vartheta)^2} = \tfrac{1}{3}\left(\cos^2(\vartheta_0) + \cos(\vartheta_0) + 1\right),$$

$$\overline{\sin(\varphi)^2} = \frac{\varphi_0 - \tfrac{1}{2}\sin(2\varphi_0)}{2\varphi_0} = \frac{1}{2}\left(1 - \frac{\sin(2\varphi_0)}{2\varphi_0}\right),$$

$$\overline{\sin(\psi)^2} = \frac{\psi_0 - \tfrac{1}{2}\sin(2\psi_0)}{2\psi_0} = \frac{1}{2}\left(1 - \frac{\sin(2\psi_0)}{2\psi_0}\right),$$

$$\overline{\sin(2\vartheta)} = \tfrac{2}{3}\left(\cos(\vartheta_0) + 1\right)\sin(\vartheta_0),$$

$$\overline{\cos(\psi)} = \frac{\sin(\psi_0)}{\psi_0},$$

- The Hamiltonian matrix is solved in the "S_z (I_z, I_+, I_-) *terms only*" approximation with a relative accuracy of 0.01%, and absolute accuracy of 0.01 mT (including the diagonals of the super-block-diagonal and sub-block diagonal; see Chapter 3, Figure 6).
- The resonant line positions are calculated from

$$\Delta E_{M_I} = h\nu \rightarrow B = \frac{h\nu - M_I\sqrt{\left(A_{zz}^{\text{eff}}(\Theta_i,\Phi_i)\right)^2 + 2\left|A_{xzyz}^{\text{eff}}(\Theta_i,\Phi_i)\right|^2}}{\mu_B\ g^{\text{eff}}(\Theta_i,\Phi_i)},$$

with additional expressions

$$A_{zz}^{\text{eff}}(\Theta_i,\Phi_i) = \Big[\tilde{A}_{xx}^{\text{avg}}\cos^2(\Phi_i)\sin^2(\Theta_i) + \tilde{A}_{yy}^{\text{avg}}\sin^2(\Phi_i)\sin^2(\Theta_i) + \tilde{A}_{zz}^{\text{avg}}\cos^2(\Theta_i)$$
$$+\tilde{A}_{xz}^{\text{avg}}\cos(\Phi_i)\sin(2\Theta_i)\Big]\ \left(g^{\text{eff}}(\Theta_i,\Phi_i)\right)^2,$$

$$A_{xzyz}^{\text{eff}}(\Theta_i,\Phi_i) = \Bigg[\frac{1}{2}\Big(\tfrac{1}{2}\Big(\tilde{A}_{xx}^{\text{avg}}\cos^2(\Phi_i) + \tilde{A}_{yy}^{\text{avg}}\sin^2(\Phi_i) - \tilde{A}_{zz}^{\text{avg}}\Big)\sin(2\Theta_i) + \tilde{A}_{xz}^{\text{avg}}\cos(\Phi_i)\cos(2\Theta_i)\Big)$$
$$+\frac{i}{2}\Big(\tfrac{1}{2}\Big(\tilde{A}_{xx}^{\text{avg}} - \tilde{A}_{yy}^{\text{avg}}\Big)\sin(2\Phi_i)\sin(\Theta_i) + \tilde{A}_{xz}^{\text{avg}}\sin(\Phi_i)\cos(\Theta_i)\Big)\Bigg]\ \left(g^{eff}(\Theta_i,\Phi_i)\right)^2,$$

$$\left(g^{\text{eff}}(\Theta_i,\Phi_i)\right)^2 = \tilde{G}_{xx}^{\text{avg}}\cos^2(\Phi_i)\sin^2(\Theta_i) + \tilde{G}_{yy}^{\text{avg}}\sin^2(\Phi_i)\sin^2(\Theta_i)$$
$$+\tilde{G}_{zz}^{\text{avg}}\cos^2(\Theta_i) + \tilde{G}_{xz}^{\text{avg}}\cos(\Phi_i)\sin(2(\Theta_i),$$

and

$$\tilde{G}_{xx}^{\text{avg}} = g_{xx}^2\left(\overline{\cos^2(\varphi)}\ \overline{\cos^2(\vartheta)}\ \overline{\cos^2(\psi)} + \overline{\sin^2(\varphi)}\ \overline{\sin^2(\psi)}\right)$$
$$+g_{yy}^2\left(\overline{\sin^2(\varphi)}\ \overline{\cos^2(\vartheta)}\ \overline{\cos^2(\psi)} + \overline{\cos^2(\varphi)}\ \overline{\sin^2(\psi)}\right) + g_{zz}^2\overline{\sin^2(\vartheta)}\ \overline{\cos^2(\psi)},$$

$$\tilde{G}_{yy}^{\text{avg}} = g_{xx}^2\left(\overline{\cos^2(\varphi)}\ \overline{\cos^2(\vartheta)}\ \overline{\sin^2(\psi)} + \overline{\sin^2(\varphi)}\ \overline{\cos^2(\psi)}\right)$$
$$+g_{yy}^2\left(\overline{\sin^2(\varphi)}\ \overline{\cos^2(\vartheta)}\ \overline{\sin^2(\psi)} + \overline{\cos^2(\varphi)}\ \overline{\cos^2(\psi)}\right) + g_{zz}^2\overline{\sin^2(\vartheta)}\ \overline{\sin^2(\psi)},$$

$$\tilde{G}_{zz}^{\text{avg}} = g_{xx}^2\overline{\cos^2(\varphi)}\ \overline{\sin^2(\vartheta)} + g_{yy}^2\overline{\sin^2(\varphi)}\ \overline{\sin^2(\vartheta)} + g_{zz}^2\overline{\cos^2(\vartheta)},$$

$$\tilde{G}_{xz}^{\text{avg}} = 0,\ \tilde{G}_{xy}^{\text{avg}} = 0,\quad \tilde{G}_{yz}^{\text{avg}} = 0,$$

$$\tilde{A}_{xx}^{avg} = \frac{A_{xx}}{g_{xx}^2}\left(\overline{\cos^2(\varphi)}\ \overline{\cos^2(\vartheta)}\ \overline{\cos^2(\psi)} + \overline{\sin^2(\varphi)}\ \overline{\sin^2(\psi)}\right)$$
$$+\frac{A_{yy}}{g_{yy}^2}\left(\overline{\sin^2(\varphi)}\ \overline{\cos^2(\vartheta)}\ \overline{\cos^2(\psi)} + \overline{\cos^2(\varphi)}\ \overline{\sin^2(\psi)}\right) + \frac{A_{zz}}{g_{zz}^2}\overline{\sin^2(\vartheta)}\ \overline{\cos^2(\psi)},$$

$$\tilde{A}_{yy}^{avg} = \frac{A_{xx}}{g_{xx}^2}\left(\overline{\cos^2(\varphi)}\ \overline{\cos^2(\vartheta)}\ \overline{\sin^2(\psi)} + \overline{\sin^2(\varphi)}\ \overline{\cos^2(\psi)}\right)$$
$$+\frac{A_{yy}}{g_{yy}^2}\left(\overline{\sin^2(\varphi)}\ \overline{\cos^2(\vartheta)}\ \overline{\sin^2(\psi)} + \overline{\cos^2(\varphi)}\ \overline{\cos^2(\psi)}\right) + \frac{A_{zz}}{g_{zz}^2}\overline{\sin^2(\vartheta)}\ \overline{\sin^2(\psi)},$$

$$\tilde{A}_{zz}^{avg} = \frac{A_{xx}}{g_{xx}^2}\overline{\cos^2(\varphi)}\ \overline{\sin^2(\vartheta)} + \frac{A_{yy}}{g_{yy}^2}\overline{\sin^2(\varphi)}\ \overline{\sin^2(\vartheta)} + \frac{A_{zz}}{g_{zz}^2}\overline{\cos^2(\vartheta)},$$

$$\tilde{A}_{xz}^{avg} = 0,\quad \tilde{A}_{xy}^{avg} = 0,\quad \tilde{A}_{yz}^{avg} = 0.$$

- The stick spectrum is calculated on *the* basis of a 2D extrapolation of the master grid over Θ and Φ (orientation relative to magnetic field), where the number of master steps in Θ and Φ dimensions corresponds to the maximal resonant field change for particular π/2 rotation at a resolution of 0.6 G (see Chapter 3, Figure 11).
- Geometrical parameter 1 (g_1) is used to define the geometry of the orientation distribution relative to the magnetic field:
  - $g_1 = -1$ is used to define an isotropic distribution
  - $g_1 = 0$ is used to define a planar distribution (perpendicular to the magnetic field)
  - $g_1 = -2$ is used to define a cylindrical distribution (symmetry axis parallel to the magnetic field)
  - $0 < g_1 < 1$ is used to define a pore distribution where g_1 represents the ratio between bilayer thickness *d* and pore outer diameter *D*; if $g_1 = 1/2$ it can represent also a neck distribution (see Chapter 3, Figure 16).
- Renormalization is done to the stick spectrum after all corrections
- A convolution of a Lorentzian derivative and stick spectrum is done up to a maximum of ten linewidths from each point

3.2.6. *Ninth Model: Spin Trap Simulations (TRP)*

The TRP simulation model can be used to simulate ESR spectra of a spin trap in an isotropic environment or any other spin system interacting with up to five different kinds of nuclei regarding hyperfine coupling. Note that the spin values as well as number of spins cannot be optimized yet (as they are not real numbers).

Table 6. Parameter Space of the Ninth Model: Spin Trap Simulations (TRP)

Parameter name	Notation	Unit	Definition interval	Default switch
Proportion of Lorentzian/ Gaussian mixture	L_w		0 ...1	*T*
Additional broadening constant	*W*	G	0.1 ... 4	*T*
Polarity correction p_g	p_g		0.9996 ... 1.0002	*T*
Spin of nucleus 1	S_1		Half integer: 0 ... 5	*F*
Number of nucleus 1	N_1		Integer: 0 ... 10	*F*
Hyperfine coupling 1 with nucleus	a_1	G	0 .. 50	*T*
Spin of nucleus 2	S_2		Half integer: 0 ... 5	*F*
Number of nucleus 2	N_2		Integer: 0 ... 10	*F*
Hyperfine coupling 2 with nucleus	a_2	G	0 ... 50	*T*
Spin of nucleus 3	S_3		Half integer: 0 ... 5	*F*
Number of nucleus 3	N_3		Integer: 0 ... 10	*F*
Hyperfine coupling 3 with nucleus	a_3	G	0 ... 50	*T*
Spin of nucleus 4	S_4		Half integer: 0 ... 5	*F*
Number of nucleus 4	N_4		Integer: 0 ... 10	*F*
Hyperfine coupling 4 with nucleus	a_4	G	0 ... 50	*T*
Spin of nucleus 5	S_5		Half integer: 0 ... 5	*F*
Number of nucleus 5	N_5		Integer: 0 .. 10	*F*
Hyperfine coupling 5 with nucleus	a_5	G	0 ... 50	*T*
Weight	*d*		0 ... 1	*T*

Here are some implementation notes about the TRP model:

- Up to 5 different nuclei can be coupled to one radical (component)
- For each group (type of nuclei) multiplet positions and amplitudes of lines are calculated using binomial coefficients and hyperfine coupling of the particular nuclei
- A convolution with a mixture of a Lorentzian and Gaussian line shape is implemented; $L_w = 0$ for a pure Gaussian line

3.4. Hybrid Evolutionary Optimization (HEA = GA Hybridized with DS)

One of the most important issues of the EPRSIM-C package is application of a robust optimization scheme given by a hybrid evolutionary optimization (HEO) based on a generational genetic algorithm (GA). It is related to genetic-like transformations of parameters of a large population of M points, where $M >> N_p$ (each point is a set or vector of spectral parameters $\{p_i\}$), where N_p is the number of optimization parameters. In our case GA is hybridized with a local search operator (like DSO) and knowledge-based operators (KBO).

A starting population is created randomly within definition intervals $[p_{i,\min}, p_{i,\max}]$.

The main loop consists of selection, real coding, and application of genetic operators in the following scheme:

- Fitness (χ^2) function evaluation for all new individuals
- Sorting
- Elitism implementation
- Fitness-proportional or tournament selection
- Genetic-operations application:
  - Multi-point/uniform crossover
  - Uniform mutation
  - Local mutation (DSO)
  - Knowledge-based operators
  - Random replacement

The stop criterion for the HEO routine used in EPRSIM C is the maximum number of generations, which is taken as 100.

Some special implementation notes are as follows:

- The definition intervals of the parameters are defined with minimal and maximal values in the MTP file ("Limits" section)
- Grids defined in the MTP file (section "Grids") are used to define the first step ("simplex")
- The generation number is 50, 100, 100, 100 for up to 8, 8–14, 14–20 parameters and above, respectively
- The population size is 130, 200, 280, 400 for up to 8, 8–14, 14–20 parameters and above, respectively
- The number of crossover sites in multipoint crossover is 1, 1, 3 and 3 for up to 8, 8–14, 14–20 parameters and above, respectively
- Other parameters (mutation probability, crossover probability, DSO probability, KBO probability, elite size) are defined in the MTP file ("Hybrid Evolutionary Algorithm" section)
- Elite cannot contain more than one copy of the same solution;
- Elite solutions replace the worst in the population every generation if the elite solutions cannot be found in the existing population (to prevent crowding of the same solution)
- The tournament selection is used to determine the pool of solutions that will be transformed by genetic operators (better solutions win for selection in random pairs)

- Random selection is used for selection of the pair of solutions from the pool of tournament-selected solutions
- Shaking is applied within the crossover with a shaking amplitude defined from the product of a Gaussian random deviation (amplitude 0.5) and grids (defined in the MTP file in the "Grids" section)
- Knowledge-based operator (KBO) 1: $\tau_{c,1} - d_1 - \tau_{c,2} - d_2$ equilibrator — multiplies smaller τ_c by 1.3 and larger by 0.7 and subtracts half of the smaller weight from the weight of the component that has larger τ_c value and adds the same amount to the other one
- Knowledge-based operator (KBO) 2: $W_1 - d_1 - W_2 - d_2$ equilibrator — multiplies smaller W by 1.3 and larger by 0.7 and subtracts half of the smaller weight from the weight of the component that has a larger W value and adds the same amount to the other one
- Knowledge-based operator (KBO) 2: polarity sorter — remixes polarity correction factors p_g of the components in the opposite order of sorted polarity correction factor p_A; note that the polarity sorter should not be used for the fifth model (MES)
- DSO is implemented as a mutation operator with a probability defined in the MTP file (section "Hybridization with DS during HEA")
- Resolution auto tune adopts the grid for the (GA) mutation operator (initially defined by the grid constants from the MTP file to standard deviations found by the covariance matrix analysis every 10 generations)
- After a maximal generation number is achieved, substitution from the elite set is done once more, the population is sorted, and a single DSO is used on solutions of the better part of population as defined in the MTP file (section "Hybridization with DS after HEA" — possible usage and part of the population should be defined)

4. EPRSIM-C PROGRAMS

4.1. EPRSIM-WIZ

The purpose of the EPRSIM-WIZ module is to handle ESR data, open spectral files, create multiple experiments in the MTP files, collect or combine ESR experiments, save files (with a corrected baseline, normalized, integrated), print the experimental data sets with simulations together, etc. Within EPRSIM-WIZ one usually defines the spectral simulation models that have to be used in the optimization procedure for each of the experimental data sets. In addition, local optimization of spectral parameters by the Downhill-Simplex optimization routine is possible for fine tuning. The graphical interface of EPRSIM-WIZ is designed to adapt to different arrangements of the experimental and simulation data in such a way that any part of the software can be easily upgraded for new models or new types of ESR experiments.

4.2. EPRSIM-BBW

The purpose of the EPRSIM-BBW module is to perform automatic optimization of a series of spectra. Multi-threading and calculation in the background are the two most important properties of this program.

EPRSIM-BBW can be run for optimization of problems defined in a single MTP file. However, EPRSIM-BBW can be run also in a "stay-resident" mode in which it is continuously checking the specified directory for any new tasks to be performed. The EPRSIM-BBW modes are triggered via various switches in command prompt. For a detailed description of switches, see the implementation notes and the manual for the EPRSIM-C software package on the accompanying CD. Here are two examples to illustrate the usage of EPRSIM-BBW:

1. To run EPRSIM BBW in the "single MTP optimization" mode, pointing to the »c:\temp\test.mtp« file containing problems for optimization, allowing at maximum 2 parallel threads and forcing 20 independent runs on each problem defined in the MTP file, one should input the following line:

   eprsimbbw.exe –p=c:\temp\test.mtp –n=2 –k=20

   When the optimization is finished, the optimized results are written in file "test1.mtp," which contains the best fits and the subdirectory POP that contains the population files for a GHOST construction. Then EPRSIM-BBW terminates.

2. To run EPRSIM BBW in "stay-resident" mode, pointing to the »c:\temp« directory that eventually contains the MTP files with problems for optimization, allowing at maximum 2 parallel threads and forcing 20 independent runs on each problem defined in the MTP file, one should input the following line:

   eprsimbbw.exe –d=c:\temp –n=2 –k=20

   After EPRSIM-BBW is started, it checks the »c:\temp« directory every 30 s for new MTP files to start optimization. When an MTP file (together with the appropriate spectral data files) is copied to the »c:\temp« directory, EPRSIM-BBW will automatically start optimization. When finished, it starts to scan the directory again.

   EPRSIM-BBW can be started directly from the command line (prompt) or from Windows Schedule to enable pure background running. EPRSIM BBW can be terminated through Task Manager or Windows Schedule (if used).

File transfer to the directory can be from anywhere (with appropriate permission) through the local network. When several new MTP files are found in the directory, the oldest one starts first.

4.3. EPR-Convert

The purpose of the EPR-Convert module is to convert Bruker binary spectral files into ASCII files in order to be used in EPRSIM-C programs or plotted with any other program. EPR-Convert can convert Bruker SPC (OS-9 file format) and DTA (Elexsys file format). Note that in case of DTA files, 2D experimental data are converted by splitting the data in several spectral files, denoted by "_slice#" with # as a slice index. EPR-Convert can also convert several spectral files into a single multi-column ASCII file. To distinguish between the data, the columns are indexed by the appropriate filenames.

4.4. GHOSTMAKER

The purpose of GHOSTMAKER is to condense the multi-run dHEO solutions into GHOSTs to be able to represent and distinguish groups of solutions. In general, one can extract at least the best solution from each run when a genetic algorithm is applied for optimization. As this kind of optimization is purely stochastic, one should compare at least the best solutions from each run to determine the uncertainty of solutions. If the proposed model complexity is large enough, the uncertainty represents the pure deviation of solutions. However, if the proposed complexity is too low and the optimization finds larger regions of the phase space to describe the spectral data, these uncertainties also represent the distribution of solutions.

The basic HEO algorithm always strongly converges to one good region, because of the strong effect of the crossover operator. Therefore, in the case when the proposed model complexity is too low, one should never extract more than only the best fit per run. However, with dHEO, with a shaking operator introduced to prevent crowding, one can extract more different solutions than just the best fit per run. In this way, the reduction in computational time for EPRSIM-BBW is very strong.

Despite the way in which solutions are generated, either by HEO or dHEO, a large number of solutions (about 200 per problem) have to be condensed, taking into account the different efficiencies of the individual runs as well as the probability with which the solution regions are found.

χ^2-filtering and **density filtering** are the two main tasks of the GHOST condensation algorithm. In the local solution density calculation as well as in the **grouping of solutions**, the principle of neighborhood of solutions is implemented. For a detailed description, see the implementation notes and the manual for the EPRSIM-C software package on the accompanying CD. In the final step of GHOST condensation, the groups of solutions have to be quantified for their proportion. First, an approximation is made according to the values condensed out from individual solutions. However, the final determination of proportions is done according to the spectral weights found by an additional run of EPRSIM-BBW after sending up to 5 descriptive solutions of GHOST condensate as spectral components to EPRSIM-BBW.

After a GHOST condensation, the results are **represented by GHOSTs** (for an example of a GHOST diagram, see Chapter 3, Figure 22). Currently, GHOST condensation is working with the MEM, MES, and LLE simulation models for which the x-axes parameter is chosen to be order parameter S_z, main cone angle ϑ, and spin-exchange rate W_{exc}, respectively, for the three models. Three of the other parameters — rotational correlation time τ_c, additional broadening W, and polarity correction p_A — are always coded by an RGB color directive, i.e., with red, green, and blue color intensity coding τ_c, W, and p_A, respectively. Additional parameters, like φ, prot, etc. are represented on the y-axes.

It should be stressed that when using GHOST condensation methodology, certain rules have to be obeyed, e.g., taking into account the signal-to-noise ratio, checking the uniformity of the run-contribution histogram, adopting the minimal threshold density, etc. A detailed description can be found in the implementation notes and the manual of the EPRSIM-C software package on the accompanying CD.

5. FURTHER READING

Filipič B, Štrancar J. 2001. Tuning EPR spectral parameters with a genetic algorithm. *Appl Soft Comp* **1**:83–90.

Filipič B, Štrancar J. 2003. Evolutionary computational support for the characterization of biological systems. In *Evolutionary computation in bioinformatics*, pp. 279–294. Ed GB Fogel, D Corne. Amsterdam: Elsevier.

Kavalenka AA, Filipič B, Hemminga MA, Štrancar J. 2005. Speeding up a genetic algorithm for EPR-based spin label characterization of biosystem complexity. *J Chem Inf Model* **45**:1628–1635.

Stopar D, Štrancar J, Spruijt RB, Hemminga MA. 2005. Exploring the local conformational space of a membrane protein by site-directed spin labeling. *J Chem Inf Model* **45**:1621–1627.

Štrancar J, Sentjurc M, Schara MV. 2000. Fast and accurate characterization of biological membranes by EPR spectra. *J Magn Reson* **142**:254–265.

Štrancar J, Koklič T, Arsov Z. 2003a. Soft picture of lateral heterogeneity in biomembranes. *J Membr Biol* **196**:135–146.

Štrancar J, Schara MV, Pečar S. 2003b. New EPR method for cellular surface characterization. *J Membr Biol* **193**:15–22.

Štrancar J, Koklič T, Arsov Z, Filipič B, Stopar D, Hemminga MA. 2005. Spin label EPR-based characterization of biosystem complexity. *J Chem Inf Model* **45**:394–406.

EPRSIM-C webpage: http://www.ijs.si/ijs/dept/epr/EPRSIMC_overview.htm

Contents of Previous Volumes

VOLUME 3

VOLUME 6

VOLUME 7

VOLUME 8

VOLUME 10

VOLUME 11

VOLUME 12

VOLUME 13

VOLUME 14

VOLUME 15

VOLUME 16

VOLUME 18

VOLUME 19

VOLUME 21

VOLUME 23

VOLUME 24

VOLUME 25

VOLUME 26

INDEX

Printed in the United States of America.